Animal Evolution

Animal Evolution
Genomes, Fossils, and Trees

EDITED BY

Maximilian J. Telford
University College London

AND

D. T. J. Littlewood
The Natural History Museum, London

Originating from a Royal Society Discussion Meeting first published in
Philosophical Transactions of the Royal Society B: Biological Sciences

OXFORD
UNIVERSITY PRESS

OXFORD

UNIVERSITY PRESS

Great Clarendon Street, Oxford OX2 6DP

Oxford University Press is a department of the University of Oxford.
It furthers the University's objective of excellence in research, scholarship,
and education by publishing worldwide in

Oxford New York

Auckland Cape Town Dar es Salaam Hong Kong Karachi
Kuala Lumpur Madrid Melbourne Mexico City Nairobi
New Delhi Shanghai Taipei Toronto

With offices in

Argentina Austria Brazil Chile Czech Republic France Greece
Guatemala Hungary Italy Japan Poland Portugal Singapore
South Korea Switzerland Thailand Turkey Ukraine Vietnam

Oxford is a registered trade mark of Oxford University Press
in the UK and in certain other countries

Published in the United States
by Oxford University Press Inc., New York

British Library Cataloguing in Publication Data
Data available

Library of Congress Cataloging in Publication Data
Data available

Typeset by Newgen Imaging Systems (P) Ltd., Chennai, India
Printed in Great Britain
on acid-free paper by
CPI Antony Rowe Ltd., Chippenham

ISBN 978–0–19–954942–9 (Hbk.)
ISBN 978–0–19–957030–0 (Pbk.)

10 9 8 7 6 5 4 3 2 1

Foreword

The subject of animal evolution has been explored and discussed for a century and a half. As long ago as the late 19th century, Ernst Haeckel, a renowned evolutionary biologist and supporter of Darwin, drew some beautiful phylogenetic trees depicting the possible course of animal evolution. Haeckel's trees included many of the major animal groups, placed on the tips of gnarled and life-like branches, each drawn complete with bark and twigs. One could be forgiven for thinking that the framework of animal evolution was completed long ago. Such a view would be grossly mistaken. The past two decades in particular have seen a revolution in our understanding of animal evolution, and revisiting this topic in 2009 is very timely. As the contents of this volume will testify, research into animal evolution is currently in its most vibrant phase ever, with novel conclusions being generated at great pace.

It is interesting to contrast the nature of current research, as described here, to the first golden age of animal evolutionary biology, the late 19th and early 20th centuries. Many of the early zoologists, including Kowalevsky, Metchnikoff, van Beneden, Lankester, Sedgwick, Jägersten, Goodrich, and T. H. Huxley were superb anatomists and embryologists, and their work has left us with a wonderful and extensive legacy of descriptive data. While invaluable for understanding animal form and function, however, these data ultimately proved of less use for discerning deep animal relationships, or understanding homologies and evolutionary transformations. Until very recently, textbooks and specialized works alike carried speculative, and often highly imaginative, scenarios of animal evolution, depicting how various body forms could be derived from others. For every question discussed,

however, there were almost as many scenarios proposed as there were scientists considering the question, and as a consequence debate was often heated, personal, and inconclusive. A comment made by the Rev. T. R. R. Stebbing at the end of a 1910 symposium discussing the origin of vertebrates reveals the lack of progress on just one of these issues: 'When we return home and our friends gleefully enquire, "What then has been decided as to the Origin of Vertebrates?", so far we seem to have no reply ready, except that the disputants agreed on one single point, namely that their opponents were all in the wrong' [Stebbing, T.R.R (1910) Discussion on the origin of vertebrates. *Proceedings of the Linnean Society of London* **122**, 9–50].

The fundamental problem plaguing the study of animal evolution a century ago was that there was no reliable way to test alternative scenarios, no objective source of data to evaluate putative homologies or proposed relationships. Every scenario was consistent with the available data, although certainly some theories were more outlandish than others! This was *the* major stumbling block to advance in the study of animal evolution, and it persisted through much of the 20th century. The problem is now clearly in focus and at least part of the solution is at hand. A major component of the solution is based on the application of molecular biology to animal evolutionary biology, in partnership with other approaches including developmental biology and palaeontology. Of course, genetics was first incorporated into evolutionary thinking in the 1920s to 1940s, when the work of Thomas Hunt Morgan, R. A. Fisher, Theodosius Dobzhansky, J. B. S. Haldane, Sewall Wright, Julian Huxley, Ernst Mayr, G. G. Simpson, and G. Ledyard Stebbins gradually merged Mendelian genetics with Darwinian evolution in a population context, and convincingly explained microevolution through changes in allele frequencies. But it still remained unclear if large anatomical differences, such as those between animal phyla, could be explained in the same way. Even if they could, as many argued, these early genetic studies did not reveal what genes were actually involved, what anatomical transformations had happened in evolution, and hence which of the multiple scenarios for animal evolution were correct. Until the 1980s, those working on evolutionary genetics had little interface with whole-organism biology, including embryology and animal diversity. Zoology was a divided field. As Frank R. Lillie, former Director of the Woods Hole Marine Biological Laboratory, commented in 1927: 'There can be no doubt, I think, that the majority of geneticists, and many physiologists certainly, hope for and expect a reunion. The spectacle of the biological sciences divided permanently into two camps is evidently for them too serious a one to be regarded with satisfaction' [Lillie, F.R. (1927) The gene and the ontogenetic process. *Science* **66**, 361].

Lillie's hoped for reunion is now upon us. In the past two decades, molecular genetics has started to have major impacts on the study of animal evolution, and in two distinct ways. First, comparison of DNA sequences is being used to trace evolutionary relationships between long-divergent animal phyla, and thus to reconstruct a fairly true genealogy of animal life spanning over half a billion years. It is not a simple exercise, and one that is fraught with methodological and analytical problems, but enormous progress is being made. It is now clear that combining data from many genes simultaneously, sometimes over 100 different genes from each species, can increase the reliability of phylogenetic inference, as can analysis of rare genome-level changes, such as inversions and rearrangements of the DNA. Examples of both approaches are discussed in this volume. These strategies have greatly benefited from recent advances in high-throughput DNA sequencing, a technology that is moving forward apace, and as such we are truly in the middle of a revolution in animal phylogenetics. Haeckel's trees are being updated with confidence.

The second application of molecular biology is closely linked with comparative embryology, the favoured approach of a century ago. Work on convenient laboratory species, such as *Drosophila*, nematodes, and mice, has produced a wealth of data on the identity and function of genes controlling specific aspects of embryonic development, from setting up of embryonic axes (dorsoventral, left–right, anteroposterior), to the establishment of tissue layers, the formation of organs, and the differentiation of specific cell types. With the discovery that many (but certainly not all) of these genes are ancient and present in highly divergent phyla, it has now become possible to trace how developmental pathways have changed in evolution, or been deployed in different ways, and relate this to the evolution of specific structures in animals. As above, there are technological difficulties and analytical pitfalls, but the results can be highly persuasive. Examples in this volume include new insights into the diversification of nervous systems, evolution of larval forms, relationships between body axes, evolution of segmentation, and the origin of novel characters.

These applications of molecular biology to zoology may have stimulated the reunion of developmental and evolutionary zoology, but they are not taking place in isolation. A holistic understanding of animal evolution is promised as insights from these methods are coupled with deeper knowledge of animal anatomy and embryology, based on old and new data, insights into divergence dates and extinct character combinations, informed through palaeontology, and refined understanding of gene functions. A revolution in understanding animal evolution is upon us.

Peter W. H. Holland
Linacre Professor of Zoology,
Department of Zoology,
University of Oxford, South Parks Road,
Oxford OX1 3PS, UK

Acknowledgements

The basis for this book was as part of the tercentenary celebrations of Linnaeus, for which we organized a Discussion Meeting at the Royal Society and a follow-on meeting at a (now sadly extinct) venue provided by the Novartis Foundation. An issue of *Philosophical Transactions of the Royal Society B: Biological Sciences* (Volume 363, Number 1496) provided the core chapters and content for this volume, and we thank OUP for their interest in publishing an updated issue as a stand alone volume. We gratefully acknowledge funding from The Royal Society, The Linnean Society of London, The Systematics Association, and the Novartis Foundation for making the original meeting *Evolution of the Animals—a Linnean Tercentenary Celebration* (June 2007) possible.

The majority of contributions have been considerably updated or even significantly changed, and we are delighted to include two entirely new chapters. We would like to thank the authors and the many reviewers for their considerable efforts.

We thank our editors at OUP, Helen Eaton and Ian Sherman, for their patience and support. We are pleased that we have been provided with the opportunity to celebrate the bicentenary of the birth of Charles Darwin with the publication of this volume.

Lastly, we dedicate this volume to those who inspired us to become zoologists and to our children, who demonstrate daily the delights and surprises of development, discovery, genotype, and phenotype.

Max Telford
Tim Littlewood

Contents

Contributors

Detlev Arendt, Developmental Biology Unit, European Molecular Biology Laboratory, Heidelberg, 69117, Germany

Patrícia Beldade, Evolutionary Biology Group, Institute of Biology, Leiden University, PO Box 9516, 2300 RA Leiden, The Netherlands *and* Instituto Gulbenkian de Ciência, Oeiras, Portugal

Jeffrey L. Boore, Genome Project Solutions, 1024 Promenade Street, Hercules, CA 94547, USA *and* University of California Berkeley, Department of Integrative Biology, 3060 Valley Life Sciences Building, Berkeley, CA 94720, USA *and* DOE Joint Genome Institute and Lawrence Berkeley National Laboratory, 2800 Mitchell Drive, Walnut Creek, CA 94598, USA

Sarah J. Bourlat, Department of Biology, University College London, Darwin Building, Gower Street, London WC1E 6BT, UK

Graham E. Budd, Department of Earth Sciences, Palaeobiology, Villavägen 16, Uppsala, Sweden SE-752 36

Richard R. Copley, Wellcome Trust Centre for Human Genetics, Roosevelt Drive, Oxford OX3 7BN, UK

James A. Cotton, School of Biological and Chemical Sciences, Queen Mary University of London, Mile End Road, London E1 4NS, UK

Mark J. Dayel, Department of Molecular and Cell Biology, 142 Life Science Addition #3200, University of California, Berkeley, CA 94720, USA

Alexandru S. Denes, Developmental Biology Unit, European Molecular Biology Laboratory, Heidelberg, 69117, Germany

Casey W. Dunn, Department of Ecology and Evolutionary Biology, Brown University, 80 Waterman Street, Providence, RI 02912, USA

Andrew Economou, Department of Biology, University College London, Darwin Building, Gower Street, London WC1E 6BT, UK

Gregory D. Edgecombe, Department of Palaeontology, The Natural History Museum, Cromwell Road, London SW7 5BD, UK

Susan I. Fuerstenberg, Genome Project Solutions, 1024 Promenade Street, Hercules, CA 94547, USA

James G. Gehling, South Australian Museum, North Terrace, Adelaide, South Australia 5000, Australia *and* Department of Geological Sciences, Queen's University, Kingston, Ontario, K7L 3N6, Canada *and* Department of Earth Sciences, Monash University, Clayton, Victoria 3168, Australia

Gonzalo Giribet, Department of Organismic and Evolutionary Biology, & Museum of Comparative Zoology, Harvard University, 26 Oxford Street, Cambridge, MA 02138, USA

Andreas Hejnol, Kewalo Marine Laboratory, PBRC, University of Hawaii, 41 Ahui Street, Honolulu, HI 96813, USA

Ronald A. Jenner, Department of Zoology, The Natural History Museum, Cromwell Road, London SW7 5BD, UK

Gáspár Jékely, Developmental Biology Unit, European Molecular Biology Laboratory, Heidelberg, 69117, Germany

Nicole King, Department of Molecular and Cell Biology, 142 Life Science Addition #3200, University of California, Berkeley, CA 94720, USA

Nicolas Lartillot, Laboratoire d'Informatique, de Robotique et de Mathématiques de Montpellier, CNRS – Université de Montpellier 2, 161, Rue Ada, 34392 Montpellier Cedex 5, France

D. Timothy J. Littlewood, Department of Zoology, The Natural History Museum, Cromwell Road, London SW7 5BD, UK

Christopher J. Lowe, Department of Organismal Biology and Anatomy, University of Chicago, Chicago, IL 60637, USA

Mark Q. Martindale, Kewalo Marine Laboratory, PBRC, University of Hawaii, 41 Ahui Street, Honolulu, HI 96813, USA

Scott A. Nichols, Department of Molecular and Cell Biology, 142 Life Science Addition #3200, University of California, Berkeley, CA 94720, USA

Daniel Papillon, Department of Biology, University College London, Darwin Building, Gower Street, London WC1E 6BT, UK

Andrew D. Peel, Institute for Molecular Biology and Biotechnology, Vassilika Vouton, 711 10 Iraklio, Crete, Greece

Hervé Philippe, Canadian Institute for Advanced Research, Département de Biochimie, Université de Montréal, Succursale Centre-Ville, Montréal, Québec H3C3J7, Canada

Davide Pisani, Laboratory of Evolutionary Biology, The National University of Ireland, Maynooth, County Kildare, Ireland

Rudolf A. Raff, Department of Biology, Indiana University, Bloomington, IN 47405, USA *and* School of Biological Sciences, University of Sydney, Sydney, NSW 2006, Australia

Omar Rota-Stabelli, Department of Biology, University College London, Darwin Building, Gower Street, London WC1E 6BT, UK

Greg W. Rouse, Scripps Institution of Oceanography, University of California San Diego, 9500 Gilman Drive #0202, La Jolla, CA 92093, USA

Suzanne V. Saenko, Evolutionary Biology Group, Institute of Biology, Leiden University, PO Box 9516, 2300 RA Leiden, The Netherlands

Andrew B. Smith, Department of Palaeontology, The Natural History Museum, Cromwell Road, London SW7 5BD, UK

Erik A. Sperling, Department of Geology and Geophysics, Yale University, PO Box 208109, New Haven, CT 06520, USA

Billie J. Swalla, Box 351800, 24 Kincaid Hall, Biology Department, Un iversity of Washington, Seattle, WA 98195–1800, USA

Maximilian J. Telford, Department of Biology, University College London, Darwin Building, Gower Street, London WC1E 6BT, UK

Kristin Tessmar-Raible, Developmental Biology Unit, European Molecular Biology Laboratory, Heidelberg, 69117, Germany

Introduction

The year 2009 is an important one for evolutionary biologists, encompassing the 200th anniversary of the publication of Lamarck's *Philosophie Zoologique*, the 200th anniversary of Darwin's birth and the 150th anniversary of the publication of *On the Origin of Species*; the Darwin bicentennial comes hot on the heels of the tercentenary of the birth of Linnaeus (2007), the original excuse for the meeting that assembled the authors of this book.

While the anniversaries of Linnaeus, Lamarck, and Darwin are clearly of particular significance to evolutionary biologists, it is not hard to identify other contemporary anniversaries marking the beginnings of research topics that have had a major impact on the content of chapters in this book. The year 2009 sees the 100th anniversary of Walcott's discovery of the Cambrian fossils of the Burgess Shales, the 25th anniversary of the discovery (twice) of the homeobox, and marks 21 years since the first analysis of metazoan phylogeny using small subunit rRNA sequences was published by Field *et al.* (1988) and, whilst the great significance of a 21st birthday might be questioned, we note that the original Field *et al.* publication appeared on 12 February—Darwin's birthday.

The passage of time, its punctuation by notable events, the significance of individuals, and the passing on of experience and knowledge are of course all notable components of evolution itself, with the important additions of selection and diversity. Our selection of authors for this volume was by no means random, as we had crossed paths with them in the past either personally (as collaborators or colleagues), or through their publications and presentations at scientific meetings.

The central questions of how animals originated and how they diverged and radiated to become the diverse forms they are today are of sufficient interest to engage a varied group of scientists using an equally broad variety of approaches. More importantly, in spite of the problems so far encountered, recent history suggests that much can be revealed about animal evolution and that the resolution of key branching points in the tree of life is indeed achievable. Our choice of authors, then, was further guided by the need to sample diversely across taxa, disciplines, and over various scales of perspective (time, level of biological organization), whilst providing an overview of the key elements that make up modern studies of animal evolution through an understanding of their genomes, fossils, and interrelationships. In this volume we chose to promote dialogue between systematists, palaeontologists, and evolutionary developmental biologists, reflecting our own interests but also, we believe, an area where collaboration is driving a greater understanding of animal evolution.

Fossils are in a unique position to provide additional characters for the resolution of phylogenies, polarization, and ordering of character transformations and provide the time and ecological background for the evolution of key novelties. Graham Budd (Chapter 1) explores the nature and beginnings of the animal fossil record, and considers in particular the recent findings of fossil embryos and other key forms, the incongruence between molecular and palaeontological estimates of the time of origin of major clades, and the nature and significance of events around the Cambrian. Finally, Graham considers the evidence implicating oxygen as a potential engine driving the Cambrian explosion of animal diversity.

Employing the latest Bayesian methods for estimating divergence times from molecular data, Kevin Peterson and colleagues (Chapter 2) consider the vagaries of estimating divergence times from the fossil record. They conclude that available data satisfy the notions of a Cambrian explosion of metazoans but indicate that the ecological and evolutionary fuses were set with the emergence of the Bilateria in the Ediacaran.

The origins of multicellularity are the focus of attention of Nicole King's team, led by Scott Nichols (Chapter 3), with a consideration of last common metazoan ancestors by means of a comparison of shared features, including patterns of gene expression, particularly in the diploblasts. Finally, they consider the emergence of the eumetazoan epithelium that, arguably, provided the means by which complex specialized organ systems subsequently evolved (Schmidt-Rhaesa, 2007).

The characteristics of the last common ancestor of the Bilateria, the so-called 'Urbilateria', are of great current interest and, in addition to the study of fossils, there are two approaches being employed to reconstruct this animal. The first is to attribute to Urbilateria the shared characteristics of the protostomes (Lophotrochozoa and Ecdysozoa) and the deuterostomes. The second approach is to look directly at the extant members of what may be an even earlier branch to consider their biology as a clue to the nature of bilaterian ancestors. Andreas Hejnol and Mark Martindale (Chapter 4) draw upon their recent studies of gene expression in the basally branching acoel *Convolutriloba longifissura* to consider gastrulation and, in particular, the relationships between the openings of the alimentary canal and the blastopore, details that may define major subdivisions of the Bilateria.

As Rudy Raff (Chapter 5) reminds us, the origins of bilaterian animal body plans are not only about adult forms. Understanding animal evolution is as much about revealing and explaining the evolution of ontogenies. Reviewing evidence from gene expression of patterning genes, phylogeny, morphology, and palaeontology, Raff argues that many larval features may have arisen independently, with new features emerging as adult bilaterian-expressed genes being co-opted for use in pelagic larvae. With the appearance of larval forms, the forces of evolution had an entirely new set of life forms (developmental stages) to act upon.

Following in the wake of recent new phylogenomic data sets involving hundreds of genes, Gonzalo Giribet and colleagues (Chapter 6) tackle the interrelationships of the Lophotrochozoa, or the Spiralia as they prefer to call them, and consider the newest of the new animal phylogenies, whilst highlighting the anomalies, inconsistencies, and persistent gaps in both gene sampling and morphological and developmental character coding.

Studies of individual organ systems provide a particular insight into evolutionary patterns and processes. Drawing from elegant comparative studies, Detlev Arendt and co-authors (Chapter 7) reveal complex similarities between the patterning of the central nervous system of a protostome (the annelid worm *Platynereis dumerilii*) and that of the chordates. They conclude that the protostome/deuterostome ancestor already had a centralized rather than diffuse nervous system patterned in this way and suggest that the diffuse nervous system in hemichordates is therefore a derived characteristic.

Charting the recent emergence of a major new clade, the Ecdysozoa, and the battles for and against accepting its validity is an exercise in understanding modern animal evolutionary studies. Max Telford and colleagues (Chapter 8) show that the Ecdysozoa are here to stay and discuss the new interpretation of morphological characters suggested by their interrelationships.

One major clade that has persisted since before the molecular revolution in phylogenetics is the Deuterostomia but its membership and their interrelationships, particularly the placement of enigmatic fossils and worms, has provided a vibrant forum for interdisciplinary studies. With the wealth of information from genomes, development, and morphology, this mixture of invertebrates and vertebrates has posed a considerable number of problems for scientists integrating independent data sets, as Andrew Smith and Billie Swalla explain (Chapter 9).

Chris Lowe (Chapter 10) describes studies comparing development in the chordates with those of

Xenambulacrarian hemichordates. Despite significant differences in morphology, the degree of conservation of gene expression patterns is striking, and Lowe alludes to the confidence and insight gleaned from establishing ancestral gene networks as a basis for understanding homology. These at least provide a sound basis for interpreting the diversity of forms that such gene networks have given rise to.

The study of animal evolution is beset by problems in all shapes and forms. From a systematic perspective, particular taxa have risen to become problematic in themselves; interpreting their biology in order to glean statements of homology, placing them in a phylogeny, or reconciling their biology with their phylogenetic placement. Such Problematica are the subject of the contribution by Ronald Jenner and Tim Littlewood (Chapter 11), who review the kinds and causes of problematic taxa amongst the invertebrates, whilst attempting to formulate some possible solutions for dealing with them in time.

Denser taxon and character (particularly gene) sampling is widely heralded as the means by which more accurate phylogenies can be resolved, yet Nicolas Lartillot and Hervé Philippe (Chapter 12) explain how the power of phylogenomics can only be harnessed properly by employing improved models of molecular evolution. Regardless of method of analysis, their contribution shows that novel relationships require a biological explanation. Of course, evolutionary signal from molecular data does not stem uniquely from nucleotides and amino acid sequences of individual genes. Jeff Boore and Susan Fuerstenberg (Chapter 13) look at the prospects for comparing entire genomes, from their constituent molecules to the biochemical and developmental pathways they control, for providing suites of new characters for phylogenetic reconstruction.

As evolutionary biologists become more involved in choosing taxa for genome characterization, or even characterizing genomes in their own laboratories thanks to second-generation sequencing technologies, it is clear that the depth of sampling needed to build and interpret the new animal phylogeny is currently thin. Richard Copley

(Chapter 14) asks where in the genomes do the phenotypic differences between animal taxa arise? Notwithstanding the paucity of current taxon sampling that requires us to consider existing model laboratory organisms as exemplars of metazoan diversity, it seems clear that the more we know about comparative genomics the more we can reveal about function across the genome. Copley argues that to understand fully the differences and similarities between genomes, it is necessary to go well beyond catalogues of shared genes. Instead, it is an understanding of the interactive components that link genotype with phenotype that will allow genomic studies to contribute to what might be construed as a return to organismal biology in its modern sense, where entire animals are viewed in a comparative evolutionary context integrating all available evidence.

Meanwhile, somewhat in contrast to the 'more genes' approach to phylogenetics, Erik Sperling and Kevin Peterson (Chapter 15) show that microRNAs, small, ubiquitous, non-coding regulatory genes, have the power to resolve phylogenies across the animal tree of life and argue that their unique properties make them the new characters of choice.

Andrew Peel (Chapter 16) addresses questions of the evolution of novelty in the insects, looking at the evolution of long- versus short-germ development in the holometabolous insects. One major conclusion is that developmental modes are not fixed in stone and have evolved both divergently and convergently in the insects. Morphology and developmental genetic networks can effectively become decoupled; one result of which is that attributing homology to developmental features based on common gene expression can be misleading.

The diversity of non-model systems is steadily increasing. Patrícia Beldade and Suzanne Saenko (Chapter 17) describe one such system, the butterfly *Bicyclus anynana*, and their approach for studying one striking aspect of these butterflies, their wing eyespots. The finding that evolutionary novelties such as wing eyespot development have involved the redeployment of genes from well-understood pathways involved in other diverse aspects of patterning in fruitflies gives one line of promise for the inclusion of many more diverse taxa as model laboratory organisms for evo-devo research.

In the final chapter we take the opportunity to consider some of the major steps made in the study of animal evolution, and the remaining hurdles that have a chance of being overcome in the next few decades. With new data come new perspectives and with new technologies comes renewed enthusiasm. Each generation has the opportunity to build on the successes and insights of those gone by and to contribute to further understanding the animal in us all.

Maximilian J. Telford
and D. Timothy J. Littlewood

PART I

Origins of animals

CHAPTER 1

The earliest fossil record of the animals and its significance

Graham E. Budd

The fossil record of the earliest animals has been enlivened in recent years by a series of spectacular discoveries, including embryos, from the Ediacaran to the Cambrian, but many issues, not least of dating and interpretation, remain controversial. In particular, aspects of the taphonomy of the earliest fossils require careful consideration before pronouncements about their affinities. Nevertheless, a reasonable case can be now made for the extension of the fossil record of at least basal animals (sponges and perhaps cnidarians) to a period of time significantly before the beginning of the Cambrian. The Cambrian explosion itself still seems to represent the arrival of the bilaterians, and many new fossils in recent years have added significant data on the origin of the three major bilaterian clades. Why animals appear so late in the fossil record is still unclear, but the recent trend to embrace rising oxygen levels as being the proximate cause remains unproven and may even involve a degree of circularity.

1.1 Introduction

The 'Cambrian explosion' is a popular term that refers to the period of profound evolutionary and environmental change that took place at the opening of the Phanerozoic some 540 million years ago (Ma). Although this set of events is multifaceted, it is associated primarily with the origin of animals in the fossil record. For over 150 years, an argument has raged about the reality of this event. Is it a genuine evolutionary event, or merely a sudden manifestation in the fossil record of evolutionary

processes that took place long before? Even if the fossil record of that time is accurately recording the unfolding of events in real time, the question of why the events took place then—and what the potential trigger was—has continued to be problematic. The Cambrian explosion itself has been much discussed (Gould, 1989; Conway Morris, 1998a, 2003a; Knoll and Carroll, 1999; Budd and Jensen, 2000; Knoll, 2003). Here I want to focus on three issues: the age of the earliest animal fossils, the continuing debate about their affinities, and finally, a critical examination of the most popular candidate for 'triggering' the explosion; the concentration of atmospheric oxygen.

Geologists as long ago as William Buckland (1784–1856) realized that a dramatic step change in the fossil record occurred at the base of what we now call the Cambrian. The apparent appearance in the fossil record of many animal groups with few or no antecedents caused Charles Darwin great trouble—indeed he devoted a substantial chapter of the *Origin* to this problem. Further insights were provided by the remarkable amount of work on North American faunas by Charles D. Walcott, who proposed that an interval of time, or the 'Lipalian', was not represented in the fossil record and/or did not preserve fossils and that the forms ancestral to the Cambrian taxa evolved during this time. However, the intense modern interest in the subject was probably sparked by the work of Whittington and colleagues in their redescriptions of the Burgess Shale (see below), together with Stephen Jay Gould's popular account of this work, *Wonderful Life*, published in 1989. In recent years, the attention paid to the youngest part of the Precambrian has led to the erection of the formal Ediacaran Period of *c.* 630–542 Ma (Knoll *et al.*, 2006), an interval that has been intensely scrutinized for its bearing on the origin of the animals.

1.2 Fossil evidence for the origin of animals: the state of play

The classical fossil evidence for the early evolution of animals consists of several sources: trace fossils, the Ediacaran biota from just before the beginning of the Cambrian (Narbonne, 2005), the conventional Cambrian fossil record (Bengtson, 1992), and the Burgess Shale fauna (Briggs *et al.*, 1995). In recent years these data sources have been enriched by further important discoveries, especially new Cambrian exceptional faunas such as the Chengjiang fauna (Hou *et al.*, 2004) and indeed very substantial new discoveries from the Burgess Shale itself (Caron *et al.*, 2006; Conway Morris and Caron, 2007); the Doushantuo fossils from the Ediacaran period of the latest Precambrian (Xiao and Knoll 2000; Xiao *et al.*, 2007a; Yin *et al.*, 2007), and more Ediacaran discoveries, such as from Namibia, Newfoundland, and the White Sea (Grazhdankin and Seilacher, 2002; Narbonne, 2004). Outside the Cambrian, the Silurian Herefordshire fauna has also yielded some remarkable fossils that have had significant bearing on the origins of various animal clades (e.g. Sutton *et al.*, 2001a, 2002; Siveter *et al.*, 2007). The volume of data that the fossil record has brought to bear on the issue of the origin of the animals has thus notably increased in recent years, explaining the exciting dynamism that currently characterizes the field. Nevertheless, even a casual observer would note that few of these new inputs have been without controversy; with high-profile publications regularly attracting published responses or critical reviews. The undeniable difficulties surrounding these data can be attributed to several causes: (1) an often incomplete understanding of the taphonomy (i.e. the complete set of preservational processes surrounding the production of the final fossil), a lack that has often led to interpretation of ambiguous fossils in a pre-conceived manner; (2) the continuing discussion of how Cambrian taxa should be classified; and (3) various dating problems.

1.2.1 The Doushantuo Formation and its taphonomy

The processes that convert a living organism into a mineralized or organically preserved fossil are far from being fully understood; nevertheless, at least some understanding of them is essential if fossils are to be successfully interpreted (Butterfield *et al.*, 2007). Nowhere has this been more important than the evaluation of the various exceptional faunas around the Precambrian/Cambrian boundary. Of particular recent interest has been the Doushantou Formation (Fm) of South China. This *c.* 250-m thick sequence of siliciclastic, phosphatic, and carbonate rocks has yielded exceptionally preserved putative examples of algae, acritarchs, and metazoan embryos and adults including sponges and a bilaterian (Chen *et al.*, 2000; Xiao and Knoll 2000; Yin *et al.*, 2001, 2007; Chen and Chi 2005; Dornbos *et al.*, 2006; P. J. Liu *et al.*, 2006; Tang *et al.*, 2006; Xiao *et al.*, 2007a). However, nearly all of these fossils have proved highly controversial. One reason for this is clear: the Doushantuo Fm has been dated to well before the beginning of the Cambrian, and thus these fossils would undoubtedly include the oldest animals in the record (but see below).

The preservation in phosphate of many Doushantuo fossils leads to problems of disentangling primary morphology from subsequent taphonomic overprints (Bengtson and Budd, 2004; Xiao *et al.*, 2000). As a result of such concerns, some of the more extravagant claims, such as that the Doushantuo biota includes representatives of bilaterians and deuterostomes, do not currently stand up to scrutiny. Nevertheless, and not withstanding attempts to provide alternative bacterial-affinity explanations (Bailey *et al.*, 2007a,b; Xiao *et al.*, 2007b), the Doushantuo fossils remain as convincing embryos. Even if the presence of phosphatized embryos is accepted though, a significant amount of disagreement over their precise dating remains, which, in its extreme, would extend the range of animals down to close to the opening of the Ediacaran at around 630 Ma, while at the other extreme the Doushantuo fossils may not significantly pre-date the oldest Ediacaran fossils at around 565 Ma.

1.2.2   Towards a chronology of the latest Precambrian

The later stages of the Precambrian are marked by glaciations of global extent that show up in the record as, for example, a series of tillites (lithified glacial sedimentary rocks of mixed composition that are formed as the result of movement by ice). These glaciations have been suggested to be evidence for the so-called 'snowball earth', i.e. intervals of time when the earth was effectively deep-frozen. The amelioration of conditions after these glaciations has been suggested to be a key factor in the rise of the animals (Runnegar, 2000), although the mechanism for such a direct causality remains largely obscure. The interval of time known informally as the 'Cryogenian', from approximately 850–630 Ma is marked in the Australian record by two distinct ice intervals: the 'Sturtian' and the 'Marinoan' (Kennedy et al., 1998). These glacial intervals can be correlated with glacial deposits elsewhere in the world, such as in China (Zhou et al., 2004). In addition, a further short-lived glacial interval, the 'Gaskiers', known primarily from Newfoundland (Eyles and Eyles, 1989), has been dated to be c. 580 Ma. Correlating Precambrian glacial intervals worldwide is difficult at best, largely because of the lack of accurate biostratigraphical control, and the task is complicated by the technical problems associated with the various types of absolute radiometric dating. As a result, a number of minority views exist, such as that the Marinoan and Gaskiers glaciations are identical (based on dating in Tasmania; Calver et al., 2004). As far as the dating of the Doushantuo Fm goes, the glacial rocks below can be dated to close to 635 Ma, and the base of the overlying Dengying Fm, has been dated to 551 Ma (Condon et al., 2005). A complicating factor is that the well-preserved fossils of the Doushantuo Fm are known not from its type locality but from the Weng'an locality, which consists of a much shorter (c. 40 m thick) section made up largely of two phosphoritic units (Dornbos et al., 2006).

An additional aid to dating comes in the form of chemostratigraphy, especially using δC^{13}, which suggests that the Doushantuo Fm is marked by three negative δC^{13} excursions: one at the base, associated with the so-called 'cap carbonates' that directly overlay the glacial deposits; one in the middle, and one near the top (Condon et al., 2005). It has often been thought that the excursion towards the top is associated with the Gaskiers glaciation, in which case the age of the Doushantuo Fm would range from about 580–635 Ma. The significance of these dates is that all of the Doushantuo fossils would pre-date the oldest of the famous Ediacaran fossils such as *Dickinsonia* etc., and thus would provide an independent record of animal life during a period of time for which no large-body fossils or trace fossils are known. Indeed, the overlying Dengying Fm does yield Ediacaran-type fossils, which could be said to support this contention. However, some recent work has questioned this view, suggesting that it is the middle δC^{13} in the Doushantuo Fm that corresponds to the Gaskiers Fm (despite the lack of other evidence for glaciation in the type area; in the Weng'an section, a definite break in the sequence at this point could be correlated with glacially related drop in sea-level). This would constrain the age of the upper Doushantuo Fm units to lie within about 551 and 580 Ma (Dornbos et al., 2006), and, as it is this interval that is thought to yield the animal fossils, these fossils could plausibly be regarded as being of a similar age to the Ediacaran assemblages. In order for this model to be correct, some of the published radiometric dates for the Doushantuo Fm would have to be incorrect (Barfod et al., 2002), but given the care required to interpret whole-rock radiometric dates, this possibility cannot simply be ruled out.

More recently, the claim has been made that at least one of the enigmatic acanthomorphic (i.e. spinose) acritarchs (see Figure 1.2), which are normally assigned to protist groups such as the green algae and dinoflagellates, are actually the hulls of diapause animal eggs (Yin et al., 2004, 2007). Although the fossil in question, *Tianzhushania*, is known to contain embryos only in the upper part of the Doushantuo Fm, it ranges down to very close to the base, and thus to 630 Ma or so. The claim would be that the oldest animal fossils of the Doushantuo Fm, dating back to just after the Nantuo glaciation (i.e. the Chinese glacial deposits normally correlated with the Marinoan) are of this age, a time that pre-dates the first Ediacaran fossils by some 60 million years, as well as the more conservative molecular clock estimates for the divergence of the bilaterians.

Despite the obvious uncertainties, the most reasonable interpretation of the data is thus that embryo-forming animals of some sort had evolved by just after Marinoan time; that sponges and presumed other animals had started to emerge by 580 Ma at the latest; and that the Ediacaran biotas are likely to be a little younger than the Doushantuo embryos. The upshot of the new data is that much more convincing evidence exists in the fossil record for an origin of the animals considerably before the Cambrian than it did 10 years ago (Budd and Jensen, 2000), with an inferred documented fossil origin of the entire clade being datable to just after 635 Ma—a significant result (see Figure 1.1 for summary).

If animals had already evolved at this time, why is it that the rest of the record does not correlate with it—why are there no macro body fossils and no (generally accepted) trace fossils? The answer to this question, which on the face of it seems to directly contradict predictions (Budd and Jensen, 2000) that no animals existed significantly before the first good trace fossils at around 555 Ma, may hinge on what sorts of organisms these embryos represent. Given their relatively unusual development, with large numbers of cell divisions taking place without any sign of gastrulation or epithelial formation, it has been suggested that they are from stem-group metazoans; i.e. from organisms more basal than any living animals including sponges (Hagadorn et al., 2006). Given that such an organism, lacking muscles and other features of the more derived bilaterians, would be unlikely readily to form either body or trace fossils, such an assignment is consistent with the hypothesis that bilaterians emerged later, close to the Precambrian–Cambrian boundary.

What is perhaps more surprising is the general lack of convincing sponge spicules from the Precambrian (Gehling and Rigby, 1996; Brasier

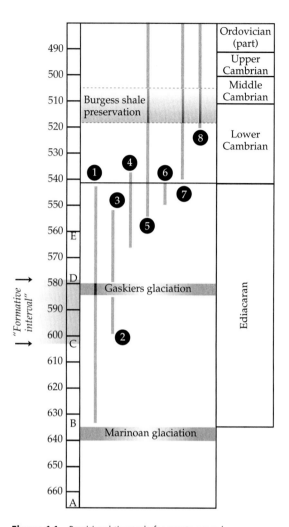

Figure 1.1 Provisional timescale for events around the Precambrian/Cambrian boundary: 1, range of large, acanthomorphic 'Ediacaran' acritarchs; a genus that contains metazoan-like embryos is found from close to the bottom of their range just above the Marinoan glaciation rocks; 2, possible range of Doushantuo embryos and cnidarian-like fossils according to Barfod et al. (2002); 3, possible range of the same according to Condon et al. (2005), which if correct is uncertain, but the former is favoured here; 4, the 'Ediacaran' biota; 5, trace fossils; 6, *Cloudina* and *Namacalathus*; 7, classical small shell fossils. The letters correspond to key dated points in metazoan evolution in Peterson and Butterfield (2005) based on minimum evolution: A, origin of crown-group Metazoa; B, total-group Eumetazoa; C, Crown-group Eumetazoa; D, crown-group Bilateria (here equivalent to Protostomia plus Deuterostomia); E, crown-group Protostomia. The 'formative interval' during which distinctive bilaterian features were assembled according to this dating is marked by arrows.

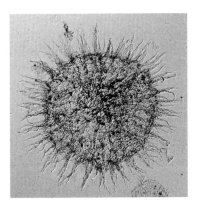

Figure 1.2 The Ediacaran acanthomorphic acritarch *Tanarium pluriprotensum* from the Tanana Formation, in the Giles 1 drillcore, Officer Basin, Australia (×75). At least some Precambrian acanthomorphic acritarchs may be the eggs of animals. Courtesy of S. Willman.

et al., 1997; Li *et al.*, 1998); given that living sponges may be paraphyletic, stem-group metazoans should be spiculate, and spicules should thus be present very early on. If this absence is genuine as opposed to taphonomic (Pisera, 2006), then the suggestion would be that crown-group metazoans did not evolve until close to the beginning of the Cambrian. Finally, the suggestion that mineralized sponge spicules are convergent within sponges and thus need not characterize basal metazoans at all (Sperling *et al.*, 2007) is one other obvious way around this impasse. In summary there seems to be no good reason as yet to place the radiation of the bilaterians significantly before the first decent trace fossils at around 555 Ma or so; although evidence for the presence of metazoans of some sort considerably before this point seems to be hardening.

1.2.3   The status of Cambrian fossils

The years in which the various exceptionally preserved fossils from the Cambrian were viewed as representing a plethora of body plans essentially unrelated to the extant phyla have now passed, closing a tradition that dates back many decades (Nursall, 1959). Nevertheless, the significance of Cambrian taxa continues to be hotly debated.

Many workers (Smith, 1984; Runnegar, 1996; Budd and Jensen, 2000) have seen the apparently more bizarre forms as lying in the stem groups of the extant phyla, thus providing a critical basis for understanding the origin of animals as we see them today. A corollary of this view is that the modern phyla, strictly considered, often do not emerge until some time after the classical Cambrian explosion, with the major radiation associated with the beginning of the succeeding Ordovician period being at least as important for the emergence of modern body plans. Conversely, other writers have seen this definition of the phyla to be overly legalistic (Knoll, 2003; Valentine, 2004; Briggs and Fortey, 2005). Part of the disagreement is a relatively uninteresting one over terminology (i.e. when does a fossil qualify to be straightforwardly called an 'echinoderm' for example), but this surface dispute conceals a more important issue, which is over the actual timing of the establishment of the extant body plans. The difficulty partly arises because it is often hard to say with confidence when the 'crown node' that subtends the crown group has been attained.

Although a crown group can be defined empirically by the results of a cladistic analysis and supported by the synapomorphies at its base, the practical issues involved in placing any particular fossil within or outside it can be difficult to resolve. In order to identify membership of the crown group, it is necessary not only to show that the fossil in question possesses the set of plesiomorphic features associated with its crown group (i.e. it lies within at least the total group), but also that it possesses at least one apomorphy of one of the clades included within it. This task is complicated by the phylogeny of the ingroups often being uncertain (with a good example being provided by the molluscs), and by the possibility of apomorphic character states for an in-group of the phylum actually being plesiomorphic, but lost in the sister group to the group that now possesses them (Budd and Jensen, 2000). Such a possibility is locally unparsimonious, but may not be globally so. The net result of these two effects is that although a taxon may look rather similar to a crown-group member of its phylum, its crown-group status cannot be confirmed. For example, there are several Cambrian taxa such as *Ottoia* that closely resemble crown-group priapulids, but nevertheless lie near the top of the stem group (Wills, 1998; Dong *et al.*, 2004). As a result, the formal origin of the crown group is pushed much later, probably to the Carboniferous. It nevertheless seems fairly clear that the basic features of the priapulids had been attained early on in their history, and the formal origin of the crown group, although strictly accurate for determining the point at which the modern-day body plan appeared (at least as measured by the fossil record), is a trivial event compared with the evolution of the basic form that took place in the Cambrian (Fortey *et al.*, 1996).

As a result of this potentially misleading application of the stem-group/crown-group distinction, the alternative idea of extending the phylum concept phylogenetically backwards to incorporate formal members of the upper stem group is favoured by several writers. While this proposal has its merits, it has obvious drawbacks too: for example, how unlike the modern phylum does a stem-group member need to be before being excluded from the group and other problems associated with the erection of subjective paraphyletic groups? Although the formal stem-group concept is of course paraphyletic, it at least has the advantage of being objectively so, with the arbitrary but empirical datum point being survivorship to the modern day.

Despite the objections to the idea then, the use of the stem-group/crown-group distinction has the advantages of providing a fixed and objective measure that is comparable across phyla for when the modern clade can be formally recognized in the fossil record; and it does not seem that any alternative proposals, which may rely on a subjective or even misleading assessment of what an 'important' character for a particular clade is, offer much of an advance.

Although I wish to continue to defend the use of the stem-group/crown-group distinction as being of phylogenetic and historical importance, the reasons for its rejection are certainly worth serious consideration. It is clear that by Burgess Shale time in the Middle Cambrian (i.e. about 507 Ma) most extant clades had appeared, and many of them had members that were, at least in a broad sense, recognizable as being similar to the crown group

itself. A far from inclusive list might include the arthropods, molluscs, priapulids, and brachiopods: Cambrian life is different, but not alien. Therefore, although the recognition that crown groups in general evolve late allowing some body-plan evolution to be 'smeared upwards' into the Palaeozoic (Budd and Jensen, 2000), the latest Proterozoic and earliest Cambrian were still highly significant periods during which the classical features of the phyla as we see them today were partly, or even largely, assembled. These include the origins of segmentation, the coelom, blood-vascular and nervous systems, and nephridia. A major unsolved question of course is whether or not these features evolved once, at the base of the bilaterians, and were then subsequently lost as the early bilaterians radiated into niches where they were functionally pointless (e.g. in the meiofauna) or whether they evolved independently several times under strong convergent pressure (Conway Morris, 2003a,b), often using a similar developmental toolkit to do so. This question would be resolvable by a much more precise phylogeny than is currently available and must be regarded as a major aim of the investigation of the origins of the animals.

1.2.4   Recent advances in basal animal palaeontology

Study of the fossil record of the oldest animals has been enlivened by the molecular evidence that the extant sponges are paraphyletic, with the Calcarea being more closely related to the Eumetazoa than the other sponges (Cavalier Smith *et al.*, 1996; Borchiellini *et al.*, 2001; Peterson and Butterfield, 2005). Such a finding gives hope of understanding the vexed issue of what sort of organism the eumetazoans (i.e. cnidarians plus bilaterians) evolved from. Indeed, the notable discovery (Botting and Butterfield 2005) that the Burgess Shale sponge *Eiffelia* (Figure 1.3a) possesses both hexaradiate spicules (characteristic of calcareans) and tetraradiate spicules (characteristic of hexactinellids), suggests that the fossil record may allow at least some insights into the earliest transitions in animal evolution; insights that complement those, not uncontroversially, already attained for other basal groups such as the ctenophores

(a)

(b)

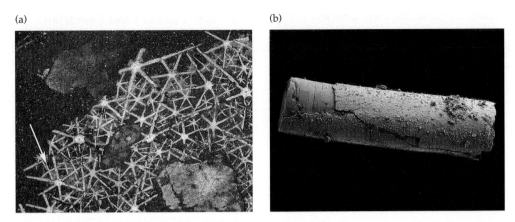

Figure 1.3 Basal metazoan fossils. (a) *Eiffelia globosa* from the Middle Cambrian Burgess Shale (ROM 57023; ×5.0) (Botting and Butterfield 2005). As well as the prominent hexaradiate spicules typical of calcarean sponges, rows of smaller, hexactinellid-like tetraradiate spicules are also visible (arrowed). Courtesy of N. J. Butterfield. (b) A section of *Sinocyclocyclicus guizhouensis* from the Ediacaran Doushantuo Formation (Xiao *et al.* 2000; Liu *et al.* 2008) (×130). This small, branching tabulate fossil has been interpreted as being a potential stem-group cnidarian. Courtesy of Shuhai Xiao.

(Conway Morris and Collins, 1996; Shu *et al.*, 2006; Chen *et al.*, 2007).

Conversely, the early record of cnidarians remains uncertain. Whilst some of the Ediacaran taxa, especially the fronds and disc-shaped fossils, have classically been interpreted as cnidarians, the interpretation of these remains in doubt, partly because of profound differences in growth patterns (Antcliffe and Brasier, 2007). On the other hand, the Doushantuo Fm has once again generated material of interest, especially the branching tabulate form *Sinocyclocyclicus* (Xiao *et al.*, 2000), material that, although potentially algal, does display a set of characters that are compatible with cnidarian affinities (Figure 1.3b). Thus, sponges, cnidarians, and potentially ctenophores are all known from Precambrian strata. These findings, and the continuing general lack of convincing evidence for bilaterians until just before the beginning of the Cambrian, all suggest that 'radiate' animals were radiating during the Ediacaran, and that the Cambrian explosion itself represents the radiation of bilaterians (Benton and Donoghue, 2007).

The status of the classical Ediacaran fossils, such as *Spriggina*, *Dickinsonia*, etc., remains highly uncertain. While new well-preserved material from, for example, Namibia (Dzik, 2002), the White Sea area

(Zhang and Reitner, 2006), and Newfoundland (Narbonne, 2004), has added information about their morphology, and has led to claims that some of these taxa can now be accommodated in the stem or crown of groups such as the ctenophores (Dzik, 2002; Shu *et al.*, 2006; Zhang and Reitner, 2006), the ever-present problem of taphonomy, particularly acute in the ediacarans, means that any claims for certain affinities must be treated with a great deal of caution. Nevertheless, given the potentially pivotal morphology, molecular development, and phylogenetic position of the ctenophores (Yamada *et al.*, 2007), the developing leitmotif of ctenophore-like morphologies in the late Ediacaran might just be pointing towards substantial advances in the area of understanding stem-group eumetazoans and bilaterians in the not too distant future.

As for the bilaterians themselves, new data continue to be generated from the major Cambrian lagerstätten such as new collections of Burgess Shale material, including a remarkable reassessment of the previously highly problematic *Odontogriphus* as a stem-group mollusc (Caron *et al.*, 2006) and other taxa claimed as stem-group lophotrochozoans, such as the 'halwaxiids' (Conway Morris and Caron, 2007). It should also be noted that advances in photographic techniques (Bengtson, 2000) have also greatly increased the ease with which data from Burgess Shale fossils can be extracted.

Persistent claims are made that members of the Ediacaran biota should be considered to be bilaterians, especially the clearly complex *Kimberella* from the White Sea area (Fedonkin and Waggoner, 1997), a claim that has been revitalized by the discovery of the molluscan affinities of the rather similar *Odontogriphus* from the Burgess Shale (Butterfield, 2006; Caron *et al.*, 2006).

The conventional record, too, continues to provide provocative material, including recent evidence that the highly enigmatic but very widespread tommotiids from the Lower Cambrian are lophophorate relatives (Holmer *et al.*, 2002, 2008; Skovsted *et al.*, 2008). Thus, the fossil record is providing important new data that might go some way to help resolving one of the most vexed problems in animal phylogeny, the relationships between the protostomes. The Chengjiang fauna has also provided material (controversially) relevant to the origins of the deuterostomes, with the vetulicolans being claimed as a new deuterostome phylum, as well as several craniates and even vertebrates that significantly extend their record back in time (Chen *et al.*, 1995,1999; Shu *et al.*, 1996b, 1999, 2001b, 2003a,b). The final major group of bilaterians, the ecdysozoans, although widely accepted, remains controversial in terms of in-group relationships (Budd, 2002; Waloszek *et al.*, 2005a, 2008). The arthropods are now largely accepted to have arisen via a rather heterogeneous group of lobopods, although the exact root is far from agreed on (Budd, 1996; Zhang and Briggs, 2007). In addition to the arthropods, the cycloneuralians have come under some scrutiny, especially since the description of stem-group scalidophoran embryos from the Lower Cambrian (Budd, 2001a; Dong *et al.*, 2004; Donoghue *et al.*, 2006a; Maas *et al.*, 2007). Nevertheless, the intriguing question of what sort of animal the last common ancestor of the ecdysozoans was like (Budd, 2001b) remains currently unanswered, at least from the fossil record, although the suspicion that the earliest lobopods such as *Aysheaia* (Whittington, 1978) are more or less priapulids on legs is not one that is easily shaken off (Dzik and Krumbiegel, 1989).

1.3 What caused the Cambrian explosion?

The age-old question of why animals evolved when they did, and not, for example, 500 million years before, continues to trouble researchers. In one sense, the question is trivial, in the same way that the question of 'why did the First World War take place in the 20th, rather than the 16th century?' is. Clearly, whenever this event took place, the same question could be asked, and the general answer of 'many other things had to happen first' is not as vacuous as it at first appears. Nevertheless, a serious point remains: is there a set of conditions that had to be in place in order to release animal evolution? When David Nicol reviewed the question 40 years ago (Nicol, 1966) he listed some of the hypotheses that had been put forward up to that point, some of which now seem quaint, for example the view that life evolved on land and only reached the sea, and thus could become readily fossilizable, in the Cambrian, or that animals adopted a more sluggish mode of life to which hard parts were appropriate—the exact opposite of the more normal 'arms race' view of the development of hard parts prevalent today (Vermeij, 1993; Bengtson, 2002). In all of these ideas a more or less constant factor has been the level of oxygen.

1.3.1 Did oxygen fuel an explosion?

Without any doubt, the most popular candidate for causing—or allowing—the Cambrian explosion is a rise in oxygen levels at the end of the Proterozoic (Nursall, 1959). In one sense, this is an excellent choice of causal agent, as no-one will ever know exactly what oxygen levels were like during that period of time. Nevertheless, the perennial debate about oxygen levels in the Proterozoic has been sharpened recently by intense interest in the subject, which has led to many more data and a clearer picture of the rise of oxygen levels in the atmosphere.

The oxygen debate is not, in this context, simply about what levels of oxygen pertained at various times in the Proterozoic, interesting and intractable though that question has proved (Lambert and Donnelly, 1991; Runnegar, 1991; Canfield and Teske, 1996; Thomas, 1997; Canfield *et al.*, 2007). It is narrowly focused on the following two questions: (1) when did oxygen levels first permanently rise high enough to permit the evolution of any sort of metazoan? and (2) did low oxygen levels limit the fossilization potential of early metazoans? The second question has widely been considered to have a positive answer, and to provide the explanation for why animal fossils do not appear in the record until just before the Cambrian, despite some evidence that they evolved hundreds of millions of years before this. It is also worth stating at the outset that the whole oxygen level debate has recently been rejuvenated and enriched by the realization that oxygen is merely one component in a multifactorial geochemical setting. In order to understand oxygen levels, one must consider other elements as well, such as sulphur (Shen *et al.*, 2002; Canfield *et al.*, 2007), as well as temperature and salinity (Knauth, 2005). Further, oxygen availability is also of importance: oxygen levels in the atmosphere, deep oceans, and shelves may all have significantly different values (Canfield, 1998; Holland, 2006).

1.3.2 Why is oxygen important?

Simply put, oxygen plays a critical role in animals for two reasons. The first is that it is necessary for certain important biosynthetic pathways; and the second is that it is used in energy production, i.e. in aerobic respiration. If it is the limiting factor in either of these roles, then low oxygen levels might have impeded animal evolution. These two cases can be called the *biosynthetic* argument and the *physiological* argument, respectively.

The biosynthetic argument: oxygen as a structural necessity
The most famous argument for the importance of oxygen in animal evolution was put forward by Towe (1970). It relies on a quirk of the genetic code that has interesting evolutionary consequences. The genetic code allows the assembly of 20 amino acids into first polypeptides and then proteins. However, some important amino acids are synthesized *after* this translation of the code. The classical example is the formation of one of these, hydroxyproline, from the encoded proline (technically proline is an imino acid). Lysine is also, on a much smaller scale,

modified in a similar way, although the roles of the two respective end products, hydroxyproline and hydroxylysine, are very different.

Although post-translational modification of proteins and indeed their component amino acids is common (for example, removing the methionine start codon; acetylation or phosphorylation), the specific hydroxylation of proline raises the question of why hydroxyproline is not included in the primary code. The obvious answer is that this amino acid was not used by organisms when the genetic code first originated. The reason is clear, because the complex—but well understood—biosynthetic pathway by which hydroxyproline is formed involves free oxygen. Simplistically, as life certainly evolved under very low-oxygen conditions, it was not possible to synthesize hydroxyproline at this time. It was only when oxygen levels had risen to a certain level that hydroxyproline synthesis was possible. The strategy for identifying the rise in oxygen levels, then, is to identify where hydroxyproline synthesis evolved in the history of life. To put it another way, when oxygen levels were lower than this critical value, hydroxyproline synthesis would have been impossible, and thus any clades that now synthesize hydroxyproline could not have existed. As usual, this simple picture needs some careful qualification. How do we know when hydroxyproline appeared phylogenetically, and could it have arisen more than once? Did hydroxyproline synthesis require free oxygen when it *first* arose? What *is* the present day phylogenetic distribution of hydroxproline?

It should be noted that hydroxyproline has long been considered of interest in the debate about animal origins because of its critical importance in one of the most important of all animal proteins, collagen, although this is not the only biochemical pathway requiring oxygen (Catling *et al.*, 2005). Collagen is an unusual protein because it is made of repeating units of a few amino acids, including hydroxyproline. The hydroxyproline is produced *in situ* by modification of proline after the basic protein structure has already formed. It seems that this process needs free oxygen levels to be about 1% of present-day atmospheric levels (PAL). The process also requires ascorbic acid (i.e. vitamin C); collagen defects are the reason behind the symptoms of scurvy. Towe's reasonable argument, therefore was that animals, all of which produce collagen, could not have evolved before oxygen reached 1% PAL (Towe, 1970). This value, which can be called the Towe limit, sets an absolute limit to the conditions in which animals could have evolved, and provides the basic mechanism by which animal evolution could have been controlled by oxygen. However, although the focus has largely been on animals, the scope of this enquiry must be broadened, because animals are not the only organisms to produce either hydroxyproline or, indeed, collagen.

The phylogeny of hydroxyproline and collagen synthesis; primitive or convergent?

Collagen itself has long been thought of as one of the (few) classical synapomorphies that uniquely unite metazoans (Conway Morris, 1998b). As a result, its discovery in fungi (Celerin *et al.*, 1996) came as a considerable surprise. The fungal collagen is considerably different from any of the many types known from animals, and may have arisen by convergence. Nevertheless, this discovery supplies intriguing evidence for an animal–fungus sister-group relationship, one that has gained some support in recent years (Wainwright *et al.*, 1993). The exciting discovery of several collagen-domain-encoding genes in the recently published genome of the choanoflagellate *Monosiga brevicollis* (King *et al.*, 2008; cf. Ruiz-Trillo *et al.*, 2008) adds further critical information for tracing the evolution of structural proteins that have until recently been thought to be metazoan autapomorphies.

Although plants do not synthesize collagen, they do produce various proteins such as extensins and pherophorins that are an important component of the cell wall; i.e. they are structural proteins similar in function and form to collagen, and are found in both the algae and higher plants (Sommer-Knudsen *et al.*, 1998; Hallmann, 2006). Further, in such molecules, hydroxyproline and hydroxy-lysine are produced in a very similar way as in collagen—by *in situ* post-translational modification of proline and lysine—and in both animals and plants the enzyme prolyl 4-hydroxylase is used for the former. The overall similarity in synthetic pathway, structure, and function suggests that collagen

and extensin synthesis grew out of a common biochemical pathway that also utilized hydroxyproline and hydroxylysine, and that this pathway may be shared by fungi. A simple survey thus suggests that these multicellular eukaryotes share hydroxyproline synthesis, and indeed synthesis of a shared family of structural proteins.

The above suggests that the multicellular eukaryotes arose in an environment that allowed the hydroxylation of proline and lysine, and was thus above the Towe limit, a view supported by modelling of the atmosphere (Canfield, 1998; Holland, 2006). More controversially, the 'fungi first' model of eukaryote relationships (Martin *et al.*, 2003) suggests that hydroxyproline synthesis was a *basal* eukaryotic feature. If true, the important result would be that there would be no level of atmospheric oxygen that would permit eukaryotic evolution in general but not animal evolution in particular; they share the same requirements. If one is searching for a general mechanism for delaying animal evolution after the appearance of eukaryotes then this appears not to be it. On these grounds alone, oxygen levels must have been at least 1% of PAL ever since the origin of the eukaryotes, which is almost certainly over a billion years ago (Butterfield *et al.*, 1990).

1.3.3 Oxygen requirements, size and shape

One of the first efforts at relating oxygen levels to the rise of animals was made by Nursall (1959), who argued that large animals, with their concomitant complex ecologies, were simply not possible in a low-oxygen environment. Not until oxygen levels had risen above a certain level would large animals be able to evolve, especially equidimensional animals such as brachiopods. For many people (Runnegar,1982c; Knoll, 2003; Shen *et al.*, 2008) this is the best reason for why the Cambrian explosion happened when it did. But does this argument hold water?

Most animals are able to generate energy using either aerobic or anaerobic metabolic pathways; with glycolytic anaerobic respiration generating about two ATP molecules, and aerobic respiration (citric acid cycle plus oxidative phosphorylation) about 36. Although the citric acid cycle does not directly rely on free oxygen, it does not take place under anaerobic conditions. As there is no free oxygen to act as the final electron acceptor, the intermediates all along the oxidative phyosphorylation chain remain in a reduced state. As a result, the chain stops functioning; and the build up of end products means (via Le Chatelier's principle) that the citric acid cycle halts. However, glycolysis can still occur, leading to a build-up of pyruvate and a small amount of ATP (two or three molecules).

So much for the basic biochemistry, the broad outline of which is extremely well known. What is less well known, however, is the presence of a variety of anaerobic respiratory pathways in metazoans. Some metazoans, for example, are able to ferment as well as produce lactic acid (from glycolysis) or opines, formed by condensing pyruvic acid with an amino acid. Simply because the yield of ATP from glycolysis is so low, some invertebrates also have pathways that avoid glycolysis. For example, some invertebrates use a fumarate electron transport system that increases the yield of ATP to up to eight molecules (Fenchel and Finlay, 1995; McMullin *et al.*, 2000; Tielens *et al.*, 2002), including some parasites such as the nematode *Ascaris*, but also free-living invertebrates such as the mussels *Mytilus* and *Geukensia* and the polychaete *Arenicola*. Whilst most of the sources of electrons in these various anaerobic pathways are organic, it is also now known that these invertebrates can switch to sulphide oxidation in hypoxic conditions, a presumed remnant of eukaryotic diversification in a high-sulphide Proterozoic ocean (Theissen *et al.*, 2003; *contra* Anbar and Knoll, 2002). Thus, respiratory mechanisms, and the mitochondria that generate them, are surprisingly diverse: as they do not fall into obvious well-defined clades, it is likely that they have been convergently derived (Tielens *et al.*, 2002).

The presence of diverse, mitochondrial based anaerobic respiratory pathways, even in metazoans, is significant because it suggests that at least some metazoans can (and could have) function well even under low-oxygen conditions, producing more energy than from mere glycolysis, thus somewhat undermining the claim that rising oxygen levels were a pre-requisite for animal evolution. Furthermore, not all organisms require the same amount of oxygen; as might be expected, mode of life is a critical variable too. Organisms that swim generally need more oxygen than those that walk, dig, or just open their valves. Floating in the water column requires least energy of all, of course (Pörtner, 2002). For some of the more 'athletic' extant organisms, such as squid, it seems that swimming takes place close to their functional and environmental limits. They manage to achieve this 'life on the edge' by living in a very stable environment, i.e. the open ocean. Although they use both aerobic and anaerobic respiratory pathways, they maximize aerobic respiration and eventually tire during anaerobic activity, as levels of free ATP drop.

For other organisms, though, a very different picture emerges. Sipunculans, for example, that typically spend their time slowly digging in low-oxygen mud, produce identical metabolites whether they work under oxygen-rich conditions or artificially induced oxygen-deficient ones, suggesting, with other evidence, that almost all muscular activity of any significance takes place anaerobically (Pörtner, 2002). In other words, low oxygen levels hardly affect such organisms because almost everything they do requires them to switch to anaerobic respiration in any case. Only resting respiration is performed aerobically, i.e. mitochondria are fuelled by oxygen when the organism is not actually doing anything. As might be expected, such organisms have an extreme tolerance to anaerobic respiration, and do not seem to tire while performing their constant but low-energy functions. Such modes of life may provide important clues to how early animal life functioned in the early Cambrian.

Despite the arguments above, a powerful case has recently been put forward that high oxygen levels are indeed necessary to sustain a complex ecology, based partly on the ability of organisms to produce a large body size and generate enough energy to sustain complex food chains (Catling *et al.*, 2005). While their calculations do not seem to take into account the possibility of fumarate-based anaerobic pathways that would generate more ATP than glycolysis, their points must be well taken, especially given the demonstrable effect on body size and mineralization that low-oxygen environments have on organisms today

(Rhoads and Morse, 1971). However, to return to the two questions asked at the beginning of the section, the real question is not whether or not, for example, hard parts could be formed under low-oxygen conditions, but rather if any sort of animals could evolve in such a regime that would generate a fossil record? Given that minute trace fossils and indeed body fossils, as in the Doushantuo Fm, can be preserved in the record, it seems that the answer must be yes.

Although animals can obviously persist in, and have distinctive adaptations for, low-oxygen environments, there can similarly be little doubt that high oxygen levels (perhaps 10% PAL) are really necessary for modern food chains and large animals to flourish. Determining when this level was first permanently achieved in the atmosphere must remain an important goal for studies of the late Precambrian and the influence of environment on animal evolution. Thus there are considerable uncertainties about Proterozoic oxygen levels and the physiological requirements of early animals; after all, recent animals living in low-oxygen environments usually possess distinct adaptations that it would be reasonable to suppose were also possessed by early animals. As a result, the current fashion for rising oxygen levels being the primary engine for the Cambrian explosion may not be as well founded as is sometimes assumed. A perfectly reasonable alternative is that the Cambrian explosion is an ecological event (Butterfield, 1997; Budd and Jensen, 2000; Marshall, 2006), consisting largely of a cascade of knock-on effects that emerged from multicellularity and mobility; although it would be misleading to identify these milestones as stand-alone 'key innovations', embedded as they are in a nexus of other morphological and ecological changes (e.g. Budd, 1998). Thus although the undoubtedly important suite of geological changes that took place during the close of the Proterozoic and opening of the Phanerozoic form the essential backdrop against which the Cambrian explosion must be viewed, it still seems reasonable to regard them as scenery rather than the major players in the Cambrian drama.

1.4 Conclusions

Although the dating of the early animal fossils remains problematic, a reasonable case for stem-group animals existing shortly after the Marinoan glaciation at around 630 Ma can be made. Nevertheless, evidence for mobile bilaterians does not appear in the record until around 555 Ma, just before the beginning of the Cambrian; a time that is no longer wildly inconsistent with some molecular clock estimates (e.g. Aris-Brosou and Yang, 2003; Peterson *et al.*

1.5 Acknowledgements

Discussions with many colleagues including Simon Conway Morris, Nick Butterfield, Sören Jensen, and Sebastian Willman are gratefully acknowledged, as are the providers of images as detailed in the figure captions. This work was supported by the Swedish Research Council (VR) and the Swedish Royal Academy of Sciences (KVA).

CHAPTER 2

The Ediacaran emergence of bilaterians: congruence between the genetic and the geological fossil records

Kevin J. Peterson, James A. Cotton, James G. Gehling, and Davide Pisani

Unravelling the timing of the metazoan radiation is crucial for elucidating the macroevolutionary processes associated with the Cambrian explosion. Because estimates of metazoan divergence times derived from molecular clocks range from quite shallow (Ediacaran) to very deep (Mesoproterozoic), it has been difficult to ascertain whether there is concordance or quite dramatic discordance between the genetic and geological fossil records. Here, using a range of molecular clock methods, we show that the major pulse of metazoan divergence times was during the Ediacaran, consistent with a synoptic reading of the Ediacaran macrobiota. These estimates are robust to changes in priors, and are returned with or without the inclusion of a palaeontologically derived maximal calibration point. The two historical records of life both suggest, therefore, that although the cradle of the Metazoa lies in the Cryogenian, and despite the explosion of ecology that occurs in the Cambrian, it is the emergence of bilaterian taxa in the Ediacaran that sets the tempo and mode of macroevolution for the remainder of geological time.

2.1 Introduction

Accurately and precisely elucidating the times of origin of the metazoan phyla is central to unravelling the causality and biological significance of the Cambrian explosion. Despite the fact that the Cambrian explosion is geologically obvious (Darwin, 1859), it has long been argued that this same geological record, because of its incompleteness, might be misleading when considering metazoan origins (Runnegar, 1982b). As Runnegar recognized, a second 'fossil record', the genetic record written in the DNA of all living organisms (Runnegar, 1986), could be used to test hypotheses about the completeness of the geological record (Peterson et al., 2007), and initial attempts at using a molecular clock strongly suggested that metazoans had a deep and cryptic Precambrian history (Runnegar, 1982a, 1986; Wray et al., 1996; reviewed recently by Conway Morris, 2006). Nonetheless, several palaeontologists have cogently argued that the fossil record provides positive evidence for the absence of early Neoproterozoic and Mesoproterozoic animals, casting doubt on the veracity of these molecular clock estimates (Budd and Jensen, 2000, 2003; Jensen et al., 2005; Conway Morris, 2006; Butterfield, 2007). Comparisons between the genetic and geological fossil records of early animal evolution, as currently understood, therefore suggest that either the geological record is woefully incomplete or there is something seriously awry with our reading of the genetic record (Bromham, 2006).

To explore the apparent incongruity between the known fossil record and the very deep estimates of metazoan diversification suggested by molecular clocks, Peterson and colleagues (Peterson et al., 2004; Peterson and Butterfield, 2005) assembled the largest novel data set yet, showing that the two records were remarkably concordant: metazoans originated at some time during the Cryogenian, and bilaterians arose during the Ediacaran. Part of the reason for the prior discrepancy concerned the use of vertebrate divergence times. Peterson *et al.* (2004) discovered that there was an approximately two-fold rate reduction across the vertebrate protein-coding genome as compared with the three invertebrate lineages examined (echinoderms, molluscs, and insects), consistent with total genome comparisons between vertebrates and dipteran insects (Zdobnov *et al.*, 2002). However, some studies using invertebrate calibrations have also inferred divergence times consistent with a cryptic Precambrian history of the Metazoa (Pisani *et al.*, 2004; Regier *et al.*, 2005), suggesting that the two-fold rate reduction across the vertebrate genome is only one of many factors influencing the estimation of divergence times (Linder *et al.*, 2005; Peterson and Butterfield, 2005).

In addition, Peterson *et al.*'s (2004) estimates and explanations were called into question by several workers, notably Blair and Hedges (2005) who argued that Peterson *et al.* (2004) used palaeontologically derived calibration points as maxima as opposed to minima, which generated spuriously shallow estimates for metazoan divergences. Although false, as Peterson *et al.* (2004) stated explicitly (see also Peterson and Butterfield, 2005), this criticism highlights an important issue surrounding the use of molecular clocks, namely the proper way to incorporate calibration points into molecular clock analyses (Benton and Donoghue, 2007). Recent experimental analyses have shown the importance of numerous, well-constrained calibration points for returning accurate and precise estimates of divergence times, and thus highlighting the need to pay particular attention to this aspect of molecular dating (Roger and Hug, 2006; Hug and Roger, 2007). Nonetheless, difficulties arise when incorporating fossils into a molecular clock analysis: unlike the establishment of a minimal divergence time for any two taxa, which is simply the first appearance of either one of the taxa, estimating the maximum divergence time is much more difficult (Benton and Donoghue, 2007). Two types of maxima have been proposed: a 'hard'

maximum proposes an absolute value for the oldest possible date of divergence, whereas a 'soft' maximum treats a divergence as having some chance of being older than a particular date, depending on a probability distribution used to describe the calibration point (Hedges and Kumar, 2004; Yang and Rannala, 2006; Benton and Donoghue, 2007).

Most modern molecular clock methods (e.g. Sanderson, 1997, 2002; Thorne *et al.*, 1998; Drummond *et al.*, 2006) allow one to constrain, as well as fix, the age of a calibration point, so that every fossil divergence can be defined using a minimum and a maximum. This is a significant improvement over older molecular clock approaches (e.g. Kumar and Hedges, 1998) because it allows the integration of palaeontological uncertainty in the estimation of divergence times. However, most existing molecular clock software including r8s (Sanderson, 2004) and Multidivtime (Thorne and Kishino, 2002), do not distinguish between hard and soft maxima, instead treating all maxima as hard. The difficulty here is that divergence times estimated with uncertain maxima treated as if they were hard can only give minimum estimates for the true divergence time, as the soft maxima might significantly underestimate the true age of the calibration points. Nonetheless, Drummond *et al.* (2006) have now implemented Bayesian relaxed molecular clock methods (in the software package BEAST) where soft maxima can be properly modelled using a probability distribution, and can thus be older than their proposed fossil date.

Here, we set out to explore the diversification of animal phyla in the Neoproterozoic using alternative relaxed molecular clock approaches while testing the stability of our results to the choice of different priors and to the deletion of palaeontologically derived maxima, and modelling soft maxima using the most appropriate probability distribution. We find that although deleting or relaxing maxima tends to push divergence times toward the past (as expected), all estimates are largely congruent between algorithms. We conclude that a synoptic reading of both the geological *and* genetic fossil records demonstrates that the Ediacaran was the time of major diversification of most higher-level animal taxa and set the stage for Phanerozoic-like macroecology and macroevolution.

2.2 Materials and methods

2.2.1 Molecular characters

All taxa are taken from Sperling *et al.*, (2007) where a concatenated alignment of seven different house-keeping genes, for a total of 2059 amino acid positions and 44 representative species (see Peterson *et al.*, 2004, and Peterson and Butterfield, 2005), was analysed using Bayesian methods (MrBayes 3.1.2; Ronquist and Huelsenbeck, 2003). See Sperling *et al.*, (2007) for further details.

2.2.2 Molecular clock calibration

Calibration points were taken from Peterson *et al.*, (2004) except for the minimum estimate for crown-group Eleutherozoa, which was adjusted from 475 to 480 million years ago (Ma) in light of the discovery of a slightly older asterozoan (Blake and Guensberg, 2005), and the minimum and maximum for crown-group Diptera was taken from Benton and Donoghue (2007). Several new maxima and minima were incorporated into this analysis. First, the maximum for the origin of crown-group echinoderms was set at 520 Ma, the first appearance of stereom in the fossil record. Because stereom is a highly distinctive skeletal material, and its presence in numerous stem-group taxa (Smith, 2005) demonstrates that stereom is a total-group echinoderm character, it must have evolved before the origin of the crown group. Second, this same time point also sets the minimum for Ambulacraria (Echinodermata + Hemichordata), as echinoderms appear before hemichordates in the rock record (Budd and Jensen, 2003). Third, because ambulacrarians are characterized by the possession of four to six coeloms in each animal (Peterson *et al.*, 2000; Smith *et al.*, 2004), and because coeloms cannot pre-date the first appearance of bilaterian traces (Budd and Jensen, 2000, 2003), the first appearance of traces sets the maximum age for crown-group Ambulacraria at approximately 555 Ma (Martin *et al.*, 2000; Jensen *et al.*, 2005). Fourth, the maximum for the origin of Gastropoda + Bivalvia is the first appearance of skeletons in the fossil record, about 542 Ma (Amthor *et al.*, 2003; Bengtson, 1994). Fifth, the maximum for the origin of crown-group

demosponges is the first appearance of demosponge-specific biomarkers (McCaffrey *et al.*, 1994; Love *et al.*, 2006; see Peterson *et al.*, 2007 for discussion) sometime after the Sturtian, *c.* 657 Ma (Kendall *et al.*, 2006). Finally, the maximum for the origin of crown-group Eumetazoa, which was only used in the BEAST analyses, is argued to be 635 Ma based on palaeoecological observations (Peterson and Butterfield, 2005).

Newly incorporated minima include the first appearance of arthropod traces 525 Ma (Budd and Jensen, 2003) as a minimum for the divergence between insects and the priapulids, the first appearance of medusozoans 500 Ma (Hagadorn *et al.*, 2002) as a minimum for the origin of the crown-group Cnidaria, and the first appearance of vertebrates 520 Ma as the minimum for the origin of crown-group chordates (Benton and Donoghue, 2007).

2.2.3 Molecular estimates of divergence times

Molecular estimates of divergence times were obtained using the Bayesian methods of Thorne *et al.*, (1998), as implemented in Multidivtime (Thorne and Kishino, 2002), and Drummond *et al.*, (2006) as implemented in BEAST version 1.4.2 (Drummond and Rambaut, 2006). All divergence times were calculated assuming the tree topology of Figure 2.1, which was derived from MrBayes (see above and Sperling *et al.*, 2007). For the Multidivtime analyses, branch lengths were estimated using the Estbranches program from the Multidivtime package, under the WAG model. For BEAST analyses, starting branch lengths were assigned arbitrarily to match the constraints imposed by the calibrations.

For the Multidivtime analyses a prior age for the root node (in our case the Fungi–Metazoa split) must be specified. We assumed a 1000 Ma prior for this node (Knoll, 1992; Douzery *et al.*, 2004) and then tested whether this choice affected our results by performing analyses in which this age was changed to 100 Ma (standard deviation (SD) = 500 Ma), 1500 Ma (SD = 500 Ma), and 2000 Ma (SD = 750 Ma). Other priors used in Multidivtime analyses include the mean and standard deviation of the prior distribution at the root node, and 'Minab'

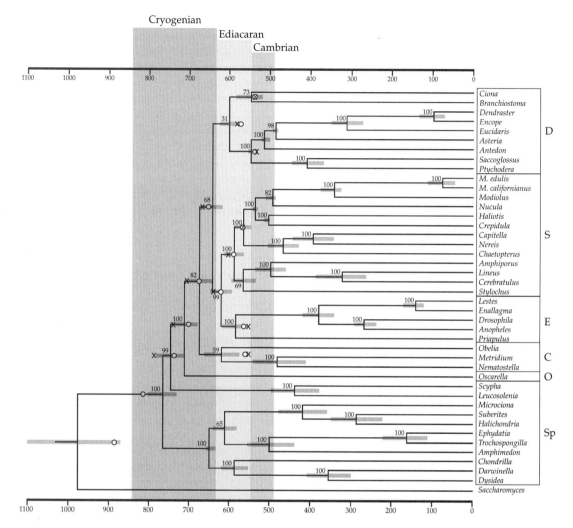

Figure 2.1 The timing of the metazoan radiation according to the molecular clock. The figure shows the phylogenetic tree for 41 metazoan taxa rooted on the yeast *Saccharomyces cerevisiae* as determined by Bayesian phylogenetic analysis (see text). The deuterostomes are shown in box D, spiralian protostomes in box S, ecdysozoan protostomes in box E, cnidarians in box C, the homoscleromorph *Oscarella* in box O, calcisponges in box C, and demosponges in box Sp. The nodes of the tree are positioned according to the optimum as determined from the Bayesian autocorrelated method of Thorne *et al.* (1998), as implemented in the software package Multidivtime (Thorne and Kishino, 2002) using a root prior of 1000 Ma (SD 500 Ma). The 95% highest-probability-density (HPD) credibility intervals are shown in brackets. The white circles are the estimates for clades with internal calibration points as determined by the Bayesian algorithm BEAST (Drummond *et al.*, 2006) using uniform priors and an exponential rate distribution. Black Xs are the estimates using exponential priors and the same rate distribution. Note that much of the metazoan diversification occurs during the Ediacaran, which lies between the Cryogenian and the Cambrian.

(parameter for the beta prior on proportional node depth). The mean and standard deviation of the prior distribution of the rate at the root node were set to 0.039, as estimated from the data following the procedure outlined in the Multidivtime manual, and the effects that 100-fold changes to this parameter had on the results were assessed. The Minab parameter affects the distribution of the nodes through time—Minab values greater than 1 will cause the nodes to repel each other, while values less than 1 will cause the nodes to attract each other. This parameter was set to 1 for our analyses, but we assessed how changing the Minab parameter from 0.6 to 1.4 affected our results.

BEAST implements uncorrelated relaxed clock methods, which assume an overall distribution of rates across branches but do not assume that the rates on adjacent branches are autocorrelated. We used both the exponential and lognormal rate distributions with two different calibration schemes: one with hard maxima, in which most calibrations were treated as uniform priors on clade ages, and a second with only soft maxima, in which all calibrations were treated as exponential priors, with 95% of their density lying between the uniform maximum and minimum. In both schemes, the maximum at 635 Ma was treated as an exponential prior, with 90% of its density lying below 635 Ma, giving a 10% prior chance that this calibration point is incorrect. All other priors and operators were kept at default settings, except that all operators that alter the tree topology were disabled.

Ninety-five per cent highest-probability-density (HPD) credibility intervals are automatically calculated by Multidivtime, and were calculated using the program Tracer for the BEAST analyses. To test whether our priors dominated the posterior distribution, all our BEAST and Multidivtime analyses were also performed without data and the results obtained in these runs were compared with those obtained when the actual data were analysed.

2.3   Results

Molecular divergence times were estimated using the topology shown in Figure 2.1. Support for Cnidaria and Deuterostomia was low (67% and 33%, respectively), probably because of long-branch artefacts (Pisani, 2004) associated with *Ciona* and *Obelia* in particular (indeed the value for Deuterostomia goes to > 90% with the removal of *Ciona*), but given the clear monophyly of the phyla Chordata and Cnidaria, constraining these nodes should not generate spurious molecular divergence estimates. Most of the other nodes were strongly supported, including Calcispongia + Eumetazoa, and Eumetazoa, supporting the results of Peterson and Butterfield (2005), and *contra* the conclusions of Rokas and colleagues (Rokas *et al.*, 2005; see also Baurain *et al.*, 2007). Indeed, within Protostomia, for example, all but one node (*Stylochus* + Nemertea) have posterior probability values above 80%, and both Lophotrochozoa and Ecdysozoa, as well as Annelida + Mollusca, have clade credibility values of 100%. In addition, we find strong support for the node Homoscleromorpha + Eumetazoa, which indicates that there are at least three independent extant sponge lineages (Sperling *et al.*, 2007).

Using this topology as a constraint tree, divergence times were estimated using the Bayesian autocorrelated method of Thorne *et al.* (1998), as implemented in the software package Multidivtime (Thorne and Kishino, 2002). These Bayesian estimates are robust to changes in the age of the root prior as the estimates are essentially the same whether the age is 100 Ma (SD 500 Ma) or 2000 Ma (SD 750 Ma) (Table 2.1), suggesting that the age of the root prior is not biasing the analyses. Also, changing the value of Minab, or the mean rate of evolution of the root node, did not change our results (not shown). Running the analyses without data confirmed that our results were not dominated by our choice of priors (not shown). Because of the suggestion that fungi diverged from animals *c.* 1000 Ma (Knoll, 1992; Douzery *et al.*, 2004), was confirmed by all our analyses that did not assume a particular age for the root node, and in our Bayesian analyses performed assuming different prior root ages (100, 1500, and 2000 Ma) we used the values derived from the 1000 Ma (SD = 500 Ma) prior in Figure 2.1.

The removal of the deeper calibration point, namely the maximum age of 657 Ma for the origin of crown-group demosponges, resulted in increasing the estimate for the age of crown-group Metazoa by *c.* 18% (from 766 Ma to 904 Ma; Table 2.1). Nonetheless, the age for both crown-group Protostomia and crown-group Deuterostomia increased by only *c.* 4–5%, suggesting that the results derived with the use of this maximum are generally robust. Given its position in the tree, the geological depth of the divergence, and the unique nature of the evidence (biomarkers), this maximum is most likely adding both accuracy and precision to the clock estimates.

We next explored these same divergence times using the models implemented in BEAST (Drummond *et al.*, 2006). In general, the estimates

Table 2.1 Optima (maxima, minima) in millions of years derived from Multidivtime (M) and BEAST (B) analyses for five key metazoan divergences

Method	Metazoa	Eumetazoa	Bilateria	Protostomia	Deuterostomia
M-1000[1]	766 (803, 731)	676 (709, 645)	643 (671, 617)	619 (648, 594)	601 (625, 579)
M-100[2]	760 (798, 725)	672 (706, 642)	641 (669, 615)	618 (645, 592)	600 (624, 578)
M-2000[3]	774 (812, 739)	679 (712, 648)	645 (674, 619)	622 (649, 595)	602 (626, 580)
M-1000-D[4]	904 (997, 825)	743 (798, 694)	686 (727, 649)	653 (689, 619)	624 (655, 596)
B-UCEX Uniform[5]	815 (1621, 625)	676 (849, 579)	652 (764, 570)	620 (692, 556)	572 (614, 537)
B-UCEX Exp[6]	1067 (2358, 612)	707 (985, 581)	669 (870, 566)	638 (784, 556)	582 (695, 529)
B-UCLN Uniform[7]	891 (995, 640)	739 (822, 607)	699 (768, 588)	660 (715, 572)	640 (706, 559)
B-UCLN Exp[8]	953 (1093, 821)	779 (869, 694)	733 (808, 663)	688 (751, 629)	677 (746, 607)

[1] Age of the root prior is 1000 Ma (SD 500 Ma).
[2] Age of the root prior is 100 Ma (SD 500 Ma).
[3] Age of the root prior is 2000 Ma (SD 750 Ma).
[4] Age of the root prior is 1000 Ma (SD 500 Ma) and estimates are derived without considering the demosponge maximum of 657 Ma.
[5] Estimates derived using an exponential rate distribution and uniform priors.
[6] Estimates derived using an exponential rate distribution and exponential priors.
[7] Estimates derived using a lognormal rate distribution and uniform priors.
[8] Estimates derived using a lognormal rate distribution and exponential priors.

derived from BEAST using an exponential rate distribution and uniform priors (white circles in Figure 2.1) are similar to those derived from Multidivtime (Table 2.1). The analyses that use exponential priors are somewhat deeper than those that use uniform priors (black Xs in Figure 2.1), and those using a lognormal rate distribution are deeper than those derived from an exponential rate distribution (Table 2.1), presumably because the exponential distribution on rates is leading to greater autocorrelation between rates. Analyses without data again confirmed that the priors were not dominating the data (results not shown).

2.4 Discussion

2.4.1 Concordance between the genetic and geological fossil records

Here we have shown, using a variety of analyses and appropriately testing for biases that may have been introduced by the use of palaeontologically derived maxima, that the genetic fossil record strongly supports the notion that the diversificat-ion of metazoans in general, and bilaterian meta-zoans in particular, occurred during the Ediacaran Period, 635–542 Ma (Knoll *et al.*, 2004, 2006). How

do these molecular estimates compare with the known geological record? Macroscopic fossils of the Ediacara biota span the upper half of the Ediacaran Period, from 575–542 Ma (Grotzinger *et al.*, 1995; Martin *et al.*, 2000; Bowring *et al.*, 2003; Condon *et al.*, 2005). Because most of these fos-sils occur as soft-bodied impressions in relatively coarse-grained siliciclastic sedimentary rocks, a comprehensive array of palaeobiological inter-pretations of the Ediacara biota has been put forth. Nonetheless, a few taxa stand out as poten-tial candidates for affinities within the Metazoa. One taxon in particular, *Kimberella*, has generated much discussion as a possible triploblastic meta-zoan. It compares well in external form to mol-luscs (Fedonkin and Waggoner, 1997) and in a few cases an everted proboscis is preserved (Gehling *et al.*, 2005) that is inferred to contain a radula-like organ given the association between specimens of *Kimberella* (Figure 2.2a, asterisk, and Plate 1) and aligned sets of paired scratch marks (Figure 2.2a, arrows) (Gehling *et al.*, 2005). These finds sug-gest that *Kimberella* was preserved in place while grazing on substrate microbial mats (Seilacher, 1999; Gehling *et al.*, 2005). Given that we estimated the divergence between annelids and molluscs to

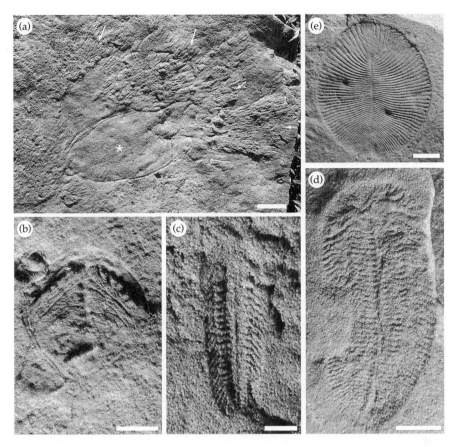

Figure 2.2 Putative ediacaran metazoans: (a) natural cast on bed base of *Kimberella* resting trace (asterisk) and *Radulichnus* radular feeding trace fans (arrows) (scale bar 1 cm); (b) *Dickinsonia costata* (scale bar 2 cm); (c) *Marywadea ovata* (scale bar 10 mm); (d) *Spriggina floundersi* (scale bar 10 mm); (e) *Parvancorina minchami* (scale bar 1 cm). See also Plate 1.

be *c.* 570 Ma (Figure 2.1), it is possible, if not probable, that *Kimberella* is allied somehow with modern molluscs.

What about other higher-level clades? Our estimates suggest that arthropods diverged from priapulids *c.* 575 Ma, suggesting that stem-group panarthropods (Nielsen 2001) should be present in Upper Ediacaran rocks. Interestingly, several taxa compare favourably with a panarthropod interpretation. For example, large specimens of *Parvancorina* show lateral structures originating on either side of the medial ridge that might be characterised as appendages (Figure 2.2b). In fact, in external form, *Parvancorina* bears a striking resemblance to the unmineralized, kite-shaped Cambrian arthropod

Skania (Lin *et al.*, 2006). *Spriggina* (Figure 2.2c) also preserves large numbers of appendage-like structures, and still others like *Marywadea* (Figure 2.2d) show apparent cephalic branching structures that resemble digestive caecae in arthropods. Importantly (see below), all of these taxa were no larger than 10 cm in maximum dimension (Gehling, 1999; Fedonkin, 2003) (see Figure 2.2), and appear simultaneously with the first demonstrable trace fossils (Jensen *et al.*, 2005). The absence of arthropod scratch marks (Seilacher, 1999), though, is not too worrisome given that such traces would demand the presence of sclerotized appendages to cut through the ubiquitously present microbial mats, a character not necessitated by the presence of stem-group panarthropods, or even deeply nested stem-group arthropods, in Ediacaran-aged sediments.

Indeed, the distinct possibility remains that this fauna preserves numerous stem-group forms ranging from basal triploblasts up through basal ecdysozoans, spiralians, and possibly even deuterostomes. Given the enigmatic nature of some very prominent taxa like *Dickinsonia* (Figure 2.2e), a taxon that appears capable of some form of limited motility (Gehling *et al.*, 2005), a position for *Dickinsonia* within total-group Eumetazoa is not out of the question. In fact, mobile but saprophytic feeding without the use of a gut would be compelling evidence that some form of ectomesoderm pre-dates the advent of endoderm.

2.4.2 Discordance between the genetic and geological fossil records

Of course, many others have addressed these questions using a similar approach, and it is worth comparing our results not only against the fossil record but also with other molecular clock estimates as well. They compares well with some molecular analyses, notably Aris-Brosou and Yang (2002, 2003), Peterson *et al.* (2004), and Peterson and Butterfield (2005), all of whom argued that the last common ancestor of protostomes and deuterostomes evolved not more than 635 Ma. However, Blair and Hedges (2005) have recently argued for much deeper divergences, based on a series of penalized likelihood (Sanderson, 2002) analyses using r8s (Sanderson, 2004) in which every calibration point was treated as a minimum. They suggested that the divergence between ambulacrarian and chordate deuterostomes was 896 Ma (with the 95% confidence interval spanning from 832 to 1022 Ma). They further argued that the divergence between hemichordates and echinoderms was 876 Ma (725, 1074 Ma), and the origin of crown-group echinoderms was 730 Ma. Finally, they estimated that the divergence between starfish and sea urchins was 580 Ma. Unfortunately, their results are most likely spurious because, as Sanderson (2004) pointed out, r8s cannot converge on a unique solution if only minima are used to calibrate penalized likelihood

analyses, which is supported by the fact that their estimate for the origin of a mineralized, coelomate taxon like crown-group Echinodermata precedes their appearance in the fossil record by some 200 million years.

Of course, neither the genetic nor the geological fossil record has a monopoly on historical accuracy, and as much as molecular evolutionists need to keep in mind the relevant palaeontological data, palaeontologists need to keep in mind estimates derived from molecular clocks (Donoghue and Benton, 2007). For example, Budd and Jensen (2000, 2003) argued that bilaterians could not have had an extensive Precambrian history, as suggested by almost all molecular clocks, as the trace fossil record, and the inferred morphology of these animals, is not consistent with an origin much before 555 Ma. They observed that possession of coelom(s) and a blood-vascular system (BVS) are inconsistent with a meiofaunal origin, as tiny organisms would have had no need for a transport system like the BVS, and are only consistent with a size large enough to be detected in the geological record. In general, we agree with their arguments, and use their insights to set a maximum age for crown-group Ambulacraria (see above).

However, the same argument cannot be extended to many other parts of the bilaterian tree. *Contra* Budd and Jensen (2000), there is no evidence for homology of coeloms either between protostomes and deuterostomes or even within both protostomes and deuterostomes. Because the coelom is, by definition, just a mesodermally lined cavity (Ruppert, 1991a; Nielsen, 2001) the possession of the space itself cannot be used as an argument for homology. Instead, topological similarity must be used, and when it is, it strongly suggests homology, for example, within Ambulacraria (Peterson *et al.*, 2000; Smith *et al.*, 2004), but not homology between any other higher taxa (Nielsen, 2001; Ruppert, 1991a). Thus, outside of Ambulacraria, the trace fossil record cannot be used to set a maximum for most bilaterian divergences. In fact, the small size of many putative Ediacaran bilaterians (Figure 2.2), and the fact that acoel flatworms are now recognized as the sister group to the remaining bilaterians (Baguñà and Riutort, 2004; Peterson *et al.*, 2005; Sempere *et al.*, 2007), is consistent with an argument that small size and absence of a coelom are primitive for Bilateria. This then removes the final obstacle to a pre-555 Ma origin for Bilateria, which is consistent with both the appearance of many different bilaterian lineages in the Ediacaran (Figure 2.2) and the molecular clock (Figure 2.1).

But despite the presence of many different stem-group taxa, the Ediacaran is still a transitional ecology, with these organisms confined to a two-dimensional mat-world. This stands in dramatic contrast to the Early Cambrian where the multitiered food webs that typify the Phanerozoic were established with the eumetazoan invasion of both the pelagos and the infaunal benthos (Butterfield, 1997, 2001; Vannier and Chen, 2000, 2005; Dzik, 2005; Peterson *et al.*, 2005; Vannier *et al.*, 2007). Hence, although the Ediacaran is an apparent quantum leap in ecological complexity as compared with the 'boring billions' that characterize earth before the Ediacaran, it is still relatively simple when compared with the Cambrian. The Cambrian was yet another quantum leap in organismal and ecological evolution, and which thus stands as the transition interval between the 'Precambrian' and the Phanerozoic (Butterfield, 2007). Whether it was triggered by the introduction of eumetazoans, as argued by Peterson and Butterfield (2005), by the introduction of mobile, macrophagous triploblasts, as is suggested by our analyses reported here (Figure 2.1), or by some other factor or combination of factors, remains to be more fully studied through continued exploration of the relevant rock sections throughout the world, and continued improvements in molecular clock methods.

2.5 Conclusions

Both the genetic and geological fossil records, each with their own inherent biases and artefacts, are largely congruent with one another, and for historical disciplines congruence of independent data sets is the strongest argument one can make for historical accuracy (Pisani *et al.*, 2007). Our analyses suggest that while the cradle of metazoan life occurred in the Cryogenian, and the explosion of metazoan ecology occurred in the Cambrian, it is the emergence of bilaterians in the Ediacaran that established the ecological and evolutionary rules that have largely governed earth's macrobiota for the remainder of geological time.

2.6 Acknowledgements

KJP was supported by the National Science Foundation; JAC was supported by an Irish Research Council for Science, Engineering and Technology Post-doctoral Fellowship; JGG is supported by the Australian Research Council Discovery Project (DG0453393), the ARC Linkage Project LP0774959 including the South Australian Museum and Beach Petroleum Pty Ltd, and the SA Museum Waterhouse Club. We would like to thank P. Donoghue (University of Bristol) for his usual perspicacity, two anonymous reviewers for their helpful comments on an earlier version of this chapter, and T. Littlewood (Natural History Museum, London) and M. Telford (University College London) for inviting us to contribute to this volume. Finally, KJP would like to thank all of the students who have come through the lab and contributed data to this project, and the South Australian Museum for a very enlightening visit.

CHAPTER 3

Genomic, phylogenetic, and cell biological insights into metazoan origins

Scott A. Nichols, Mark J. Dayel, and Nicole King

Over 600 million years ago (Ma), the first multicellular metazoans evolved from their single-celled ancestors. Although not recorded in the fossil record, the earliest events in metazoan evolution can be inferred by integrating findings from phylogenetics, genomics, and cell biology. Comparisons of choanoflagellates (microeukaryote relatives of metazoans) with sponges (the earliest known metazoans) reveal genetic innovations associated with metazoan origins. Among these are the evolution of the gene families required for cell adhesion and cell signalling, the presence of which catalysed the evolution of multicellularity and the functions of which have since been elaborated to regulate cell differentiation, developmental patterning, morphogenesis, and the functional integration of tissues. The most ancient tissues—differentiated epithelia—are found in sponges and evolved before the origin and diversification of modern phyla.

3.1 Introduction

Metazoans are one of evolution's most dramatic experiments with multicellularity, and yet we know surprisingly little about their origins. The fossil record provides no insight into the biology of the unicellular ancestors of metazoans. Indeed, the relatively abrupt appearance of fossils attributable to modern metazoan phyla over the c. 80 million year span of the Cambrian radiation obscures the sequence of metazoan phylogenesis. Nonetheless, by merging phylogenetics and comparative genomics with comparative cell biology, we can infer some of the earliest events in metazoan evolution.

Metazoan origins required at least two innovations: the evolution of simple colonies of equipotent cells followed by the organization and integration of cell function and behaviour within an 'individualized' organism (Pfeiffer and Bonhoeffer, 2003; King, 2004; Michod, 2007). Both of these phenomena required regulated cell signalling, adhesion, and differentiation mechanisms, the origins of which directly address fundamental questions about the evolution of multicellularity.

Metazoan multicellularity evolved independently from that of all other macroscopic lineages. In fact, although unicellular life predominates in all considerations of total biomass and biodiversity, at least 16 separate transitions to multicellularity have occurred during the history of eukaryotic life (King, 2004). The imprint of these separate origins can be seen at the level of phylogenetics, comparative genomics, and comparative cell biology. In the following discussion, we review how insights from choanoflagellates and sponges have begun to illuminate some of the earliest events in metazoan history, the origin of multicellularity, and the differentiation of epithelial tissues.

3.2 Phylogenetics: are there any 'living models' of early metazoan ancestors?

3.2.1 The case for choanoflagellates

Choanoflagellates and sponges have classically been thought to straddle the evolutionary divide between metazoans and their unicellular ancestors. Choanoflagellates, a group of heterotrophic microeukaryotes, originally captured the attention of cell biologists for their striking similarity to the 'feeding cells' (choanocytes) of sponges (James-Clark, 1866; Saville-Kent, 1880–82; see Figure 3.1). This resemblance was first noted by Henry James-Clark in 1866, prompting one of two interpretations: either that sponges and choanoflagellates are derived from an ancestral species that used choanoflagellate-like cells to capture bacterial prey, or that these cell types are only superficially similar and have evolved independently. Subsequent molecular phylogenetic analyses and comparative genomic data have

firmly established that sponges are metazoans, that metazoans are monophyletic, and that choanoflagellates are sister to metazoans (Burger *et al.*, 2003; Medina *et al.*, 2003; Steenkamp *et al.*, 2006; Moreira *et al.*, 2007; King *et al.*, 2008; Ruiz-Trillo *et al.*, 2008; see Figure 3.2). Furthermore, mitochondrial genome data and species-rich phylogenetic analyses demonstrate that choanoflagellates are not derived from metazoans, but instead represent a distinct lineage that evolved before the origin and diversification of metazoans (Lavrov *et al.*, 2005; Steenkamp *et al.*, 2006; Rokas *et al.*, 2005; Jimenez-Guri *et al.*, 2007).

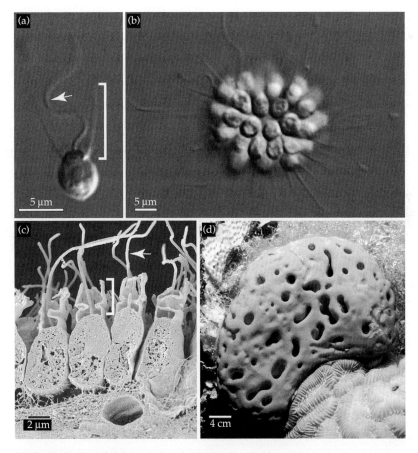

Figure 3.1 Similarities between choanoflagellates and sponge choanocytes. Choanoflagellates are heterotrophic microeukaryotes that use an apical flagellum to swim and to generate water flow, thus trapping bacterial prey on an actin-filled microvillar collar. Some choanoflagellates, like the species of *Proterospongia* shown here, have both unicellular (a) and colonial (b) life-history stages. The ultrastructural and functional characteristics of choanoflagellates are conserved in the feeding cells of sponges, choanocytes (c, adapted from Leys and Eerkes-Medrano, 2006), despite vast differences in overall organismal morphology (d). Arrows indicate flagellum and braces indicate the collar of individual cells.

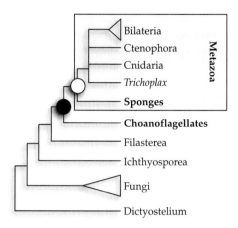

Figure 3.2 Phylogenetic relationships among metazoans and their close relatives. The preponderance of available evidence supports sponges as the earliest branching metazoan lineage and choanoflagellates as the closest living relatives of the Metazoa. As such, comparisons with these lineages can uniquely inform us about the nature of the last common metazoan ancestor (white circle) and the last unicellular ancestor of Metazoa (black circle). Other close unicellular relatives of Metazoa, such as Filasterea and Ichthyosporea are poorly understood, but ongoing genome projects for members of these lineages promise to feature prominently in future studies of metazoan origins.

Increasing numbers of molecular phylogenetic analyses of diverse microeukaryotes have recently revealed a collection of taxa (including *Filasterea, Ministeria, Capsaspora owczarzaki,* and *Ichthyosporea*) in the internode between metazoans and fungi (Ruiz-Trillo *et al.,* 2004, 2008; Steenkamp *et al.,* 2006; Shalchian-Tabrizi *et al.,* 2008). Choanoflagellates remain the closest known relatives of Metazoa, and the cell morphology of these other diverse microeukaryotes does not provide an obvious link to choanoflagellates and metazoans. Nonetheless, molecular phylogenetic analyses reveal that metazoans count diverse single-celled and colony-forming lineages, in addition to choanoflagellates, among their close relatives (Medina *et al.,* 2001; Ruiz-Trillo *et al.,* 2004, 2007, 2008; Steenkamp *et al.* 2006). A phylogenetically informed comparison of genomes from diverse microeukaryotes with those of metazoans and choanoflagellates promises to further refine our understanding of pre-metazoan genome evolution.

3.2.2 The case for sponges

Arguments in support of sponges as useful 'living models' of the last common metazoan ancestor (LCMA) stem from their cytological similarities with choanoflagellates, their phylogenetic position, and the antiquity of their fossil record. Of these arguments the evolutionary link between choanoflagellates and sponge choanocytes is perhaps the most compelling. The strength of this argument lies in its proven predictive power; it was the hypothesized homology of sponge choanocytes with choanoflagellates that first suggested an evolutionary relationship between choanoflagellates and metazoans (to the exclusion of countless other eukaryotes). As discussed above, this predicted relationship has since been independently borne out in phylogenetic analyses.

In addition to the observation that the sponge body plan is organized around the most ancient metazoan cell type, a preponderance of phylogenetic analyses based upon both morphological and molecular data sets place sponges as the 'earliest branching' metazoans (i.e. all other metazoans are more closely related to each other than to sponges: Collins, 1998; Borchiellini *et al.,* 2001; Medina *et al.,* 2001, 2003; Eernisse and Peterson, 2004; Peterson and Butterfield, 2005; da Silva *et al.,* 2007; Jimenez-Guri *et al.,* 2007; Sperling *et al.,* 2007; Ruiz-Trillo *et al.,* 2008). With this perspective, we can begin to reconcile the 'primitive' nature of the modern sponge body plan with the fact that they, like most metazoans, are the product of at least 600 million years of independent evolution. Chance, key innovations, or (more likely) both, resulted in drastically different evolutionary outcomes in sponges compared with other metazoans. Only after other metazoans diverged from sponges did traits such as nerves, muscles, tissues, and a digestive gut arise.

The fossil record is consistent with the hypothesis that the sponge body plan has remained nearly unchanged since the late Neoproterozoic (reviewed in Carrera and Botting, 2008). Specifically, sponge fossils from between 750 Ma (Reitner and Wörheide, 2002) and 580 Ma (Li *et al.,* 1998) represent the earliest known metazoan body fossils. By the time of the Cambrian, sponge diversity was high, with spicules from most major sponge groups forming an abundant component of the Cambrian fossil record globally (Gehling and Rigby, 1996).

A second, and less well established, phylogenetic result that has emerged is the possibility of sponge paraphyly. Under this scenario, some sponge lineages (e.g. calcareous and homoscleromorph sponges) might be more closely related to eumetazoans than to other, earlier branching sponge lineages (Collins, 1998; Borchiellini *et al.*, 2001; Medina *et al.*, 2001; Peterson and Butterfield, 2005; Sperling *et al.*, 2007). The evolutionary implications of sponge paraphyly have been thoroughly explored and can be distilled into an argument that all extant metazoans are derived sponges (Sperling *et al.*, 2007; Nielsen, 2008). However, the proposition of sponge paraphyly remains tenuous, in part because analyses that include expressed sequence tag (EST) and mitochondrial genome data from the homoscleromorph species *Oscarella carmela* strongly support sponge monophyly (Jimenez-Guri *et al.*, 2007; Lartillot and Philippe, 2008; Lavrov *et al.*, 2008; Ruiz-Trillo *et al.*, 2008; Wang and Lavrov, 2008).

3.2.3 The controversy

Despite the weight of evidence supporting the placement of sponges at the base of the metazoan tree, placozoans (Dellaporta *et al.*, 2006) and, more recently, ctenophores have also been posited as the earliest branching metazoan phylum (Dunn *et al.*, 2008). The case for placozoans derives from an analysis of the mitochondrial genome from the only characterized species, *Trichoplax adhaerens*. This analysis can be distilled into three arguments: (1) like choanoflagellates and unlike most metazoans the mitochondrial genome of *T. adhaerens* is large (*c.* 43 kb compared with the 15–24 kb genomes typical of metazoans); (2) it contains an assortment of introns, intergenic spacers, and genes that are lacking from all other sequenced metazoan mitochondrial genomes (albeit, also without orthologues in choanoflagellates or other non-metazoans); and (3) phylogenetic analyses of predicted mitochondrial proteins support *T. adhaerens* as the earliest branching lineage in an unprecedented clade that also contains sponges, ctenophores, and cnidarians. The

existence of this clade is contradicted by numerous independent analyses and can be explained by accelerated rates of evolution within Bilateria (Dellaporta *et al.*, 2006; Wang and Lavrov, 2008). More recently, a genome-scale analysis of predicted proteins from single-copy loci in the draft genomes of the sponge *Amphimedon queenslandica* and *T. adhaerens* strongly supported placozoans as an independent lineage that branches after sponges, and before cnidarians (Srivastava *et al.*, 2008). This result and others (Collins, 1998; da Silva *et al.*, 2007) cast doubt on the hypothesis that *T. adhaerens* is the earliest branching metazoan.

Recently, Dunn *et al.* (2008) published a phylogeny based upon 150 EST-derived genes that supports ctenophores as branching before two sampled sponge species. This finding would imply one of two unlikely evolutionary scenarios: that the LCMA was much more complex than previously predicted (e.g. it had nerves, muscles, and a digestive gut) or that the ctenophore lineage and other eumetazoans underwent extensive convergent evolution (Giribet *et al.*, 2007). The former scenario is not supported by the fossil record—sponges would have had to undergo morphological simplification before their appearance as the first recognizable metazoan fossils—and the latter explanation would require the improbable, independent evolution of nerves, muscles, and a gut in the ctenophore and cnidarian/bilaterian lineages. Instead, the weight of evidence places choanoflagellates as the closest living metazoan outgroup, sponges as the earliest branching metazoan phylum, and argues that the choanocyte-based feeding strategy of sponges is ancestral to all Metazoa.

3.3 Reconstructing the genetic toolkit for cell–cell interactions

Choanoflagellates and sponges, by virtue of their positions on the tree of life, bracket metazoan origins and are well situated to help us understand the genetic innovations associated with the transition to multicellularity. Indeed, a wealth of genomic data have begun to pour out from representatives of both of these groups. The single-celled choanoflagellate *Monosiga brevicollis* is the subject of a recently completed genome project (King *et al.*, 2008), and genome sequencing projects are under way for the freshwater choanoflagellate *Monosiga ovata* and a colony-forming choanoflagellate, *Proterospongia* sp. (Ruiz-Trillo *et al.*, 2007). Likewise, sponges have received increasing attention from a genomics perspective. Pilot EST projects have been completed for the sponges *O. carmela* and *Suberites domuncula*, and genome-scale data are available for the sponge *A. queenslandica* (Nichols *et al.*, 2006; Perina *et al.*, 2006; Adamska *et al.*, 2007a). The juxtaposition of sponge and eumetazoan sequences with those from choanoflagellates is beginning to reveal the catalogue of genes present in their common ancestor, thus permitting the construction of hypotheses about genomic innovations underlying metazoan origins.

A prediction from the field of evo-devo is that genes involved in regulating development play important roles in morphological evolution. One such class of genes includes those involved in the conserved signalling pathways that transduce extracellular cues in diverse metazoans. Although all cellular organisms engage in cell signalling, the pathways required for metazoan development are more elaborate than those of unicellular organisms and distinct from those found in other multicellular lineages (e.g. fungi and plants). Traditionally, seven intercellular signaling pathways are considered unique to and abundant in Metazoa: nuclear hormone receptor, WNT, TGF-β, Jak/STAT, Notch/Delta, Hedgehog, and RTK (Gerhart, 1998; Barolo and Posakony, 2002; Pires-daSilva and Sommer, 2003). At least six of these seven pathways (Wnt, TGF-β, RTK, Notch, Hedgehog, and Jak-STAT) have conserved components that are expressed in sponges and thus were present in the LCMA (Adell *et al.*, 2003; Nichols *et al.*, 2006; Adamska *et al.*, 2007a,b; Adell *et al.*, 2007).

In contrast with sponges, only two of the major metazoan signalling pathways, the RTK pathway and components of the Hedgehog signalling pathway, are present in the genome of the choanoflagellate *M. brevicollis* (King *et al.*, 2008). Despite early suggestions that RTK signalling might represent a key innovation in the evolution of metazoans from their single-celled ancestors (Hunter, 2000), components of the pathway are abundant in the choanoflagellate genome (including *c.* 120 tyrosine kinase domains, *c.* 30 tyrosine phosphatases, and

c. 80 SH2 domains: King *et al.*, 2008; Manning *et al.*, 2008, Pincus *et al.*, 2008). In addition, two choanoflagellates (*M. brevicollis* and *M. ovata*) contain homologues of the proto-oncogene Src and biochemical analyses reveal these homologues to conserve most of the regulatory interactions associated with metazoan Srcs (Segawa *et al.*, 2006; Li *et al.*, 2008). These observations establish the presence of bona fide tyrosine kinase signalling during the pre-metazoan era.

With the accumulation of genome-scale data from early branching metazoans and their close outgroups, an emerging theme is that the functional protein domains found in developmentally important metazoan signalling and adhesion genes have histories and, presumably, ancestral functions independent of their roles in metazoan proteins. In other words, these protein domains evolved prior to metazoan origins and only later, as a product of domain or exon shuffling (see Patthy, 1999), were linked in the combinations found in the canonical signalling and adhesion proteins of modern metazoans.

One example of this is the case of the secreted ligand Hedgehog. In bilaterians, the Hedgehog signalling pathway is involved in developmental patterning events as diverse as segment polarity in *Drosophila* and brain, bone, muscle, and gut patterning in vertebrates. The canonical Hedgehog ligand is composed of two protein domains, an N-terminal signalling domain that is released through autoproteolytic cleavage by a linked C-terminal intein domain (reviewed in Perler, 1998). Analyses of the choanoflagellate, sponge, and cnidarian genomes reveal that the two functional domains known from the bilaterian Hedgehog family evolved independently and were subsequently coupled through domain shuffling early in metazoan evolution (Figure 3.3). Specifically, the genomes of the choanoflagellate *M. brevicollis* and the sponge *A. queenslandica* encode the Hedgehog N-terminal and C-terminal domains on separate, unrelated proteins, whereas the cnidarian *Nematostella vectensis* has orthologues of these proteins in addition to true Hedgehog proteins typical of bilaterians (Adamska *et al.*, 2007a; King *et al.*, 2008; Matus *et al.*, 2008). This pattern suggests that the Hedgehog gene family evolved through domain shuffling after the divergence of sponges from other metazoans and

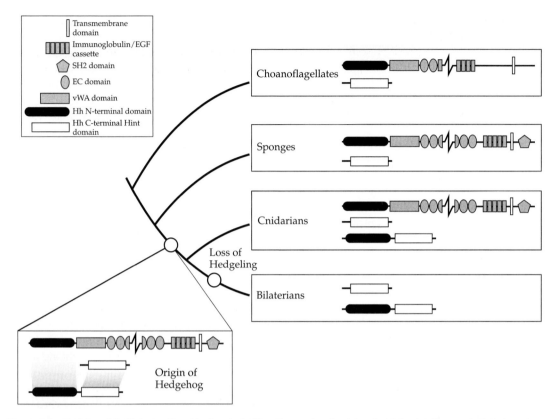

Figure 3.3 Evolution of the Hedgehog ligand by domain shuffling. The two functional domains of the signalling protein Hedgehog, the N-terminal signal domain (black), and the C-terminal Hint domain (white), evolved on separate proteins in the ancestors of choanoflagellates and animals. One of these ancient proteins, Hedgeling (that links the N-terminal signal peptide to extracellular cadherin domains on a transmembrane protein), has homologues in sponges and cnidarians but was lost in the ancestors of bilaterians. A second ancestral protein, Hoglet, containing only the Hint domain, has been conserved in choanoflagellates, sponges, cnidarians, and bilaterians. Hedgehog, a ubiquitous signalling ligand among bilaterians, is also found in cnidarians and evolved by domain shuffling after the divergence of sponges and eumetazoans (Snell *et al.*, 2006; Adamska *et al.*, 2007a; King *et al.*, 2008; Matus *et al.*, 2008).

that a heretofore unrecognized gene family containing the Hedgehog signal peptide linked to extracellular cadherin domains and a transmembrane domain was in place in the last common ancestor of choanoflagellates and metazoans.

In addition to the Hedgehog protein itself, other components of the Hedgehog signalling pathway can be traced back to the last common ancestor of choanoflagellates and metazoans. For example, sponges and choanoflagellates encode orthologues of the upstream protein that once dispatched releases the Hedgehog protein from signalling cells, and the receptor patched that localizes to the surface of downstream Hedgehog signalling targets (Nichols *et al.*, 2006; King *et al.*, 2008). Additionally,

the sponge *O. carmela* expresses a gene with similarity to the negative regulator, suppressor of fused (Nichols *et al.*, 2006).

The case of *hedgehog* evolution is illustrative of the evolution of metazoan signalling systems because it demonstrates how seemingly emergent metazoan cell signalling machinery might have been assembled piecemeal through domain shuffling and the co-option of genes with different (if related) ancestral functions. The most exciting work lies ahead and will entail the exploration of how these genes function in choanoflagellates, sponges, and cnidarians, and how their functions were altered as they were recruited for their roles in regulating bilaterian development.

3.4 From single cells to epithelia

The general architecture of choanoflagellates and sponge choanocytes extends to eumetazoan planar epithelial cells, with the central flagellum of choanoflagellates and the primary cilium of epithelial cells probably sharing a common ancestry (Singla and Reiter, 2006). Neither sponges nor choanoflagellates have epithelial tissues as classically defined, but they do form structures that exhibit the rudiments of epithelial tissue architecture and epithelial cellular machinery. The absence from sponges of abundant intercellular junctions and a basement membrane, two features that contribute to the mechanical and absorptive/transport properties of eumetazoan epithelial cells, has led to their characterization as lacking epithelia. Instead they are considered to be a somewhat loose association of cells in which the internal environment (i.e. mesohyl) is, at least in terms of ionic homeostasis, undifferentiated from the external environment (e.g. seawater: de Ceccatty, 1974; Cereijido *et al.*, 2004). Nonetheless, sponges have cell layers such as those formed by choanocytes (i.e. the choanoderm) that are specialized for filter-feeding. In addition, these cell layers are packed closely together (particularly in embryos) with highly regular paracellular spaces suggesting that they have the same kind of direct intercellular interactions that characterize eumetazoan epithelia and are not simply embedded in a common extracellular matrix (*sensu* Harwood and Coates, 2004; Figure 3.4).

Contrary to the dogma that sponges lack functional epithelia is the observation that the cell layers that line their various body cavities are differentiated. For example, in addition to the choanoderm, the body surface, internal water canals, and spermatic cysts are lined by T-shaped pinacocyte cells, whereas oocytes and embryos are often encased in a layer of large, cuboidal follicle cells. Furthermore, in some species the basal epithelium is uniquely differentiated and larvae develop an outer presumptive epithelium composed of columnar cells more than 15 μm high and *c.* 2 μm in diameter (e.g. *O. carmela*; SAN, personal observation). The morphological differences between as many as five presumptive sponge epithelial tissues plausibly reflect functional differences.

Amongst choanoflagellates that form multicelled colonies, colony architecture varies between species, with cells typically connected at their lateral surfaces to form two-dimensional 'chains' or by the bundling of secreted extracellular pedicels to form rosettes (Figure 3.5). Additionally, colonies of the genus *Proterospongia* form spherical clusters of polarized cells in direct contact with each other. Cell contacts in colonies from *Codosiga botrytis* and *Desmarella moniliformis* comprise cytoplasmic bridges (Hibberd, 1975; Karpov and Coupe, 1998) and therefore differ from the protein-plaque-based

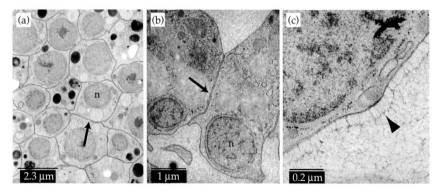

Figure 3.4 Transmission electron micrographs of larval and adult epithelial tissues in the sponge, *Oscarella carmela*. Like other sponges, the larval (a) and adult (b) epithelial tissues of *O. carmela* are characterized by closely apposed membranes that have very small, uniform paracellular spaces (arrows). However, in contrast to other sponges, only homoscleromorphs are reported to have a loose, ladder-like basement membrane composed of type IV collagen (arrowhead; Boute *et al.*, 1996). Scale bars are shown; n, nucleus.

intercellular junctions typical of eumetazoan epithelial tissues. Furthermore, with the exception of an uncorroborated account by William Saville-Kent (1880–82), no choanoflagellate colony is known to display cell differentiation.

Epithelial tissues therefore represent a metazoan innovation that evolved *de novo* after the evolution of multicellularity. Aspects of choanoflagellate biology hint at the types of cell biological phenomena that might have laid the foundation for epithelial origins. The capacity of most choanoflagellates to adhere to surfaces suggests the presence of a ubiquitous adhesion mechanism that might have been co-opted to support intercellular adhesion in diverse choanoflagellate lineages and in the lineage leading to Metazoa. This is consistent with the discovery of more than 23 cadherin genes and a diversity of predicted proteins with C-type lectin,

immunoglobulin, α-integrin, collagen, fibronectin, and laminin adhesion domains encoded by the genome of the exclusively unicellular species, *M. brevicollis* (Abedin and King, 2008; King *et al.*, 2008).

Only one group of sponges, the homoscleromorphs, has been argued to have a bona fide epithelium, complete with intercellular junctions in the larva and a basement membrane underlying larval and adult tissues (Boute *et al.*, 1996; Boury-Esnault *et al.*, 2003). However, due to uncertainty about the phylogenetic position of this group (see Section 3.2), it is unclear if other sponges have lost these epithelial features or if homoscleromorphs are more closely allied with eumetazoans. Another possibility is that the molecular machinery characteristic of intercellular junctions and the basement membrane in eumetazoan epithelia

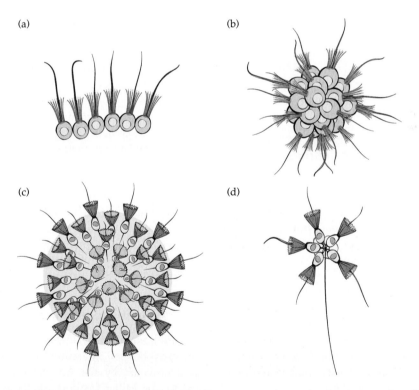

Figure 3.5 Diversity of colony morphology in choanoflagellates. All known species of choanoflagellates have unicellular life-history stages, but some species are also capable of forming simple evidently undifferentiated colonies. Among these species, colony morphology is diverse and suggestive of independent evolutionary origins. For example, colony morphology can vary from 'chains' (a) or 'balls' (b) of athecate cells in direct contact with each other to thecate cells embedding in a common gelatinous matrix (c) or sharing a common stalk (d).

is in place in all sponges, but is not sufficiently concentrated to be detected as an electron dense structure under transmission electron microscopy. Indeed, diverse adhesion gene families, and other cell junction components, are widely conserved in sponges (reviewed in Bowers-Morrow *et al.*, 2004). Furthermore, in *Ephydatia muelleri*—a species that lacks electron-dense epithelial cell junctions and a basement membrane—immunofluorescent staining reveals that actin plaques characteristic of adherens junctions (Yap *et al.*, 1997) are present at points of cell contact in the pinacoderm (Elliot and Leys, 2007).

Epithelial tissues were the first metazoan tissue type to evolve and are thought to have been required for body plan diversification by allowing early metazoans to compartmentalize and regulate physiological homeostasis within and between body compartments (Tyler, 2003). Sponges and choanoflagellates exclusively share the ancient characteristic of collared cells, yet when viewed at the cell biological level the gulf between their morphology and that of other metazoans is not as wide as it may seem. For example, it seems that the fundamental characteristics of epithelial tissues can be traced to the rudimentary epithelia of sponges and, to some extent, to the simple, undifferentiated colonies of choanoflagellates.

3.5 The biology of the earliest metazoan ancestors

The study of metazoan origins is still in its infancy, despite recent advances in our understanding catalysed by genome-scale data from choanoflagellates, sponges, and other early branching metazoan phyla. It is premature to assume that any one (or few) species is(are) representative of each phylum—the diversity within these groups is high and phylogenetic divergences are deep—so an immediate goal is to acquire genomic data from a more representative sampling. Nevertheless, from the available data we can begin to reconstruct the minimal genomic, cell, and developmental characteristics of the first metazoans.

The cell biology of the last common ancestors of choanoflagellates and metazoans probably resembled that of modern choanoflagellates. However,

we do not know whether the last common ancestor of choanoflagellates and metazoans was capable of forming simple multicelled colonies like those formed by some choanoflagellate species and other microeukaryote relatives of metazoans (Leadbeater, 1983; Ruiz-Trillo *et al.*, 2007). A key to addressing this question may be to determine the molecular mechanisms underlying cell interactions in diverse colony-forming choanoflagellates species.

If choanoflagellate-like cells are the most ancient metazoan characteristic and the LCMA was a bacterivorous filter-feeder, was the LCMA more like a choanoflagellate or a sponge? Sponges and other metazoans share many developmental and genomic characteristics that choanoflagellates lack. Specifically, sexual reproduction in sponges is typical of other metazoans in that it involves the fusion of a large, nutritive egg with a small, motile sperm to produce an embryo that undergoes programmed patterns of cleavage, cell differentiation, and morphogenetic patterning. Also, many components of the molecular machinery that regulate development in other metazoans are conserved in sponges and, in some cases, expressed during development (Nichols *et al.*, 2006; Adamska *et al.*, 2007a,b). In contrast, sex is undocumented (though likely) in choanoflagellates, and there is no record of gametic differentiation or of development beyond the formation of simple colonies in some species. The genome of the choanoflagellate *M. brevicollis* also encodes few genes with homology to those that regulate development in metazoans. Integrating these data provides an early impression of a sponge-like LCMA, and suggests a series of specific developmental and cell biological innovations that separate modern metazoans from their single-celled ancestors

3.6 Acknowledgements

We are grateful to R. Howson for contributing to the artwork used in Figure 3.5. Financial support was provided by the American Cancer Society (SAN), the Miller Institute for Basic Research in Science (MJD), Richard Melmon and the Gordon and Betty Moore Foundation Marine Microbiology Initiative (NK).

CHAPTER 4

The mouth, the anus, and the blastopore—open questions about questionable openings

Andreas Hejnol and Mark Q. Martindale

Gastrulation is one of the major events during the embryogenesis of an animal. In addition to the formation of the germ layers it is often the time when the future axial properties and digestive openings become apparent, and it is not surprising that this event plays an important role in hypotheses regarding metazoan evolution. A major difference between these theories concerns the structure of the alimentary canal and the relationship of its openings to the blastopore of the last common bilaterian ancestor. Here we review competing theories of bilaterian evolution and evaluate their plausibility in the light of recent insights into metazoan phylogeny and development.

4.1 Gastrulation as a process determining multiple body plan characteristics

The evolution of an internal germ layer enabled the compartmentalization of the body of multicellular animals (Metazoa) into a digestive layer (endoderm) and an outer layer, the integument (ectoderm). The developmental process that separates the inner from the outer cell populations is called gastrulation. During gastrulation, cells or cell layers are internalized and later form the digestive epithelium and often also germ cells. Bilaterians have in addition a third germ layer, the mesoderm, which either separates from the endoderm or ingresses as independent precursors. Gastrulation not only generates distinct cell types, it also establishes the organismal axes from a pre-existing animal–vegetal embryonic axis. One can thus picture gastrulation as

a fundamental event in every organism as it determines the main body features during development. Consequently, gastrulation plays a major role in hypotheses regarding the transition from a radially symmetrical diploblastic animal to a bilaterally symmetrical triploblastic organism with mesoderm, the last common bilaterian ancestor.

4.2 The blastopore as the site of internalization

In most animals the site of internalization of gastrulating cells is limited to a specific area, the blastopore. The fate of this site is often not only the area of germ layer specification, it sometimes becomes the connection of the endodermal digestive cavity to the ectoderm and thus to the animal's environment. This connection is usually called either the 'mouth' if the organism has only a single opening to the gut or if it is the anterior opening of a through gut, or the 'anus' when it corresponds to the posterior opening and functions as a site of excretion. Grobben (1908) subdivided the bilaterian clade into the taxa Protostomia and Deuterostomia based on the fate of the blastopore becoming either the mouth or the anus. Grobben claimed that in most protostomes ('first mouth') the blastopore becomes the mouth, and the anus is formed secondarily at a different site, later during embryogenesis. In contrast to protostomes, in deuterostome embryos the site of gastrulation becomes the anus while the mouth is formed at a different site in the animal hemisphere of the embryo (Figure 4.1). The blastopore becoming the anus appears to be ancestral for the

I. II.

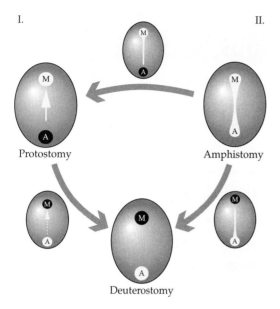

Figure 4.1 Protostomy, deuterostomy, and amphistomy and evolutionary scenarios about the direction of change. In protostomy, the blastopore (bright in all figures) is displaced from posterior to antero-ventral by morphogenetic movements (bright arrow), and gives rise to the mouth (M). The anus (A), if present, is formed at a different site at the posterior end of the embryo. In deuterostomy the blastopore stays at the posterior site of the embryo and either closes or gives rise to the anus. The mouth is formed at a different site from the former blastopore in the anterior of the embryo. In amphistomy, the blastopore elongates and closes laterally with both ends giving rise to the mouth and anus. One scenario (I) for the evolution of deuterostomy is using the protostomy as ancestral, proposing the abbreviation of the anterior movement of the blastopore (dotted bright arrow) by a molecular separation of the mouth determination from the blastopore (intermediate stage small). The other scenario (II) claims the amphistomy as being ancestral, from which deuterostomy and protostomy evolved by closure of the blastopore from the posterior to anterior direction (protostomy) or from the anterior to posterior direction (deuterostomy).

deuterostomes, but the situation in the protostomes is unclear due to its extreme variability in different forms (Table 4.1) and because the phylogenetic relationships of several key taxa are not known. In some cases a large maternal investment of yolk (e.g. hexapods, cephalopods, and onychophorans) influences the gastrulation process, similar to what is found in most amniote embryos.

However, even in embryos with small quantities of yolk, the fate of the blastopore is highly variable (see Table 4.1) and the reconstruction of the ancestral type of gastrulation is further hindered by intrataxon variation within larger clades such as annelids, nematodes, arthropods, and nemerteans (Table 4.1). The variability of the relationship between the blastopore and its possible future fates—mouth or anus—has led Ulrich (1951) to dismiss the term Protostomia and rename it Gastroneuralia, based on the ventral localization of nerve cords in these taxa. However, as it is not clear if a ventrally centralized nervous system is part of the ground pattern of the Protostomia, it is premature to rename this clade.

4.3 The fate of the blastopore and its role in scenarios of bilaterian evolution

The significance of transitions from the cnidarians to the Bilateria, which possess an antero-posterior and a dorsoventral axis, might be one of the biggest controversies in zoology over the last centuries. In cnidarians and ctenophores, two groups that diverged prior to the origin of the Bilateria, the blastopore gives rise to a single opening (mouth–anus) of the gastric digestive cavity, while in many bilaterians two openings are present, both of which can be formed independently from the blastopore. The scenarios under debate for explaining this transition differ in various details, such as the presence of a coelom or a larval stage. However, the formation of a through gut and the fate of the blastopore to the openings are central to understanding bilaterian body plan evolution.

4.3.1 The gastraea theory and the ancestral lateral closure of a slit like blastopore

The gastraea theory (Haeckel, 1872, 1874) proposes a hypothetical metazoan ancestor similar to the gastrula stage of recent animals. A fundamental assumption in the future variations of Haeckel's theme of the gastraea, for example the bilaterogastraea theory (Jägersten, 1955) and the trochaea theory (Nielsen and Nørrevang, 1985), is the extension of the ventral blastopore along the antero-posterior body axis. In these theories, a lateral closure of this elongated blastopore, which stays open at its ends,

Table 4.1 Blastopore fates in bilaterian taxa

Taxon	Fate of blastopore	References
Gastrotricha	Protostomy (*Lepidodermella, Turbanella*)	*Lepidodermella* (Sacks, 1955), *Turbanella* (Teuchert, 1968)
Nematoda	Protostomy (*Tobrilus, Ascaris*), blastopore closure (*Tylenchida*)	*Tobrilus* (Schierenberg, 2005), *Ascaris* (Boveri, 1899), *Tylenchida* (Malakhov, 1994)
Nematomorpha	Deuterostomy (*Paragordius*)	*Paragordius* (Montgomery, 1904)
Mollusca	Protostomy (*Crepidula, Patella*), deuterostomy (*Viviparus*)	*Crepidula* (Conklin, 1897; Hejnol *et al.*, 2007), *Patella* (Dictus and Damen, 1997), *Viviparus* (Dautert, 1929)
Priapulida	Blastopore closure?	*Priapulus* (Wennberg *et al.*, 2008)
Kinorhyncha	?	
Loricifera	?	
Platyhelminthes	Protostomy (*Planocera, Hoploplana*)	*Planocera* (Surface, 1907), *Hoploplana* (Boyer *et al.*, 1998)
Gnathostomulida	?	
Rotifera	Protostomy (*Asplanchna, Calidina, Philodina*)	*Asplanchna* (Lechner, 1966), *Calidina* (Zelinka, 1891), *Philodina* (AH, unpublished)
Entoprocta	Protostomy (*Pedicellina*)	*Pedicellina* (Marcus, 1939)
Nemertea	Protostomy (*Procephalotrix*), deuterostomy (*Drepanophorus*)	Procephalotrix (Iwata, 1985), *Depranophorus* (Friedrich, 1979)
Annelida	Protostomy (*Polygordius*), deuterostomy (*Eunice*), blastopore closure (*Capitella*)	*Polygordius* (Woltereck, 1904), *Eunice* (Åkesson, 1967), *Capitella* (Eisig, 1898)
Sipunculida	Protostomy (*Phascolosoma*)	*Phascolosoma* (Gerould, 1906)
Cycliophora	?	
Phoronida	Protostomy (*Phoronopsis*)	*Phoronopsis* (Rattenbury, 1954)
Brachiopoda	Protostomy (*Terebratulina*)	*Terebratulina* (Conklin, 1902)
Onychophora	Mouth and anus form at different site of blastopore (*Peripatopsis*)	*Peripatopsis* (Manton, 1949)
Tardigrada	Protostomy (*Thulinia*)	*Thulinia* (Hejnol and Schnabel, 2005)
Arthropoda	Protostomy (*Cyprideis*), deuterostomy (*Meganyctiphanes*)	*Cyprideis* (Weygoldt, 1960), *Meganyctiphanes* (Alwes and Scholtz, 2004)
Chaetognatha	Deuterostomy (after blastopore closure)	*Sagitta* (Hertwig, 1880)
Xenoturbella	?	
Hemichordata	Deuterostomy (*Balanoglossus*)	*Balanoglossus* (Heider, 1909)
Echinodermata	Deuterostomy (*Synapta*)	*Synapta* (Selenka, 1876)
Cephalochordata	Deuterostomy (*Branchiostoma*)	*Branchiostoma* (Cerfontaine, 1906)

? Indicates unknown

gives rise to a mouth and anus in the bilaterally symmetrical ancestor with a through gut. This process has been termed 'amphistomy' by Arendt and Nübler-Jung (1997). The same concept was the foundation of Remane's enterocoely theory (Remane, 1950), which begins with a cnidarian polyp transforming into an 'ur-bilaterian' by stretching along its directive axis. Remane's theory assumes the simultaneous evolution of the mouth and anus, and predicts that the coeloms of this hypothetical ancestor are formed from the common gastric pouches of an anthozoan-like polyp (Remane, 1950). These

theories thus predict a rather morphologically complex bilaterian ancestor, which is consistent with what has also been proposed on the basis of similarities of the expression of some developmental genes (Carroll *et al.*, 2001; Arendt, 2004). The two common themes of these theories are: (1) the simultaneous evolution of the mouth and anus, and (2) that the blastopore gives rise to both openings in the common ancestor. Many authors suggest that a slit-like blastopore is ancestral for the Bilateria and argue that a variety of extant animals such as onychophorans, polychaetes, insects and some nematodes appear to show this kind of gastrulation (Sedgwick, 1884; Arendt and Nübler-Jung, 1997; Nielsen, 2001; Malakhov, 2004). The most elaborate explanation of the evolution of gastrulation processes which supports the gastraea-based theories is delivered by Arendt and Nübler-Jung (1997). These authors explain the evolution of deuterostomy by a closure of the slit-like blastopore from anterior to posterior, leaving an opening, which becomes the anus (Figure 4.1). Accordingly, protostomy evolved by a closure of the blastopore from posterior to anterior leaving a mouth open (Figure 4.1). Indeed, some bilaterians show a ventrally elongated blastopore that follows this pattern (e.g. *Polygordius*).

4.3.2 The acoeloid-planuloid theory and the ancestrality of protostomy

Competing with 'gastraea'-based theories is a different view that does not require the simultaneous evolution of mouth and anus to establish bilaterality. The starting points of these hypotheses were pioneered by von Graff (1891), who proposed a paedomorphic planula larva, similar to that of recent cnidarians, which adopted a benthic lifestyle and flattened along one body axis, giving rise to the bilateral symmetry of the Bilateria (Hyman, 1951; Salvini-Plawen, 1978). The authors suggest that the former posterior blastopore—which gives rise to the mouth—is shifted to the ventral body side, to facilitate uptake of food from the ventral surface. This condition is represented by extant acoel and nemertodermatid flatworms, which have only a ventral opening to their digestive cavity. A posterior position of the mouth is found in the nearly radially symmetrical acoel *Diopisthoporus*, which is thought to reflect the ancestral planula-like condition (Beklemishev, 1969; Reisinger, 1970) of some feeding anthozoan planula larvae (Widersten, 1973).

According to these acoeloid-planuloid theories, the last common bilaterian ancestor was a rather simple benthic, probably meiofaunal, worm lacking a through-gut, coeloms, and excretory organs (Hejnol and Martindale, 2008b).

How do proponents of the acoeloid-planuloid theory explain the variation of the gastrulation types in the Bilateria and how this is related to the

evolution of the anus? The most thorough thoughts about this problem are presented by Salvini-Plawen (1978, 1980). He points out that in many spiralian embryos the vegetal (=posterior) blastopore gets displaced into the antero-ventral direction (e.g. in molluscs, annelids, nemerteans, and polyclad flatworms), and thus recapitulates the evolutionary process (protostomy). The anus is thought to have evolved independently from the blastopore in multiple lineages, which is also reflected by the late developmental formation of the anus in many protostomes. The deuterostome condition is explained with an evolutionary 'abbreviation' of the anterior movement of the blastopore (Figure 4.1). The mouth, instead of moving anteriorly in the form of the blastopore, is immediately formed at its final location (Beklemishev, 1969; Salvini-Plawen, 1980) and the blastopore either closes completely, such as in chaetognaths and nemerteans, or stays open and forms the anus, as in nematomorphs, deuterostomes, and several crustaceans (Table 4.1).

4.4 Recent progress in molecular systematics and developmental biology and their impact on the problem

Both competing scenarios differ in their assumption about which type of gastrulation is ancestral for the Bilateria. In gastraea-based theories the lateral closure of the blastopore—or 'amphistomy'—with the simultaneous evolution of the orifices would deliver the state from which the diversity of gastrulation can be derived. In the acoeloid-planuloid theory, protostomy is ancestral, including an independent evolution of an anal opening. Since both hypotheses have their roots in a time when the metazoan phylogeny was speculative, a proper phylogenetic framework is required to determine the direction of evolutionary change.

4.5 A new animal phylogeny

Recent progress in molecular biology, computer technology and the development of new phylogenetic reconstruction algorithms have improved the ability to determine animal relationships with the use of molecular data (Philippe and Telford, 2006; Dunn *et al.*, 2008). In addition to the seminal publication of Aguinaldo *et al.*, (1997) which established the subdivision of the Bilateria into three large clades, Ecdysozoa, Lophotrochozoa, and Deuterostomia, further resolution of metazoan relationships has been accomplished by increased taxon sampling and the use of phylogenomic approaches (Dunn *et al.*, 2008). A major result pertinent to understanding the role of gastrulation in body plan evolution is the placement of the acoels as the sister group to the remaining Bilateria (Ruiz-Trillo *et al.*, 1999; Baguñà and Riutort, 2004). Their position has been corroborated by multiple independent molecular approaches (see Telford *et al.*, 2003, for example). Our current understanding places the nemertodermatids as sister to Bilateria, and Acoela as sister to that group, thus breaking the monophyly of the Acoelomorpha (Jondelius *et al.*, 2002; Wallberg *et al.*, 2007). Having the acoel flatworms at the base of the Bilateria has important implications for our understanding of the evolution of organ systems (Hejnol and Martindale, 2008b). The similarity of the body plan of acoels and nemertodermatids, both possessing only one opening to their digestive system and an orthogonal nervous system and lacking a through gut and nephridia, clearly support a simple acoeloid bilaterian ancestor, which was previously proposed on the basis of morphological data (Hyman, 1951; Salvini-Plawen, 1978; Haszprunar, 1996a). These data do not support the gastraea theory of Haeckel or the transformation of a coelom bearing ur-bilaterian from a sessile cnidarian polyp (Remane, 1950).

4.6 'Amphistomy'—a common theme in bilaterian development?

While molecular phylogenetic results support the acoeloid-planuloid theory, the recently improved resolution of the metazoan relationships with the use of the phylogenomic approach has implications for accepting the 'amphistomy' hypothesis (Dunn *et al.*, 2008). If one assumes gastrulation with a slit-like blastopore is in the ground pattern of the protostomes and deuterostomes (together named Nephrozoa after Jondelius *et al.*, 2002), one would expect a broad distribution of a lateral closure of a slit-like blastopore in the Bilateria (Arendt and Nübler-Jung, 1997). The cases for which amphistomy are most commonly cited are onychophorans, the polychaete *Polygordius*, and the nematode *Pontonema* (e.g. Arendt and Nübler-Jung, 1997; Nielsen, 2001). In drawings from the early research on the onychophoran *Peripatus capensis* (Figure 4.2a), it indeed seems that a large extended blastopore closes laterally and both ends stay open and give rise to the mouth and anus (Balfour, 1883; Sedgwick, 1885). In contrast, Kennel (1885) draws a different picture for *Peripatus edwardsii*, showing that the opening that gives rise to the mouth and anus is separate from the blastopore, which is positioned more posteriorly (Figure 4.2a). The most thorough analysis of onychophoran gastrulation (Manton, 1949) corroborates Kennel's findings for several onychophoran species and describes the immigration of the mesoderm and germ cells at the posterior blastopore which is never in contact with either mouth or anus. The syncytial development of the yolky onychophoran embryos seems to be a rather derived adaptation to their terrestrial lifestyle, as is the case in other terrestrial arthropods (e.g. hexapods and myriapods), and does not represent an example of an 'amphistomic' type of gastrulation. Another taxon often referred to as being 'amphistomic' is the polychaete annelids. The traditional example is the description of the development of *Polygordius* (Woltereck, 1904). A close examination of the original work shows that the extended blastopore first closes laterally, but instead of leaving both ends open, only the anterior edge gives rise to the mouth (Figure 4.2b). The anus is formed later in development one cell row posterior (Figure 4.2b) to the former 'seam' of the blastopore (Woltereck, 1904). Thus, the development of *Polygordius* follows a protostomic pattern rather than amphistomy. Even if a polychaete can be shown to possess a prototypical amphistomy form of gastrulation, it is difficult to be sure that it is a plesiomorphic character, since we observe a high variation in gastrulation patterns in polychaete annelids. Both protostomy (Mead, 1897) and deuterostomy (Åkesson, 1967), have been described in polychaetes as well as in other trochozoan taxa including nemerteans (Friedrich, 1979; Iwata, 1985) and molluscs (Fioroni, 1979; see Table 4.1).

A similar variation of gastrulation is found in the third taxon for which amphistomy has been

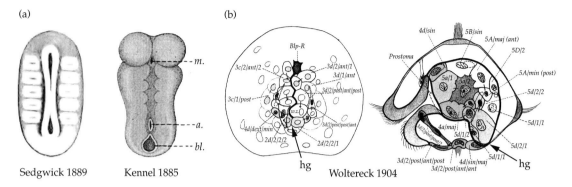

Figure 4.2 Original drawings of the development of the onychophoran *Peripatus* and the annelid *Polygordius*. (a) The drawing of Sedgwick (left) shows the supposed 'amphistomic' gastrulation in the onychophoran *Peripatus capensis*. On the right the drawing from the study of Kennel on *Peripatus edwarsii*, which shows the blastopore (bl.) separate from the secondary opening, which gives rise to the mouth (m.) and anus (a.). (b) Original drawings of the gastrulation of *Polygordius*, a supposed amphistomic polychaete. The drawing on the left shows the closure of the blastopore (dark black line in the original) leaving the mouth open (protostomy). The hindgut (hg) forms at a different site from the blastopore (arrow).

described, the Nematoda (Malakhov, 1994). Apart from a clear case of blastocoelic protostomy in *Tobrilus* (Schierenberg, 2005) the site of immigration of the two entoderm cells becomes the future mouth in most nematodes, which is separate from the later immigration site of the mesodermal precursors which are descendants from different lineages (Schierenberg, 2006).

The developmental stage for which amphistomic gastrulation has been described in *Pontonema* is much later than the immigration of the E precursors, which form the endoderm—a process which is usually referred to as gastrulation in nematodes (Schnabel *et al.*, 1997; Sulston *et al.*, 1983; Voronov and Panchin, 1998).

Again, even if the cursory study of *Pontonema* is correct, it is not clear what type of gastrulation the last common ancestor of nematodes had, because the sister group of nematodes, the Nematomorpha (Dunn *et al.*, 2008) gastrulate by deuterostomy (Montgomery, 1904; Inoue, 1958). Tardigrades, which form the sister group to the Arthropoda and Onychophora (Dunn *et al.*, 2008), gastrulate by immigration of mesodermal and endodermal precursor cells and show protostomy (Hejnol and Schnabel, 2005), with no evidence of enterocoely as has been assumed by early investigators (Marcus, 1929).

If amphistomy was an ancestral pattern, giving rise to both deuterostomy and protostomy, it must have been lost independently in nearly every larger protostome clade. Taken thus, the developmental diversity of bilaterians described today gives little support for either amphistomic gastrulation or Haeckel's gastraea. Both concepts seem to deliver a feasible evolutionary scenario for the human mind but are not represented in living organisms.

4.7 'Abbreviated' protostomy as a model for the variability of gastrulation?

Attempts to explain variations in bilaterian gastrulation patterns from the point of view of the acoeloid-planuloid theory are based on the movement of the vegetal blastopore in the antero-ventral direction, which then gives rise to the mouth in protostomic animals (Figure 4.1). In the case of deuterostomy, the mouth forms at a separate site and no such movement can be observed. Salvini-Plawen (1980) explains the multiple independent origins of deuterostomy in several animal lineages with an evolutionary 'abbreviation' of the antero-ventral movement by means of the mouth forming directly in the anterior part, the former animal hemisphere of the embryo. This includes a spatial separation of the molecular mechanisms determining the mouth from the site of gastrulation. It is clear from the developmental studies on ctenophores and cnidarians that the mouth and the blastopore have a common origin (Goldstein and Freeman, 1997). In both animals the blastopore gives rise to the single opening of the digestive cavity. Thus, it is not surprising that many genes which have been assigned to gastrulation and foregut development, for example *brachyury*, *goosecoid*, and *forkhead*, are expressed at the cnidarian and ctenophore blastopore (Scholz and Technau, 2003; Martindale *et al.*, 2004; Matus *et al.*, 2006a; Yamada *et al.*, 2007). It is difficult, however, to dissect the role of these genes, since the blastopore has overlapping functions in mouth formation, axis determination, and germ layer specification.

It is important to point out that the site of gastrulation has changed along the animal–vegetal egg axis in the stem lineage of the Bilateria. While cnidarians and ctenophores gastrulate at the animal pole, bilaterians, including acoels, gastrulate at the vegetal pole. Since bilaterians form their mouths at the animal hemisphere, this indicates an ancient separation of the determining factors of mouth formation and site of germ layer specification. The molecular separation of signalling centres might explain the variation of the relationships between mouth and blastopore in the Bilateria by a facilitation of movements of signalling centres. The absence of molecular data from a broader range of taxa limits detailed conclusions at this time, but in animals in which the mouth is formed at a separate site from the blastopore (deuterostomy) (including asteroids, echinoids, hemichordates, and chaetognaths), the gene *brachyury* is expressed in both locations, indicating the spatial separation of a former common expression domain at the blastopore (Peterson *et al.*, 1999; Shoguchi *et al.*, 1999; Takada *et al.*, 2002). Furthermore, recent work indicates that the mouth in protostomes and deuterostomes—although formed by variable developmental processes—is homologous, based on the shared arrangement of the expression domains of *goosecoid* and *brachyury* (Arendt *et al.*, 2001). A specific mouth signalling system which would enable a detailed explanation of how the position of the mouth becomes specified in the Bilateria has not

yet been identified, but the dissociation of the blastopore from the position of the mouth already occurred in the bilaterian stem lineage.

The fate map of the acoel *Neochildia fusca* shows that the mouth is formed at a site different from the blastopore (Henry *et al.*, 2000) and the mouth of *Isodiametra pulchra* is formed long after blastopore closure (Ladurner and Rieger, 2000). Our own studies of *brachyury* and *goosecoid* expression in the acoel *Convolutriloba longifissura* supports the homology of the acoel mouth with the protostome and deuterostome mouth (Hejnol and Martindale, 2008a). The fate map of the acoel *Neochildia fusca* furthermore shows that the vegetal part of the embryo gets shifted in an antero-ventral direction by the increased proliferation of descendants of micromere 1a versus those of its ventral counterpart micromere 1b (Henry *et al.*, 2000). This mirrors the observations made by fate map studies in spiralian taxa, like polyclads (Boyer *et al.*, 1998), molluscs (Dictus and Damen, 1997; Hejnol *et al.*, 2007), annelids (Ackermann *et al.*, 2005), and nemerteans (Maslakova *et al.*, 2004a) in which the dorsal side of the embryo proliferates more than the ventral regions (see also discussion in van den Biggelaar *et al.*, 2002). This shifting of the vegetal part of the embryo by proliferation of dorsal cells might be an ancestral bilaterian feature which was lost in the deuterostome lineage where the position of the mouth is specified independently from the blastopore. Unfortunately, the relationship of the blastopore to the primary egg axis has not been investigated in a large number of ecdysozoan taxa and appears to be highly variable and obscured by yolk content.

4.8   Has the anal orifice evolved several times convergently?

If the mouth is homologous in all animals and was the earliest opening to the digestive cavity, when did the anal opening evolve and do all anal openings share a common ancestry? The outgroups of the Nephrozoa—ctenophores, cnidarians, acoels, and nemertodermatids—as well as several other key bilaterian taxa lack an anal opening. All Platyhelminthes possess only a mouth and the presence of one or more dorsal anal pores in the branched gut system of some polyclads is a derived character (Ehlers, 1985). Gnathostomulids and some rotifers (e.g. *Asplanchna*) lack an anus and *Xenoturbella*, which in the current view forms the sister taxon to the Ambulacraria or to all remaining Deuterostomia (Bourlat *et al.*, 2003, 2006; Perseke *et al.*, 2007; Dunn *et al.*, 2008), also possesses only one opening to the digestive cavity (Westblad, 1949).

This lack of an anus, however, can be interpreted as either loss from a stem species with a through gut (as seems to be the case for the brachiopods and many parasitic forms) or that the anus has evolved later and independently in several lineages. Although it is not parsimonious that the anus evolved multiple times, its functional advantage and the differences in development and morphology could be evidence for an independent evolutionary origin (Beklemishev, 1969; Schmidt-Rhaesa, 2007).

The anus is morphologically very diverse between the protostome taxa. For example, gastrotrichs do not possess an ectodermal hindgut like most bilaterians; instead the anus is formed by a direct and often temporary connection of the endoderm to the outside (Ruppert, 1991b). Such a temporary anus is also present in Micrognathozoa (Kristensen and Funch, 2000) and in the gnathostomulid *Haplognathia* (Knauss, 1979). Another variation of anal morphology found in many lineages is a combined opening of the anus with the gonopore, a so-called cloaca. Cloacas are present in the ecdysozoan nematodes, nematomorphs, tardigrades, and rotifers. Acoels and nemertodermatids lack an anus but have a male gonopore at the posterior end of the body formed by an involution of the ectoderm. *Brachyury* is expressed in the hatchling of the acoel *Convolutriloba longifissura* in the posterior ectoderm, which later gives rise to the adult male gonopore (Hejnol and Martindale, 2008a). This might indicate that the anus of some taxa, as in the Platyzoa and Ecdysozoa, might have been derived from a connection between the endoderm and the ectoderm of the gonoduct. Thus the last common nephrozoan ancestor might have had only an antero-ventral mouth and a posterior male gonopore similar to what is found in acoels and nemertodermatids (Hejnol and Martindale, 2008a). Of the larger animal clades, the Trochozoa show the highest morphological similarity of their anal openings: most of them possess an ectodermal hindgut, which ends in an opening that is separate from the gonoducts. In most trochozoans the hindgut is formed after the mouth as a secondary ectodermal involution at a site separate from the blastopore. Despite this evidence for the homology of the trochozoan anus, it remains unclear if the anus is homologous in all bilaterians and thus part of the nephrozoan ground pattern. The independent evolution of the anus might have been a result of the extension of the body length in several lineages, since a blind gut is mostly present in smaller animals and parasitic forms. However, it is too early to draw a conclusion because important nodes in metazoan phylogeny are still not resolved. It will be important to see if the Platyzoa, a taxon comprising Rotifera, Gnathostomulida, Platyhelminthes, and Gastrotricha, turns out to be a true monophyletic group or a long-branch artefact (see Chapter 6).

4.9 Conclusion

We are far from understanding the evolution of gastrulation mechanisms in the different animal lineages. As in phylogenetic systematics where broader taxon sampling helps to reconstruct phylogeny, it is necessary to follow the same approach in comparative developmental biology to answer the questions raised from studies of a handful of arbitrarily selected animal species. The new molecular approaches to deliver the necessary phylogenetic framework, as well as detailed analyses of the development of more animal taxa using modern cell lineage studies and molecular developmental approaches, will deliver insights into the evolution of the important organ systems.

4.10 Acknowledgements

We thank Tim Littlewood and Max Telford for the invitation to write this chapter and for organizing the Royal Society meeting in London. Kevin Pang critically read the manuscript. The research was supported by the NSF grant AToL Protostome to MQM and by a grant of the German Science Foundation to AH.

PART II

The Bilateria

CHAPTER 5

Origins of metazoan body plans: the larval revolution

Rudolf A. Raff

The origins of bilaterian animal body plans are generally thought about in terms of adult forms. However, most animals have larvae with body plans, ontogenies, and ecologies distinct from their adult forms. The first of two primary hypotheses for larval origins suggests that the earliest animals were small pelagic forms similar to modern larvae, with adult bilaterian body plans evolved subsequently. The second suggests that adult bilaterian body plans evolved first and that larval body plans arose by interpolations of features into direct-developing ontogenies. The two hypotheses have different consequences for understanding parsimony in the evolution of larvae and of developmental genetic mechanisms. If primitive metazoans were like modern larvae and distinct adult forms evolved independently, there should be little commonality of patterning genes among adult body plans. However, sharing of patterning genes in adults is observed. If larvae arose by co-option of adult bilaterian-expressed genes into independently evolved larval forms, larvae may show morphological convergence but with distinct patterning genes, as is observed. Thus, comparative studies of gene expression support independent origins of larval features. Precambrian and Cambrian embryonic fossils are also consistent with direct development of the adult as primitive, with planktonic larval forms arising during the Cambrian. Larvae have continued to co-opt genes and evolve new features, allowing study of developmental evolution.

5.1 Evolutionary biology has primarily been about the study of adults

It is striking that studies of evolutionary histories are nearly all about the evolution of adults. Palaeontologists, having only a few fossil larval forms, perforce have had to study adults, which make up most of the fossil record. Transitions that can be studied are almost inevitably those of adult character states. In popular representations this translates into computer animations where fins transform into legs, dinosaurs morph into birds, or apes into hominids—beguiling but misleading images. The bias extends to phylogeny, as most available characters are adult ones. Our definitions of the body plans of phyla are of adult body plans. This bias persists in evo-devo, which largely focuses on the evolution of novel adult features, for example the loss of legs in snakes and the origin of the turtle shell. These examples are approached by studies that combine morphological, palaeontological, and gene regulatory data (Cohn and Tickle, 1999; Gilbert *et al.*, 2001). Developmental biology also focuses primarily of the development of adults. This is largely dictated by interest in major body parts, for example insect wings or tetrapod legs. Another source of the bias arises from our genetic and developmental model systems, limited to a few chosen for ease of laboratory use (Bolker, 1995; Jenner and Wills, 2007). Essentially all major genetic model systems are direct developers, where the adult body plan of the phylum is generated progressively in development, even if some form of metamorphosis occurs. This is true of *Caenorhabditis elegans*, a nematode, *Drosophila melanogaster*, an arthropod, or zebra fish, frogs, or mice—all vertebrates. The evolution of adult bias does not mean that early development is ignored, but that it is largely the study of early development leading to adult characters.

5.2 Most phyla have a second body plan

Not only has our focus been on the origins of adult body plans, but, of the phyla examined, vertebrates and arthropods have received the bulk of attention in studies of evolution among Bilateria. Both phyla have been highly successful in terrestrial as well as marine environments and are primitively direct developers (Jenner, 2000; Sly *et al.*, 2003). The result of the focus on these phyla is that the second largest episode of metazoan body plan evolution, that of larval body plans, has been less appreciated. The majority of the 33 or so bilaterian phyla are primarily or exclusively marine and exhibit indirect development in which a larval form with a body plan distinct from the adult is present (Figure 5.1). A radical metamorphic event finally releases the adult form at the end of larval development. These phyla thus have a distinct second life-history stage—that of their larval forms. These larvae differ greatly from the adults in ecology, and are generally planktonic filter feeders whereas their adults are benthic and often effectively sessile. Such larvae have historically been called 'primary larvae', on the basis of the historical idea that larval forms represent the primitive body plans of ancestral metazoans.

5.3 Did adult or larval body plans arise first?

How did two distinct kinds of body plans evolve? The classic view, which derives from Haeckel's recapitulation theory, is that the first metazoans were similar to living larvae. Jägersten (1972) summarized it like this: '...the two phases of the life cycle arose when the adult of the primeval ancestor of the metazoans, viz, the holopelagic, radially symmetrical *Blastaea*, descended to life on the bottom (and became bilateral), while *its juvenile stage remained in the pelagic zone*'. Nielsen and Nørrevang (1985) and Nielsen (1995) in the same vein suggested that a pelagic gastraea animal evolved into a pelagic trochaea animal (that is, resembling a particular type of living feeding larva), which was ancestral to protostome and deuterostome phyla. Nielsen (2008) has linked this scenario to a foundation in sponge larvae.

This hypothesis was incorporated, in the developmental-genetic era, to mesh with inferences about gene regulatory systems (Davidson *et al.*, 1995). Gene regulatory systems of ancestral planktonic animals were hypothesized to resemble those found in living marine larvae (Figure 5.2). Bilaterian adults were suggested to have evolved through the innovation of imaginal 'set aside' cells distinct from the majority of differentiated larval cells. The imaginal cells gave rise to tissues of a new adult stage, and metamorphosis evolved to complete the transition. The new adults evolved a novel gene regulatory system similar to those of living adult bilaterians, including novel use of *Hox* genes to pattern the antero-posterior body axis.

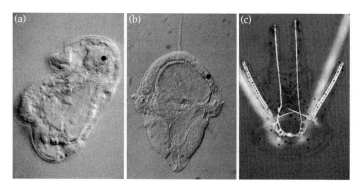

Figure 5.1 Examples of indirect-developing planktotrophic larval forms: two lophotrochozoans and a deuterostome: (a) Müller's larva of a platyhelminth flatworm; (b) trochophore of an annelid; (c) pluteus larva of a sea urchin. All are feeding larvae with guts and ciliary bands of various types. Parts (a) and (b) courtesy of G. Rouse.

This scheme explains the lack of an early metazoan body or trace fossil record as all evolution would have taken place in tiny planktonic adults. It ties larval forms into a phylogenetic scheme in which larval forms provide accessible proxies for the unfossilizable ancestors, and gives a developmental twist to the Cambrian radiation—the first fossil animals resulted from the appearance of new body plans.

There are difficulties for this interlinked suite of hypotheses (Sly *et al.*, 2003; Peterson 2005; Peterson *et al.*, 2005). Notably the larva-first hypothesis requires a vast number of convergent events, accounting for the massive molecular similarities in use of *Hox* and other regulatory genes in supposedly independently evolved descendant clades with benthic body plans. Further, somehow a selective role for set aside cells has to be accounted for before a new bilateral and benthic adult stage has evolved, which requires selection for novel developmental elements prior to need.

The planktonic metazoan ancestor has little evidence supporting it beyond analogies between the ontogeny of living larval forms and evolution of hypothetical ancestors. There is a second and more plausible evolutionary possibility, that the first bilaterians were just that, small benthic bilaterally symmetric triploblastic animals similar in complexity to living acoel flatworms (Figure 5.2). Molecular phylogenetic studies suggest that acoels may be the most basal living bilaterians (Ruiz-Trillo *et al.*, 2004; Sempere *et al.*, 2007), although this deep position is debated (Dunn *et al.*, 2008). Developmental data also support the position of acoels as basal bilaterians (Hejnol and Martindale, 2008a). Acoels are direct developers, and possess anterior, middle, and posterior group *Hox* genes (Baguñà and Riutort, 2004; Ramachandra *et al.*, 2002). The last common ancestor of protostomes plus deuterostomes (PD ancestor) was probably somewhat more complex than acoels, and possessed the genetic machinery basic to eye development, nephridia, heart, and other mesodermal tissues (Erwin and Davidson, 2002; Erwin, 2006; Baguñà *et al.*, 2008). This does not mean that these structures were present in derived states as seen in living protostomes or deuterostomes. It does mean that acquisition of bilaterian features was stepwise, with some features attained between the split from cnidarians to the acoelomorph grade, and further acquisitions from there to the PD ancestor. Further evolution of features characterizing the stem groups of phyla would have represented a third stage in evolution of features (Baguñà and Riutort, 2004).

The proposal of an ancestral benthic bilaterian ancestor requires a hypothesis for the secondary

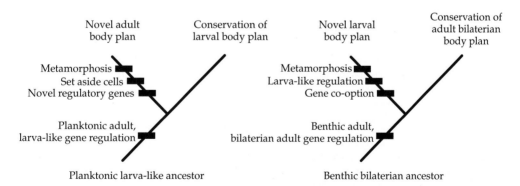

Figure 5.2 Conflicting larva-first and adult-first hypotheses of bilaterian origins. The hypotheses posit amounts of evolutionary change along branches leading to more derived developmental changes. In the larva-like ancestor hypothesis (left), most evolution of developmental characters lies on the branch leading to the benthic adult, with the larva retaining ancestral features. In the benthic bilaterian ancestor case (right), most evolution lies in the line to the planktonic larva, with the adult retaining ancestral features. Both hypotheses illustrate single lineages, but in the metazoan radiation, numerous lineages evolved in parallel. A large degree of homoplasy results in either case. The amount of convergence required to evolve planktonic larvae with their relatively simple organization is substantially less than evolving the entire basic suite of adult bilaterian features in 33 or so lineages.

evolution of the indirectly developing planktonic larvae, in place of the ancestral larval hypothesis. The inference of an ancestor lacking a larva has led to the intercalation model of larval origins (Valentine and Collins, 2000; Sly *et al.*, 2003). In this hypothesis, ancestral bilaterians are hypothesized to be small worm-like creatures, perhaps part of an acoelomorph radiation. These ancestral bilaterians were direct developers, and had evolved the basic developmental gene regulatory systems of bilaterian development. With the opening of the Cambrian radiation, the evolution of more divergent bilaterians accelerated, and produced the basal clades that gave rise to modern phyla (Budd and Jensen, 2000), but planktonic larvae and their body plans evolved secondarily.

The requirements for a planktonic larva are simpler than for the larger benthic reproductive adult. Table 5.1 separates the characters of the benthic PD ancestor from those selected for in the evolution of a planktonic larva. Larvae require ciliary bands for swimming and capture of microscopic prey. A mouth and gut are needed to process prey. Simple neural systems allow some control of muscle-cell contraction, for example in the pharynx. Other sensory information allows avoidance responses and ultimately detection of signals from the substrate biofilm to induce metamorphosis. For the development of a coherent larval symmetry, systems for the determination of the larval axes (animal–vegetal, dorsoventral, and left–right) are needed. In order for the switch from larval to adult development, a developmental switch that controls cellular fates has to be assembled from existing signalling systems in more primitive metazoans (Matus *et al.*, 2006a). Finally, a system for metamorphosis evolves, which probably initially involves transformation of most larval cells and tissue into adult tissues. However, slow metamorphosis increases vulnerability, and selection should favour evolution of a more rapid and efficient system using imaginal cells set aside as adult precursors within the larva to ensure rapid metamorphosis (Hadfield, 2000).

Sly *et al.*, (2003) predicted that some portion of genes required for adult development and life history would have been co-opted to direct the acquisition of a set of features involved in the simpler larval ontogeny required to produce a new life-history stage of an indirect-developing feeding larva. The acquisition of features would have involved step-wise intercalation of genes already used in the adult to generate features of the larva. The most basic requirement for feeding structures was probably met by the use of some of the adult gut programme. We have found evidence to support this idea in the common expression of genes in the gut of the sea urchin pluteus larva and in adult gut (Love *et al.*, 2008). Other features, for example the apical plate with its ciliary tuft, have co-opted unrelated sets of regulatory genes in sea urchin versus mollusc larvae (Dunn *et al.*, 2007). Larval evolution is suggested to have been a sequential assembly of features that would have diverted the ancestral course of development into two temporally distinct streams, one that first produced a feeding larva and a second stream that, from larval tissue, developed the juvenile adult. Imaginal cells and a discrete metamorphosis would have more sharply separated the two ontogenetic trajectories.

The second consequence of the intercalation hypothesis is that different metazoan lineages would simultaneously have evolved planktonic larvae. Convergence would have been highly prevalent as the rise of feeding larvae followed in time the splitting of metazoan phyla or their precursor lineages. These evolving lineages would have evolved planktonic larvae with features noted in Table 5.1,

Table 5.1 Characters required to evolve a planktonic feeding larva from a benthic bilaterian.

Characters required in larvae	Adult characters not required in larvae
Ciliary bands	Locomotory appendages
Gut	Respiratory system
Mouth	Reproductive organs
Simple neural/sensory system	Brain
Axial determination	Strongly expressed anteroposterior axis
Developmental switch to adult feature ontogeny	Nephridia
Metamorphosis	Eyes
	Circulatory system
	Skeleton

gained by co-option of different suits of regulatory genes to accomplish control of the development of broadly similar larval morphological structures. None the less, the convergence required would have been far less profound than that needed to independently evolve many lineages of bilaterians with the more complex features of the PD ancestor (Table 5.1).

5.4 Phylogeny and hypotheses of larval origins

The two hypotheses have distinct phylogenetic consequences with respect to mapping of developmental features onto evolutionary history. The scheme with a larva-like plan first is difficult to reconcile with recent phylogenies of bilaterian metazoan clades. First, molecular phylogenetic analyses do not support a metazoan phylogeny in which basal clades are indirect developers (Dunn et al., 2008). Jenner (2000) noted that the strongest data allowing a decision on the primitive developmental mode would come from phylogenetic studies in which a wide range of 'minor' non-coelomate phyla were included. He tested the occurrence of indirect versus direct modes of development using a phylogenetic tree on which minor as well as major phyla were mapped. Figure 5.3 shows an analogous tree. Direct development appears primitive in bilaterians and indirect-developing planktonic larvae have arisen independently in lophotrochozoans among the protostomes and in the echinoderm + hemichordate clade of deuterostomes. The other deuterostome clade, the chordates, is direct-developing. The echinoderms and hemichordates share a planktonic larval form, but the highly diverse lophotrochozoan clades (molluscs, annelids, brachiopods, bryozoans, nemertines, platyhelminths) have diverse larvae indicating a more complex history of multiple planktonic larval origins (Rouse, 2000; Peterson, 2005). Other protostome clades, notably the ecdysozoans (which includes arthropods, nematodes, and others) are direct-developing. Finally, the basal acoels and other minor clades (not shown) are direct developers. The mapping of the presence of planktonic larvae supports direct development as primitive in bilaterians, with separate origins of planktonic larvae in the echinoderm + hemichordate clade and in the lophotrochozoans.

5.5 Evidence from gene expression patterns

One potentially strong discriminator for homologous features is patterns of expression of developmental regulatory genes. This approach has had mixed success, because there has been extensive co-option of genes in evolution. There have been a small number of comparisons of gene expression patterns of putatively homologous features of protostome trochophore larvae (annelids and molluscs) with deuterostome diplurula larvae (echinoderms + hemichordates) to test for possible homologues at the level of gene deployment (Arendt et al., 2001; Dunn et al., 2007). A few genes show similar expression patterns. Others do not. The collection of genes is small and the sampling incomplete. The case of *nodal* illustrates the uncertainties. *Nodal* is involved in left–right

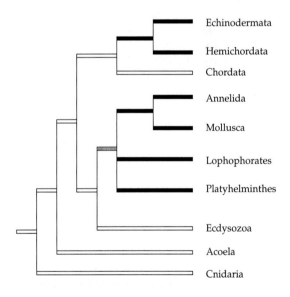

Figure 5.3 Developmental modes plotted on a metazoan phylogenetic tree. Open bars: direct development. Stippled box: ambiguous developmental mode. Filled boxes: planktotrophic indirect development. After Jenner (2000) and Peterson et al., (2005).

determination in echinoderms and vertebrates. However, it operates in a different domain (right side in echinoderms, left side in chordates), and interpretations of axial homologies are not yet possible (Duboc and Lepage, 2006). *Nodal* appears to have no role in *Drosophila*, an ecdysozoan. It has been reported from lophotrochozoans (Grande and Patel, 2009).

The small sample of cross-phylum comparisons of larval gene expression indicates homoplasy in larval gene expression (Raff, 2008). That could arise from a case of homology between the trochophore and the dipleurula, but taken with the phylogenetic considerations it appears more likely to represent a convergence in evolution of larval features accompanied by convergence in gene regulation. Convergence is likely because the structure of larvae is simpler than the structure of adult bilaterians, and because co-option of genes may have been related to shared adult and larval functions. Thus, the patterns of expression of *brachyury*, *Gsc*, and *Otx* might represent co-option of gene expression in adult oral developmental into development of similar larval oral structures, a sort of serial homology. Fully defining phylogenies and comparative gene expression will be advanced by genomic data. Most genome sequencing has concentrated on model or medically significant vertebrates, arthropods, and nematodes. A sea urchin genome has now been sequenced (Sodergren *et al.*, 2006), and the genomes of lophotrochozoans, especially planktotrophic marine annelids and molluscs, are still needed.

5.6   Hunting for the larval revolution in the fossil record

We have good fossil time markers for the visible appearance of diverse complex bilaterians in the Cambrian fossil record, which starts 542 million years ago (Ma), and especially in the famous mid-Cambrian Chengjiang and Burgess Shale faunas (520–505 Ma). The origin of bilaterians lies in the late Precambrian: recent estimates suggest somewhere between 580 and 600 Ma (Peterson *et al.*, 2005), although there is still a considerable range of uncertainty. An estimate of the timing for evolution of planktonic larvae of about 500 Ma is emerging,

which if correct puts the origin of these second body plans as much as 100 million years later than the divergence of the basal bilaterian benthic adult. Signor and Vermeij (1994) noted that the Cambrian fossil record showed relatively few benthic suspension feeders or planktonic forms. They suggested that the evolution of planktonic feeding larvae took place in the late Cambrian to early Ordovician, driven by an expansion of plankton and sanctuary from predation—a point reinforced by Peterson (2005). The radiation of indirect-developing feeding larval in the late Cambrian was probably driven by a number of selective forces on larval traits as the Cambrian radiation produced a marine trophic organization similar to that of recent times (Dunne *et al.*, 2008). Table 5.2 presents a number of evolutionary considerations affecting larvae, including ecological factors as well as effects of developmental and population features. These include egg size and provisioning (Allen and Pernet, 2007; McEdward, 2000; Moran, 2004), which dictate egg numbers and feeding or non-feeding larval forms. In addition, fully indirect development requires rapid metamorphosis and co-option of signal systems (Hadfield, 2000). Finally, predation selects for a shorter planktonic larval life, but the length of planktonic life affects features from genetic differentiation within populations of a species to the distance that a species can disperse (Shulman and Bermingham, 1995).

The list in Table 5.2 is one of interacting characters. That is, if initial evolution of feeding larvae was driven by ecological conditions favouring

Table 5.2   Likely selective forces acting on evolution of planktonic larvae in Cambrian seas.

Trait	Advantage
Larval feeding	Lower investment in each egg
Lower investment in individual egg	More eggs produced
Higher egg numbers	Increase number surviving
Planktonic swimming	Escape benthic predators
Planktonic feeding	Exploit new niche
Motility and time in water column	Increase 'dispersibility'
Character displacement vs adult	Lower ecological competition
Metamorphosis	Rapid developmental shift in body form

feeding on plankton, selection is likely to have initially favoured co-option of adult features and genes that introduced motility and feeding structures into initial stages of larval development. Selection would then have favoured rapid development and metamorphosis to reduce vulnerability to predation. Other potentially advantageous traits such as dispersibility might conflict and in turn favour selection of the production of large numbers of small eggs to reduce the effects of predation. The rapid evolution of diverse modes of development among related living marine species shows that a variety of selective domains exist and continue to influence larval evolution (Raff and Byrne, 2006).

Direct fossil evidence for larval evolution comes from exquisite phosphoritic preservation of late Proterozoic cleavage stage embryos of unknown taxa, notably the embryos of the Doushantuo fauna (Xiao *et al.*, 1998, Hagadorn *et al.*, 2006). Early to mid Cambrian developmental series have been reported that include larval forms of cnidarians and small nemathelminth ecdysozoans (Bengtson and Zhao, 1997; Budd, 2004; Donoghue *et al.*, 2006b; Maas *et al.*, 2007). An understanding of how embryos can be preserved for mineralization is emerging (Briggs, 2003; Raff *et al.*, 2006, 2008; Gosling *et al.*, 2008). The early fossil embryos so far described are large, ranging from 350–1100 mm (Xiao and Knoll, 2000) for late Precambrian embryos to 350–750 mm for early to mid Cambrian embryos (Steiner *et al.*, 2004; Donoghue *et al.*, 2006b). There are biases in the record, notably low taxonomic diversity (Donoghue *et al.*, 2006b). The possibility that small embryos typical of indirect-developing marine animals (50–200 mm) exist has been checked by Donoghue *et al.*, (2006b), but does not appear to be the case. Fossil embryo evidence for the time of appearance of indirect developing forms is still scarce. Nützel *et al.*, (2006) have observed that Cambrian larval mollusc shells are larger than those of the Ordovician and Silurian, consistent with a shift from direct to indirect development by the end of the Cambrian.

5.7 Fossils, larvae, and Linnaeus

Linnaeus created a systematic approach that created a static hierarchical system of classification, which has lent itself to evolutionary interpretation, and ultimately to modern phylogenetics. The fossil record has supplied crucial information for phylogenetics and evo-devo (Raff, 2007). Larval and adult characters have produced homoplasies that yield some contradictory phylogenetic inferences among some of the deepest Linnean taxa. Thus, the trochophore larvae of annelids and molluscs carry a different phylogenetic signal from their adult body plan features. Rather than seeing these characters as conflicting, a better knowledge of the Cambrian fossil record of clades basal to living phyla allows us to dissect more finely the timing of evolution of both adult and larval body plans. Halwaxiids and their kin are sclerite-bearing middle Cambrian animals that lie somewhere among basal forms in a clade that includes molluscs, annelids, and brachiopods (Conway Morris and Caron, 2007).

The characters of larval forms show some linkages between phyla obscured by changes in adult morphology, and in fact agree with phylogenetic inferences based on gene sequence data. Thus, the trochophore shared by annelids and molluscs belies segmentation and paired appendages shared by annelids and arthropods, the so-called Articulata. The existence of these forms suggests that the primitive trochophoran larva may have its origin in a Cambrian clade living before the split of the lophotrochozoan phyla. This would move the time of larval origin to earlier in the Cambrian. This might suggest that the earliest planktotrophic larvae have not yet been detected, or that the full suite of planktonic feeding features were acquired slowly, and included convergences among related lineages (Rouse, 2000). Similarly, the dipleurula larva links the pentameral echinoderms with the bilaterian worm-like hemichordates, indicating that the origins of this larval form occurred after the split of this clade from chordates. Basal chordates and echinoderms are present in mid Cambrian strata.

5.8 Gene co-option continues to occur in larval evolution

Larvae did not cease evolving in the Cambrian. First, novel features evolved in planktonic larvae after the initial evolution of larval body plans. This kind of evolution has been inferred by Rouse (2000) for downstream feeding in trochophore-like larvae by analysis of the distribution of features in a phylogeny of lophotrochozoan clades. Among deuterostomes, we have analysed the arms of the sea urchin pluteus larva (Figure 5.4a). This is an indirect-developing feeding planktonic larva derived from the basal dipleurula-type larva of echinoderms. The pluteus has, since the split of sea urchins from other crown echinoderm classes about 450 Ma, evolved long arms that contain a novel rigid calcium carbonate skeleton and which bear the circumoral ciliary band (Bottjer *et al.*, 2006). These arms evolved somewhere between the late Ordovician and the Permian and thus followed the initial evolution of the dipleurula. The pluteus arm is a novel larval organ (Love *et al.*, 2007). The arms consist of an ectoderm bearing a ciliary band and an underlying mesoderm consisting of skeletogenic mesenchyme cells. Expression of particular genes occurs in the tips of the growing arms (e.g. tetraspanin in ectoderm and advillin and carbonic anhydrase in mesenchyme). These genes also are expressed in various adult tissues. Their role in larval arms indicates that they have been recruited for expression in these structures following the origin of the dipleurula. This recruitment serves as an accessible proxy for the more remote events of the Cambrian.

A second type of larval evolution is that of various non-planktotrophic derivatives of larvae in various clades, for example snails (Collin, 2004) and starfish and sea urchins (Raff and Byrne, 2006). In many taxa, planktonic feeding larvae have given rise to non-feeding, direct-developing planktonic or brooded larvae, and even viviparous larvae. These modified larvae rapidly evolve distinct morphologies, as seen in the larvae of the congeneric sea urchins *Heliocidaris tuberculata* and *Heliocidaris erythrogramma* (Figure 5.4), which diverged about 4 Ma (Zigler *et al.*, 2003). *Heliocidaris tuberculata* takes about 6 weeks of feeding in the water column to reach metamorphosis. *Heliocidaris erythrogramma* takes 3 days, and does not feed. The *H. erythrogramma* egg is 100 times the volume of that of indirect-developing sea urchins and

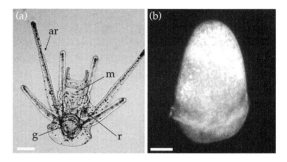

Figure 5.4   Rapid evolution of larvae shown by two congeneric sea urchins, diverged for 4 million years. (a) Planktotrophic pluteus larva of the indirect developer *Heliocidaris tuberculata*. The notable features are the arms (ar), each supported by a skeletal rod, and bearing a ciliary band; the large gut (g); the mouth (m); and the developing adult rudiment (r) that will grow to become the juvenile sea urchin released at metamorphosis (about 6 weeks' post-fertilization). (b) Non-feeding direct-developing larva of *Heliocidaris erythrogramma*. All internal features are those of the developing adult. Metamorphosis is 3–4 days' post-fertilization. Scale bar in both parts = 100 μm.

supports development through post-metamorphic development of the adult mouth.

At first glance it would appear that *H. erythrogramma* is simplified by the loss of larval features but retains adult ontogeny. Some feeding structures such as the larval arms and gut are lost, but developmental features retain a high degree of complexity, and dramatic novel features have appeared. These include changes in oogenesis and spermatogenesis, in maternal embryonic axis determination, in cleavage pattern, in embryonic cell lineages, and in heterochronies in larval gene expression and morphogenetic events (Raff and Byrne, 2006). Rapid and profound evolutionary changes in larval development occur frequently, with for example several clades of sea urchins having independently evolved larvae similar to that of *H. erythrogramma* (Sly *et al.*, 2003). The evolutionary lability of larvae suggests that evolution of primary larval features would have been rapid in the face of selection under the new ecological regime of the late Cambrian and early Ordovician. It is also likely that the developmental regulatory features of living larval clades give us strong clues to those of early larval forms.

5.9 Developmental innovations and the metazoan radiation

The origin of the ancestral benthic bilaterian body plan was an immense evolutionary developmental innovation that produced a shift from the cnidarian frond-dominated world of the late Proterozoic to the diversified bilaterian-dominated world of the Cambrian. However, evolution of novel developmental features depends both on the appearance of variation in development and on selection acting on developmental stages and processes. Developmental features of early metazoans may have been less constrained by a looseness of ecological fit, resulting in more experimentation with body plans, i.e. adaptive peaks were present, but in a fairly flat landscape where few deep valleys of low fitness were yet found. The rapid diversification of basal taxa related to living phyla was probably the result of ecological pressures and opportunities that selected for the development of novel morphologies among bilaterians of relatively simple morphology. Acoelomorph bilaterian ancestors would have possessed a large suite of genes regulating development that could be recruited for the evolution of new structures. The possibilities for body plan innovation in acoelomorph-grade animals would, in many respects, have been easier than for proposed schemes that suggest divergence from more derived ancestors. Thus, the dorsal–ventral inversion of organs of protostomes and deuterostome would have been of little consequence at the acoelomorph grade of organization, but could have become a fixed element of body plan later. Segmentation, another feature of importance, may also be a product of convergence in emerging lineages (Seaver, 2003).

The evolution of planktonic larvae followed the origins of basal bilaterian phyla by about 100 million years. Again, it less likely that developmental novelties *per se* drove this evolutionary innovation. Instead, larvae bearing features arising from the novel expression of genes used in adults were selected as agents of exploitation of greater ranges of ecological possibility, such as increasing planktonic food resources, escape from benthic filter-feeding predators, and a vastly improved dispersal than that offered by large direct-developing embryos. The evolutionary flexibility of larval development allowed diverse and rapid responses to selection. Selection on expression of existing genes in new contexts may underlie much of the evolution of novelties in development.

CHAPTER 6

Assembling the spiralian tree of life

Gonzalo Giribet, Casey W. Dunn, Gregory D. Edgecombe,
Andreas Hejnol, Mark Q. Martindale, and Greg W. Rouse

The advent of numerical methods for analysing phylogenetic relationships, along with the study of morphology and molecular data, have driven our understanding of animal relationships for the past three decades. Within the protostome branch of the animal tree of life these data have sufficed to establish two major clades—Ecdysozoa, a clade of animals that all moult, and Spiralia (often called Lophotrochozoa), a clade whose most recent common ancestor had spiral cleavage. In this chapter we outline the current knowledge of protostome relationships and discuss future perspectives and strategies to increase our understanding of relationships within the main spiralian clades. Novel approaches to coding morphological characters are a pressing concern, best dealt with by scoring real observations on species selected as terminals. Methodological issues, such as the treatment of inapplicable characters and the coding of absences, may require novel algorithmic developments. Taxon sampling is another pressing issue, as terminals within phyla should include enough species to represent their span of anatomical disparity. Furthermore, key fossil taxa that can contribute novel character state combinations, such as the so-called 'stem-group lophotrochozoans', should not be neglected. In the molecular forum, expressed sequence tag (EST)-based phylogenomics is playing an increasingly important role in elucidating animal relationships. Large-scale sequencing has recently exploded for Spiralia, and phylogenomic data are lacking from only a few phyla, including the three most recently discovered animal phyla (Cycliophora, Loricifera, and Micrognathozoa).

While the relationships between many groups now find strong support, others require additional information to be positioned with confidence. Novel morphological observations and phylogenomic data will be critical to resolving these remaining questions. Recent EST-based analyses underpin a new taxonomic proposal, Kryptrochozoa (the least inclusive clade containing the Brachiopoda and Nemertea).

6.1 Introduction

The protostomes consist of Chaetognatha, a relatively small group of uncertain affinity, and two megadiverse clades—Ecdysozoa and Spiralia (the latter also sometimes referred to as Lophotrochozoa). Spiralia, which includes many kinds of worms, flatworms, molluscs, and related animal groups, comprises a greater number of animal phyla than any other non-overlapping metazoan clade. Specifically, these are Annelida (subsuming several former phyla: Echiura, Pogonophora, Sipuncula, Vestimentifera, and perhaps Myzostomida), Brachiopoda, Bryozoa, Cycliophora, Entoprocta, Gastrotricha, Gnathostomulida, Micrognathozoa, Mollusca, Nemertea, Phoronida, Platyhelminthes, and Rotifera (including the former phylum Acanthocephala) (Giribet, 2002, 2008; Halanych, 2004; Matus et al., 2006b; Dunn et al., 2008). This amount of phyletic diversity within Spiralia adds up to about half of the traditional extant animal phyla and, in terms of species numbers, includes the second largest phylum (Mollusca) as well as the two phyla with some of the largest body plan disparity (Mollusca and Annelida). Although the relationships among these phyla have remained controversial—Ecdysozoa versus Articulata issues aside (e.g. see reviews in Giribet, 2003; Scholtz, 2002)—recent phylogenomic analyses have shed light on the subject. The goals of this chapter are to review the relationships of the spiralian phyla and the techniques for studying such relationships, to establish a current working framework for the main divisions within the clade, and to formalize the supraphyletic classification of Spiralia following current phylogenetic views.

6.1.1   Protostome groups and affinities

Developmental characters such as the fate of the blastopore—which often becomes the adult mouth (see Chapter 4)—and the mode of formation of the mesoderm are typically cited features for supporting a protostome clade (Nielsen, 2001). Depending on composition, the protostomes have sometimes been considered a paraphyletic assemblage of worm-like animals characterized by the presence of a dorsal (or circumesophageal) brain connected to a ventral longitudinal nerve cord, often paired. The proposal that acoels and nemertodermatids are basal bilaterian animals (Ruiz-Trillo *et al.*, 1999, 2002; Jondelius *et al.*, 2002) rather than Platyhelminthes, renders the traditionally formulated Protostomia (Nielsen, 2001) paraphyletic, since the deuterostomes are then closer to the remaining protostomes. Neither nemertodermatids nor acoels have a ventral centralized nerve cord (e.g. Raikova *et al.*, 2004a,b). From this evidence, and the current phylogenetic framework for metazoans, acoels and nemertodermatids are not considered part of Protostomia in the following discussion, rendering Protostomia monophyletic given our current understanding of metazoan phylogeny.

Ecdysozoa is currently recognized as monophyletic in most analyses. Some recent genome-wide analyses have questioned the validity of the clade (Blair *et al.*, 2002; Dopazo *et al.*, 2004; Philip *et al.*, 2005; Rogozin *et al.*, 2007b; Wolf *et al.*, 2004)—a result that now appears to be due to poor taxon sampling (Philippe *et al.*, 2005a; Dunn *et al.*, 2008). Ecdysozoa is discussed in detail by Telford and colleagues in Chapter 8 of this volume (see also

Telford *et al.*, 2008). The name Spiralia was first coined by Schleip (1929) because of the stereotypical spiral development that occurs only within this clade (Nielsen, 2001; Maslakova *et al.*, 2004a). Spiralia contains all animals with spiral development. This character, like many others within Metazoa, shows homoplasy, apparently in the form of secondary reduction—but never as convergence outside the clade. This indicates that any animal with spiral development is an unambiguous member of the clade Spiralia, although the absence of this type of development does not discriminate against its membership.

In one of the first uses of phylogenetic nomenclature, Lophotrochozoa was defined by Halanych *et al.*, (1995) as a node-based name, defined as the 'last common ancestor of the three traditional lophophorate taxa, the mollusks, and the annelids, and all of the descendants of that common ancestor'. This referred to a subgroup of Spiralia, perhaps being synonymous with Trochozoa, the exact scope of which depends on the placement of lophophorates. Aguinaldo *et al.*, (1997) later emended Lophotrochozoa by listing a set of taxa that they viewed as being part of this taxon, namely 'annelids, molluscs, rotifers, phoronids, brachiopods, bryozoans, platyhelminths and related phyla', thus including all or almost all non-ecdysozoan protostomes. This delineation approach to a taxon is in conflict with the original phylogenetic definition of Lophotrochozoa and has resulted in considerable confusion in the literature ever since. The clade names Lophotrochozoa and Spiralia have tended to be used as synonyms, based on Aguinaldo *et al.*, (1997), while in other cases the original phylogenetic definition of Lophotrochozoa is used and it is shown as a subtaxon of Spiralia. We prefer to employ the name Lophotrochozoa in the spirit of how it was originally defined, though Halanych *et al.*, (1995) were not clear as to whether or not they included Entoprocta as part of Bryozoa, as proposed by Nielsen (2001). There is some recent evidence that Bryozoa should include Entoprocta (Hausdorf *et al.*, 2007), in which case this would be moot. One primary reason why neither Lophotrochozoa nor Spiralia has stabilized in usage is that the relationships among the specifying taxa defining Lophotrochozoa (lophophorate taxa, molluscs, and annelids) have yet to be resolved. This is particularly problematic with regard to the lophophorate taxa. The first phylogenomic studies that include lophophorates indicate that Lophotrochozoa is nested within a clade of animals with spiral cleavage (Helmkampf *et al.*, 2008a,b) or perhaps equivalent to Spiralia (Figure 6.1) (Dunn *et al.*, 2008). In the case of synonymy, Lophotrochozoa is the junior synonym, since Spiralia has precedence. We, like many other current authors (e.g. Hausdorf *et al.*, 2007; Henry *et al.*, 2007; von Döhren and Bartolomaeus, 2007; Helmkampf *et al.*, 2008a), therefore use Spiralia rather than Lophotrochozoa, both based on the fact that it has precedence and because it refers to a clear synapomorphy.

Spiralia has been suggested to comprise two putative clades, Platyzoa (Cavalier Smith, 1998) and Trochozoa (Roule, 1891). Trochozoa is preferred (see Rouse 1999; Giribet *et al.*, 2000) over the more recently coined Eutrochozoa (Ghiselin, 1988) used by some authors (Eernisse *et al.*, 1992; Valentine, 2004). Whether or not Platyzoa is a clade (e.g. Zrzavý, 2003; Glenner *et al.*, 2004; Todaro *et al.*, 2006; Dunn *et al.*, 2008) remains unclear, as many

analyses provide low support for the exact position of the 'platyzoan' phyla, although most tend to group them in a clade or in a grade giving rise to Trochozoa.

The identification by Dunn *et al.*, (2008) of a core set of stable taxa whose relationships are well supported provides a more detailed picture of Platyzoa. The only stable taxon putatively assigned to Platyzoa was Platyhelminthes (Dunn *et al.*, 2008, their Figure 2), which was found to be sister to Trochozoa with strong support in analyses restricted to stable taxa. All other platyzoans were unstable in these analyses (Dunn *et al.*, 2008, their Figure 1 and Supplement), and their position could not be resolved with confidence. All platyzoan taxa investigated to date have relatively long branches, which has led some authors to suspect that support for the group is a systematic error (Telford, 2008). The well-supported position of Platyhelminthes as sister to Trochozoa (which cannot be a result of long branch attraction since Trochozoa does not contain taxa with long branches) may serve as an anchor that is spuriously attracting other long-branch taxa whose placement does not have

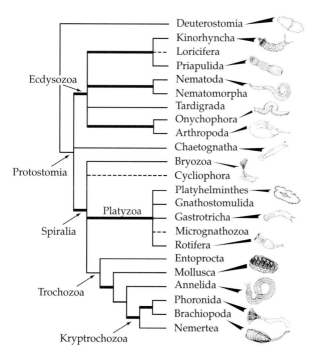

Figure 6.1 Hypothesis of protostome relationships mostly based on recent phylogenomic analysis and morphology. This tree has not been generated by a consensus or other numerical technique. Protostomes are divided into two sister clades, Ecdysozoa and Spiralia, the latter divided into Platyzoa and Trochozoa. Phyla with dashed lines lack phylogenomic data and are placed on the tree based mostly on morphology. The thickness of internal branches reflects support. Kryptrochozoa is a new supraphyletic taxon proposed here.

strong signal. The resolution of this problem (including tests of this specific hypothesis) will require a two-pronged strategy of greatly improved taxon sampling within putative platyzoan groups and detailed investigations of the inference of their position designed specifically to identify systematic error.

An unequivocal apomorphy for Platyzoa (Figure 6.2a,b and Plate 2) is hard to delineate morphologically. The putative clade contains a series of acoelomate or pseudocoelomate animals—no coelomates belong to this group—some with special types of jaws formed of cuticularised rods, or Gnathifera (Gnathostomulida, Micrognathozoa, and Rotifera) (Kristensen and Funch, 2000; Sørensen, 2003; see Figure 6.2b). Platyzoa also includes other types of flatworms, including Platyhelminthes (Figure 6.2a) and Gastrotricha (but see Zrzavý, 2003, for an alternative position of gastrotrichs as sister to Ecdysozoa). With the exception of polyclad flatworms, and the parasitic acanthocephalan rotifers, platyzoans are strict direct developers. The possible membership of Cycliophora within Platyzoa remains a contentious issue (see Giribet et al., 2004).

Trochozoa (Figure 6.2d–f; see also 6.4a–d) contains those groups with a typical trochophore larva, namely Annelida (Figure 6.2g), Mollusca (Figure 6.2f), and Entoprocta (Figure 6.2e), and also includes some lophophorates. Recently, some nemertean larvae have been interpreted as modified trochophores, with a vestigial prototroch (Maslakova et al., 2004b). Trochophores do not occur outside the clade, but several ingroup members do not develop through a trochophore. Such is the case, for instance, for Brachiopoda and Phoronida (see Figure 6.4a,c). Even within groups where a trochophore is widespread there are clear cases of it being lost (e.g. clitellate annelids and cephalopod molluscs).

The position of the chaetognaths (Figure 6.2h), also known as 'arrow worms', has been debated for a long time, but recent studies based on morphology (Harzsch and Müller 2007) and phylogenomics (Marlétaz et al., 2006; Matus et al., 2006b; Dunn et al. 2008) strongly suggest that they are early diverging relatives of protostome taxa despite having deuterostome-like development. Beyond

this, though, there is little clarity as to their exact position. Depending on character and taxon sampling, they have been placed as sister to or within Spiralia (Matus et al., 2006b; Dunn et al., 2008, Helmkampf et al., 2008b), within Ecdysozoa (Matus et al., 2006b; Helmkampf et al., 2008a,b), or as sister to Spiralia + Ecdysozoa (Marlétaz et al., 2006; Matus et al., 2006b; Dunn et al., 2008). The nervous system of chaetognaths has recently been found to be similar to that of other protostomes in having a typical circumoral arrangement of the anterior CNS (Harzsch and Müller, 2007). Resolving the placement of Chaetognatha is critical to the reconstruction of some of the most basic developmental characters, including cleavage mode and the fate of the blastopore.

Cycliophoran affinities with entoprocts (Figure 6.2e) are still under consideration (Giribet et al., 2004), and one of the cycliophoran larval forms, the chordoid larva, has been interpreted as a modified trochophore (Funch, 1996). Trochozoa includes, in general, larger protostomes than Platyzoa. Many trochozoans are coelomates with large body cavities and metanephridia-based excretory organs in the adult phase. Some, however, are functionally acoelomate, with protonephridial excretory systems as adults (Nemertea), or acoelomates, and therefore with protonephridia as excretory organs.

6.1.2 Problematica

One of the most problematic protostome groups, not just in terms of its phylogenetic placement, is the symbiotic group Myzostomida (Eeckhaut and Lanterbecq, 2005). From its original description, *Myzostoma cirriferum* (Figure 6.2c) was considered a trematode (Leuckart, 1827). Since then, myzostomids have been associated with crustaceans, tardigrades, pentastomids, and with polychaete annelids. Jägersten (1940) grouped Myzostomida and Annelida (as two separate classes) into a protostome group called Chaetophora. Based on the shared unusual ultrastructure of the sperm in myzostomes and acanthocephalan rotifers, with a pulling (instead of pushing) flagellum, both groups were classified into the phylum Procoelomata (Mattei and Marchand, 1987). However, due to the

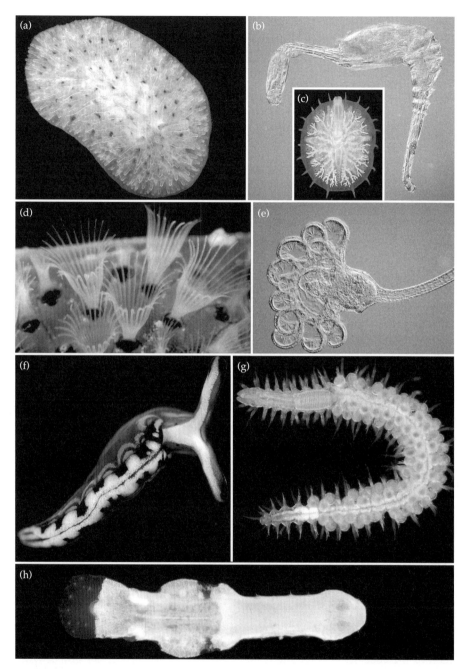

Figure 6.2   Examples of some spiralian taxa: (a), (b) Platyzoa; (c), (d), (h) uncertain; (e)–(g) Trochozoa. (a) The free-living platyhelminth *Hoploplana californica*. (b) An undescribed species of seisonid rotifer *Paraseison* taken from its crustacean host *Nebalia*. (c) A myzostomid *Myzostoma cirriferum* taken from its crinoid host. (d) Several zooids of a bryozoan colony. (e) Anterior end of entoproct *Pedicellina* sp. (f) Dorsal view of the sacoglossan mollusc *Thuridella picta*. (g) A syllid polychaete annelid brooding embryos on its dorsum. (h) A benthic spadellid chaetognath, *Spadella*, All photographs by G. W. Rouse. (See also Plate 2.)

presence of segmentation, parapodia-like structures with chaetae and aciculae, and an apparent trochophore larva, many modern authors consider myzostomes as polychaete annelids (e.g. Brusca and Brusca, 2003; Nielsen, 2001; Rouse and Fauchald, 1997; Rouse and Pleijel, 2001; but see Haszprunar, 1996b). Molecular phylogenetic studies of myzostomids have suggested rather contradictory hypotheses, with some analyses suggesting a relationship to rotifers (including acanthocephalans) and cycliophorans (Zrzavý *et al.*, 2001)—a somehow expanded 'Procoelomata' of Mattei and Marchand (1987)—or as closer to Platyhelminthes than any other spiralians (Eeckhaut *et al.*, 2000). Both cases presuppose a platyzoan affinity of myzostomes, as opposed to their more traditional trochozoan kinship. However, a recent molecular analysis using mitochondrial genome data and multiple nuclear genes suggests that myzostomes should be placed back among the annelids (Bleidorn *et al.*, 2007). This debate is probably not yet settled, as EST data do not currently support an annelid affinity for myzostomids (Dunn *et al.*, 2008).

The position of several other spiralian clades has remained elusive even in recent phylogenomic analyses using large numbers of genes (Dunn *et al.*, 2008). Such is the case for bryozoans, entoprocts, gastrotrichs, gnathostomulids, and rotifers. Despite most of these taxa forming part of Platyzoa, their position was unstable across analytical methods or model selection, and nodal support for their exact position was not conclusive. Results of those analyses also differed with respect to the positions of bryozoans (Figure 6.2d) and entoprocts (Figure 6.2e) compared with another recent phylogenomic analysis (Hausdorf *et al.*, 2007), which suggested a sister-group relationship between these two phyla, although with a much smaller taxon sampling. A reason for the instability found in Dunn *et al.*, (2008), at least for some of these phyla, was a poor EST representation due to shallow library examination (e.g. only *c.* 1000 clones sequenced, as in the case of myzostomids) or poor library quality (as in the case of entoprocts). This leaves a clear strategy for improving our understanding of the relationships of these phyla—sequencing additional clones or developing additional libraries.

6.1.3 Working framework for spiralian relationships

All the previous work, and especially the recent resolution obtained based on EST data for broad taxon sampling, has contributed to a roadmap for resolving metazoan relationships (e.g. compare the resolution of the spiralian trees in the recent reviews by Giribet *et al.*, 2007, or Giribet, 2008, with the phylogenomic results of Dunn *et al.*, 2008). Focusing on the spiralians—the topic of this chapter—previous work has delimited two putative main clades and their core membership. The affinities of bryozoans, chaetognaths, cycliophorans, and myzostomids may still be debatable, but it seems that the remaining protostome phyla (following the definition provided above) either belong to Platyzoa (Gastrotricha, Gnathostomulida, Platyhelminthes, Rotifera) or Trochozoa (Annelida, Brachiopoda, Entoprocta, Mollusca, Nemertea, Phoronida) (Figures 6.1, 6.2, and 6.4). Interestingly, only Trochozoa includes members with chaetae (Annelida, Brachiopoda) (Hausen, 2005; Lüter, 2000). Other chaeta-like structures in chitons (Leise and Cloney, 1982), juvenile octopods (Brocco *et al.*, 1974), and the gizzard-teeth of a bryozoan (Gordon, 1975) are often considered to be convergent (Hausen, 2005), but could instead be plesiomorphic for trochozoans. Resolution of the placement of myzostomids, with their unequivocal chaetae (Lanterbecq *et al.*, 2008), may provide further clarification as to how this feature has evolved.

Relationships among the phyla that constitute Platyzoa are not well established (e.g. Giribet *et al.*, 2004; Dunn *et al.*, 2008), although good apomorphies exist for one of its subclades, Gnathifera (Kristensen and Funch, 2000; Sørensen 2001, 2003). The internal phylogenies of several of these phyla are well understood as total evidence and multilocus analyses have been published for Gnathostomulida (Sørensen *et al.*, 2006), Platyhelminthes (Figure 6.2a) (Littlewood *et al.*, 1999), Rotifera (Figure 6.2b) (Sørensen and Giribet, 2006), and Gastrotricha (Zrzavý, 2003).

Many trochozoan internal relationships have been recently resolved with strong support. Monophyly of annelids (Figure 6.2g) and its membership have been corroborated only in recent phylogenomic analyses (Hausdorf *et al.*, 2007; Dunn *et al.*, 2008). There now seems to be a consensus about the affinities of the former phyla Echiura, Sipuncula, and Pogonophora/Vestimentifera as highly modified annelid subtaxa (McHugh, 1997; Hessling and Westheide, 2002; Hausdorf *et al.*, 2007; Struck *et al.*, 2007; Dunn *et al.*, 2008). In spite of this progress, resolution of relationships within Annelida, especially as to the placement of the 'root' of the annelid tree, is far from agreed upon. Molecular analyses (e.g. Bleidorn *et al.*, 2003; Rousset *et al.*, 2004, 2007; Colgan *et al.*, 2006; Struck *et al.*, 2007) radically contrast with the most comprehensive analyses of annelid relationships based on morphology (Rouse and Fauchald, 1995, 1997; Rouse and Pleijel, 2001; see a recent review in Rouse and Pleijel, 2007). Without doubt, an important leap is needed in the number of data to be incorporated into annelid phylogenetic studies.

Internal relationships of the other large trochozoan phylum, Mollusca (Figure 6.2f), do not present a much brighter picture. Relationships based on morphology (e.g. Salvini-Plawen and Steiner, 1996; Haszprunar, 2000) and molecules (e.g. Passamaneck *et al.*, 2004; Giribet *et al.*, 2006) are still at odds for relationships within this group, and molecular analyses have traditionally had trouble recovering molluscan monophyly until the incorporation of phylogenomic data (Hausdorf *et al.*, 2007; Dunn *et al.*, 2008). Only a recent multilocus analysis of molluscan relationships was able to recover monophyly, although with low clade support (Giribet *et al.*, 2006), and the internal relationships among the molluscan classes were for the most part not resolved.

Internal sipunculan phylogeny has been addressed in recent times, based on both morphology and molecular analyses (e.g. Maxmen *et al.*, 2003; Schulze *et al.*, 2005, 2007), and it is now becoming clear that Sipuncula is affiliated with Annelida (Struck *et al.*, 2007; Dunn *et al.*, 2008), though their closest annelid relatives have yet to be established. The phylogeny of Brachiopoda has also been assessed in a range of analyses (Carlson, 1995; Cohen *et al.*, 1998; Cohen, 2000; Cohen and Weydmann, 2005; the latter two also including several phoronid species), and some authors have proposed that Phoronida may be a subgroup of Brachiopoda

(Cohen, 2000; Cohen and Weydmann, 2005; but see Bourlat *et al.*, 2008; Helmkampf *et al.*, 2008a). Although several higher-level morphological and molecular analyses exist within the phylum Nemertea (e.g. Sundberg *et al.*, 2001; Thollesson and Norenburg, 2003), results still depend on few markers and are not integrated with morphological analyses. Little synthetic phylogenetic work has been published on the phylogeny of entoprocts or bryozoans.

6.1.4 Controversial fossils: molluscs, annelids, brachiopods, or stem-group Spiralia?

The Palaeozoic fossil record is rich in protostome taxa (Budd and Jensen, 2000; Valentine, 2004), and in general it is thought that most metazoan phyla were already present in the Cambrian. Recent discoveries of Lower Cambrian sipunculans (Huang *et al.*, 2004) and chaetognaths (Szaniawski, 2005; Vannier *et al.*, 2007) reduce the number of animal phyla missing from the Palaeozoic record. The most conspicuous phylogenetic gap in the Palaeozoic record is for Platyzoa (Figure 6.1), no members of which have yet been found (Giribet, 2008).

Although considerable advances have recently been made in understanding the morphology and phylogenetic context of several potentially pivotal spiralian fossils, disagreement about the interpretation of structures relative to extant phyla have left the picture clouded. Especially relevant fossils are the sclerotome-bearing *Wiwaxia* (Butterfield, 1990; Conway Morris, 1985; Eibye-Jacobsen, 2004) (Figure 6.3a and Plate 3), *Orthrozanclus* (Conway Morris and Caron, 2007) (Figure 6.3c), and *Halkieria* (Conway Morris and Peel, 1995; Vinther and Nielsen, 2005) (Figure 6.3b), and the unarmoured *Odontogriphus* (Caron *et al.*, 2006) (Figure 6.3e). Much of the controversy about these fossils—which are at least uniformly recognized as spiralians but then variably assigned to either Annelida, Mollusca, or Brachiopoda—is encapsulated in a debate over whether a clearly homologous feeding apparatus in *Odontogriphus* and *Wiwaxia* is (Scheltema *et al.*, 2003; Caron *et al.*, 2006, 2007), or is not (Butterfield, 2006, 2008), a radula, and whether these animals are (Butterfield, 2006), or are not (Eibye-Jacobsen, 2004; Caron *et al.*, 2007), segmented.

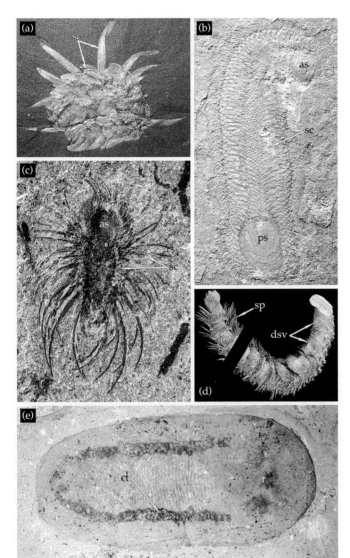

Figure 6.3 Exceptionally preserved Palaeozoic spiralian fossils. (a) *Wiwaxia corrugata* (Middle Cambrian, photo courtesy of Jean-Bernard Caron). (b) *Halkieria evangelista*, sclerites (sc) anterior shell (as) and posterior shell (ps) (Lower Cambrian, photo courtesy of Jakob Vinther). (c) *Orthrozanclus reburrus*, anterior shell (as), sclerites (sc) (Middle Cambrian, photo courtesy of Jean-Bernard Caron). (d) *Acaenoplax hayae*, dorsal shell plates (dsv), spines (sp) (Silurian, digital reconstruction courtesy of Mark Sutton). (e) *Odontogriphus omalus*, radula (r) and ctenidia (ct) (Middle Cambrian, photo courtesy of Jean-Bernard Caron). (See also Plate 3.)

Wiwaxia and *Halkieria* have long been associated, based on similar sclerite morphology and left–right sclerite zones (Bengston and Conway Morris, 1984), and some recent studies unite them (with *Orthrozanclus* and several other fossil 'coeloscleritophorans') in a putative clade, Halwaxiida (Conway Morris and Caron, 2007). Explicit cladistic analyses have resolved the halwaxiids as a clade (Conway Morris and Caron, 2007; Sigwart and Sutton, 2007) or a grade (Vinther *et al.*, 2008) in the mollusc stem group. Mollusc affinities for *Halkieria* have been advanced based on similarities to chitons in particular (Vinther and Nielsen, 2005). Arguments for a halkieriid origin of brachiopods (Holmer *et al.*, 2002) have been weakened by the discovery that the supposed intermediate tannuolinids are not in fact scleritome-bearing but are sessile, bivalved organisms with brachiopod-like ultrastructure (Holmer *et al.*, 2008).

Halwaxiid monophyly is disputed by Butterfield (2006), largely on the basis of *Wiwaxia* sharing putative autapomorphies of Annelida, especially microvillar setae that are histologically identical to annelid chaetae, and specifically similar to the flattened notochaetae of chrysopetalid polychaetes (Butterfield, 1990). Others have accepted the homology of these chaetae (as do we, as a primary homology statement), but excluded *Wiwaxia* from Annelida based on its lack of parapodia and segmentation (Eibye-Jacobsen, 2004). Butterfield (2006) attempted to retain *Wiwaxia* in Annelida by arguing that a dorsal chaetal scleritome is sufficient to identify *Wiwaxia* as an annelid. We are unconvinced, because although the relevant chaetae of the fossil polychaetes (*Canadia*) that Butterfield (2006) refers to are notochaetal and hence dorsal, *Wiwaxia* has no evidence for parapodia or segmentation whatsoever. Butterfield's (2006) evocation of a special style of creeping that transforms segmentation/parapodia/neuropodia in *Wiwaxia* beyond recognition is decidedly *ad hoc*. Likewise, assuming that *Wiwaxia* is segmented because *Odontogriphus* is supposed to be segmented (Butterfield, 2006), despite a lack of convincing evidence from the fossils (Caron *et al.*, 2007), is unnecessary. The structural similarity of *Wiwaxia* and halkieriid sclerites (Bengston, 2006) and their sclerite zones (Conway Morris and Peel, 1995) is not so easily dismissed, and relegating their similarity to convergence or inheritance from the spiralian stem lineage is uncompelling.

A recurring theme in the spiralian fossil record is the variable assignment of certain fossils to either Annelida or Mollusca (indeed these character conjunctions in fossils are an argument in favour of spiralian monophyly). For example, the Silurian *Acaeonoplax* (Figure 6.3d) has generally been accepted as an aplacophoran mollusc (Sutton *et al.*, 2001b, 2004; Sigwart and Sutton, 2007), an interpretation with which we concur, not least based on its calcareous spicules and seriated shell plates. Note that the mollusc interpretation had been challenged by Steiner and Salvini-Plawen (2001) who argue for annelid features being present in the fossils. The discovery of soft-part preservation associated with the typical shell plates permits a confident assignment of the long-problematic Ordovician–Permian Machaeridia to Annelida (Vinther *et al.*, 2008), though this finding contradicts the recent placement of machaeridians within Mollusca (Sigwart and Sutton, 2007). Machaeridians throw up an unexpected character combination in Spiralia—calcareous shell plates (with marginal growth) in animals that have parapodial chaetae. Such unique character combinations underscore the utility of including fossil terminals in cladistic analyses of morphology. All of the fossil spiralians discussed above are coded as terminals in our matrix for the *Assembling the Protostome Tree of Life* project (authors' work in progress).

In the primary reference phylogeny of Dunn *et al.*, (2008, their Figure 2) a clade labelled Clade C was proposed, including a set of spiralian animals primitively with a trochophore larva. Clade C was not considered a synonym of Trochozoa because the most current phylogenetic hypotheses (Hausdorf *et al.*, 2007; Dunn *et al.*, 2008, their Figure 2) did not include Entoprocta, or Entoprocta was considered the sister group to Clade C (Dunn *et al.*, 2008, their Figure 1). Entoprocta have spiral development and a trochophore larva, and therefore, until its exact position is resolved, we prefer not to name Clade C formally. The monophyly of Clade C is consistent with a homology between chaetae in annelids and brachiopods and spicules in molluscs. Both kinds of structure are epidermal extracellular formations whose secretory cells develop into a cup or a follicle with microvilli at their base. This homology has been anticipated by palaeontologists (Conway Morris and Peel, 1995, their Figure 50) who have proposed a common origin of mollusc spicules and annelid/brachiopod chaetae as modifications of sclerites as developed in the scleritome of various Cambrian fossil taxa that have subsequently been assigned to the Halwaxiida (Conway Morris and Caron, 2007). The hollow sclerites of the Cambrian fossils (sharing a suite of morphological details encompassed under the 'coelosclerite' concept of Bengston 2006) are variably organic (e.g. *Wiwaxia*) or aragonitic (e.g. *Halkieria*). The character delimitation of a mollusc/halwaxiid sclerite applied by Vinther *et al.*, (2008), i.e. an ectodermal element secreted to a finite size by a basal epithelium, applies to chaetae as well.

6.2 Novel approaches in morphology

Morphology and development, including early cleavage patterns, have played fundamental roles in shaping our understanding of animal relationships. A major shift within the past two decades towards using molecular evidence has resulted in some major rearrangements in the tree of animal life. Morphological analysis still has its role in continuing to decipher animal relationships and in interpreting the results derived from molecular studies. Animal morphological analyses underwent a first revolution with the advent of cladistic techniques, which led to the proposal of numerous phylogenetic hypotheses based on numerical (parsimony-based or maximum-likelihood) analyses of explicit data matrices (e.g. Eernisse et al., 1992; Nielsen et al., 1996; Zrzavý et al., 1998; Sørensen et al., 2000; Peterson and Eernisse, 2001; Jenner and Scholtz, 2005). Morphological matrices are not exempt from arbitrary decisions in inclusion (or exclusion) of characters, along with their definition and the identification of character states. Common problems with these previous approaches to metazoan morphological data matrices are the uncritical recycling of characters (see Jenner, 2001) and the assignment of homology to absences of a given character state (Jenner, 2002) (see Jenner, 2004a, for a general discussion). Another major problem is the decision of what character state is assigned to higher (supraspecific) taxa, as all metazoan morphological phylogenies published so far rely on coding. This could lead to the arbitrary choice of character states—often in a hypothesis-driven manner (Jenner, 2001). An alternative to some of these problems is to code real observations for a selected number of species instead of supraspecific taxa (Yeates, 1995; Prendini, 2001) in the same fashion that species are used for molecular analyses. This solution is not only appealing from an operational perspective, but also philosophically, since it allows for a stricter test of monophyly than previous strategies.

No metazoan-wide morphological matrix has yet been produced using species as terminals, although we are currently working on such a matrix (G. Edgecombe et al., work in progress). This strategy is not without difficulties. It requires two principal conditions: incorporating multiple species per phylum—ideally a collection of species that represent the phyletic morphological disparity and requiring careful species choice; and coding

observations for many species. Certain characters, especially those of development and ultrastructure are unobserved (or not described) for many terminals and filling the matrix requires a considerable amount of work. Filling all cells with observations for a metazoan matrix of hundreds of taxa and hundreds of characters requires substantial effort by a much larger group of researchers than the team assembled. It is therefore necessary to perfect collaborative software that allows data matrices to be updated by a group of authors over the web (see, for example, http://morphobank.geongrid.org/ or http://www.mesquiteproject.org/) and requires what we call 'coding parties', where experts in taxa or characters meet periodically to discuss characters, character states, and specific coding of taxa.

Defining the characters and their states remains difficult (e.g. Jenner, 2004a,b). Hence some researchers believe that such efforts should be done explicitly and discussed by the scientific community in large, developing specific ontologies (e.g. Ramírez et al., 2007; see also https://www.morphdbase.de/). For example, characters that have played fundamental roles in shaping animal relationships through time, such as the fate of the blastopore, segmentation, or the origins of body cavities, are still poorly understood and their states need further research and discussion. Problems with assigning a homologous state to the lack of a feature have received little attention in the literature. The same is true for inapplicable characters and their specific treatment by computer algorithms (e.g. Pleijel, 1995; Lee and Bryant, 1999; Strong and Lipscomb, 1999).

One possible solution to these issues is the addition of as many states as there are non-homologous absences, although this would incur the necessity of adding complex Sankoff characters with more than the 10 states allowed by some software implementations. Another is to have characters treated in a hierarchical and integrative way, such that less inclusive characters are not considered in an analysis until the more general characters that encompass them are. These less inclusive characters are only applied in parts of the tree that are applicable. Another possibility would be the use of a dynamic approach to morphology (Schulmeister and Wheeler, 2004; Ramírez, 2007), analogous to the direct optimization of molecular characters.

Although this option is still in its infancy, it could deal with the problem of absences and inapplicable character states in a completely different way.

The impact of fossils in numerical analyses of morphology (e.g. Cobbett *et al.*, 2007), or when combined with molecules (e.g. Wheeler *et al.*, 2004) should not be underestimated when attempting to reconstruct spiralian relationships, especially due to the presence of so many fossils that show intermediate morphologies of extant animals (see above). It is therefore imperative to incorporate data from key, exceptionally preserved fossils into these data matrices if we truly want to understand the evolution of spiralian animals, even though not all clades are equally represented in the fossil record.

6.3   Phylogenomic approaches

Since the publication of the earliest metazoan analyses based on molecular data (Field *et al.*, 1988; Lake, 1990), molecular phylogenetics has revolutionized our understanding of metazoan relationships. Novel concepts, now widely accepted by the community, such as Ecdysozoa (Aguinaldo *et al.*, 1997), Spiralia (Halanych *et al.*, 1995), and the more controversial Platyzoa (Giribet *et al.*, 2000) are rooted on molecular analyses of ribosomal RNA sequence data. The specific position of many 'odd' taxa have also benefited from molecular techniques. Salient examples are those of acoels (Ruiz-Trillo *et al.*, 1999), nemertodermatids (Jondelius *et al.* 2002), xenoturbellids (Bourlat *et al.*, 2003), nemerteans (Turbeville *et al.*, 1992), and gnathostomulids (Giribet *et al.*, 2000). However, the positions of other phyla such as loriciferans, micrognathozoans, or cyclophorans are not well resolved in molecular analyses (Giribet *et al.*, 2004; Park *et al.*, 2006). Most of these studies were based on one or at most a few targeted molecular markers.

New strategies that consider molecular data from many genes rather than just a few have emerged within the past few years, and are collectively designated with the catch-all label of 'phylogenomics' (Delsuc *et al.*, 2005). The first such analyses culled a subset of widespread genes from the few complete eukaryotic genomes to tackle metazoan relationships and test hypotheses such

as Ecdysozoa or Coelomata (Blair *et al.*, 2002; Dopazo *et al.*, 2004; Philip *et al.*, 2005; Wolf *et al.*, 2004). These studies, although broad in the number of genetic data included, suffer from one of the most crucial phylogenetic biases—deficient taxon sampling. Not surprisingly, analyses including a broader taxon sampling followed immediately, focusing not only on whole-genome approaches, but taking advantage of EST projects (Philippe and Telford, 2006). Such studies have provided insights into the relationships of several protostome phyla (Hausdorf *et al.*, 2007; Roeding *et al.*, 2007; Dunn *et al.*, 2008), allowing resolution of long-standing questions such as the sister-group relationships of Arthropoda–Onychophora (Roeding *et al.*, 2007; Dunn *et al.*, 2008), or the overall topology among the trochozoan phyla (Dunn *et al.*, 2008) (see Figure 6.1).

Several current research groups, especially those funded under the US National Science Foundation AToL and the German Deep Phylogeny programmes have focused on closing the gap in missing protostome diversity using targeted EST studies (see Hausdorf *et al.*, 2007; Roeding *et al.*, 2007; Dunn *et al.*, 2008). These novel data already encompass nearly all animal phyla, and just a few phyla of small-sized animals are missing (currently there are EST/genomic data missing for Cycliophora, Loricifera, Micrognathozoa, and Nemertodermatida). Continued improvements in sequencing technologies and computational tools will soon make it possible for far more taxa within critical groups to be incorporated, and will make phylogenomic approaches more cost-effective than traditional directed-PCR approaches for a greater number of problems. Phylogenomic approaches will also become truly genomic with the final transition to systematics labs sequencing complete genomes on a routine basis (ESTs are just a stopgap until that time).

6.4   Conclusions: the future of spiralian phylogeny

Much recent progress has been made on the phylogeny of Spiralia. There is now strong support for the existence and many internal relationships of Trochozoa, though its complete composition remains uncertain due to persistent problems placing Bryozoa and Entoprocta (Dunn *et al.*, 2008). Multiple sources of evidence place the remaining spiralian taxa in the clade Platyzoa, though there are lingering questions as to whether or not support for this clade is due to systematic error. Resolving these two questions will be important priorities for moving forward. However, key questions such as the affinities of bryozoans and cycliophorans, or the exact position of gastrotrichs, myzostomids, and micrognathozoans—to mention just a few—lack a convincing answer. Two qualitative changes taking place in the study of animal relationships may contribute towards an even more resolved picture of the spiralian tree. First is the fresh study of animal morphology and development, translated into a data matrix where observations (instead of inferences) and species (instead of supraspecific taxa)—including fossils—are coded. Second is the widespread use of phylogenomic techniques, now beginning to span a much greater swath of spiralian diversity. Once hundreds (or thousands) of genes become available for a wide sampling of protostome species, relationships may finally be established with great support. We will then be able to proceed to the even more fundamental task of attempting to explain the origins of morphological disparity and taxonomic diversity.

6.5 A new taxonomic proposal

Given recent progress in resolving several nodes of the spiralian tree of life, we formalize a taxonomic proposal derived from the analyses recently published by Dunn *et al.* (2008):

Kryptrochozoa Dunn, Edgecombe, Giribet, Hejnol, Martindale, Rouse new taxon.

• **Definition**: the least inclusive clade containing the Brachiopoda and Nemertea (Figure 6.4 and Plate 4).

• **Intention of the name**: this name is intended to refer to a clade comprised of Nemertea and Brachiopoda.

• **Etymology**: compounding the Greek *kryptos* (hidden) and Trochozoa with reference to the modification of the trochophore larvae (Figure 6.4).

• **Reference phylogeny**: in the primary reference phylogeny (Dunn *et al.*, 2008; Figure 6.2) this clade was labelled Clade A.

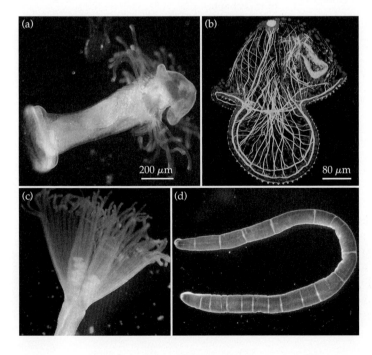

Figure 6.4 Examples of taxa and larval forms in Kryptotrochozoa, a new subtaxon of Trochozoa. (a) An actinotroch larva of an unidentified phoronid species. (b) Fluorescently labelled pilidium larva of the nemertean *Cerebatulus lacteus* (photograph by Patricia Lee and Dave Matus). (c) Anterior end of phoronid brachiopod *Phoronis hippocrepis* (photograph by G. W. Rouse). (d) Dorsal view of the nemertean *Micrura* sp. (photograph by G. W. Rouse). (See also Plate 4.)

• **Discussion**: the clade includes two phyla (note Phoronida is accepted here as part of Brachiopoda, following Cohen, 2000, and Cohen and Weydmann, 2005; though see Helmkampf *et al.*, 2008a, and Bourlat *et al.*, 2008) that have modified trochophores, in the case of some heteronemerteans (Maslakova *et al.*, 2004a,b), or larvae that do not show evident homologies to trochophore larvae in other nemerteans and brachiopods (including phoronids). Examples of these are the pilidium larva of some derived Nemertea (Figure 6.4b) and the actinotroch larvae found in Phoronida (Figure 6.4a).

• **Remarks**: this clade was proposed in the phylogenomic analysis of Dunn *et al.* (2008). This clade has never been proposed previously, though it has since been recovered in other molecular analyses (Bourlat *et al.*, 2008; Helmkampf *et al.*, 2008b). There are no obvious morphological apomorphies. Bootstrap support for Kryptrochozoa was much lower for independent analyses of non-ribosomal (14%) and ribosomal genes (15%) than it was in the combined 150-gene analyses (Dunn *et al.*, 2008, their Supplementary Figure 10a). This may help explain why it has not been recovered in previous analyses; support requires that many genes be analysed in combination, and Dunn *et al.* (2008) is the only phylogenomic analysis to date to include the relevant taxa.

6.6 Acknowledgements

We wish to acknowledge Tim Littlewood and Max Telford for organizing the Linnean Tercentenary Symposium on Animal Evolution at the Royal Society. The Royal Society and the Novartis Foundation made the symposium possible. Ron Jenner, an anonymous reviewer, and the editors provided insightful comments that improved earlier versions of this article. Errors and biases remain our responsibility. All the collaborators on the protostome AToL project (Jessica Baker, Noemí Guil, Reinhardt Møbjerg Kristensen, Dave Matus, Akiko Okusu, Joey Pakes, Elaine Seaver, Martin Sørensen, Ward Wheeler, Katrine Worsaae), without whom the research could have not been conducted, provided stimulating discussions that have led to the writing of this chapter. Illustrations of fossils were kindly provided by Jean-Bernard Caron, Mark Sutton, and Jakob Vinther, and thanks to Patricia Lee and Dave Matus for the labelled pilidium image. This material is based upon work supported by the National Science Foundation AToL programme under grant nos 0334932, 0531757, 0531558. GG was recipient of a fellowship from the Ministerio de Educación y Ciencia (Spain) for a sabbatical stay at the Centre d'Estudis Avançats de Blanes during the writing of this chapter.

CHAPTER 7

The evolution of nervous system centralization

Detlev Arendt, Alexandru S. Denes, Gáspár Jékely, and Kristin Tessmar-Raible

It is currently unknown when and in what form the central nervous system (CNS) in Bilateria first appeared, and how it further evolved in the different bilaterian phyla. To find out, a series of recent molecular studies have compared neurodevelopment in slowly evolving deuterostome and protostome invertebrates such as the enteropneust hemichordate *Saccoglossus* and the polychaete annelid *Platynereis*. These studies focus on the spatially different activation and, when accessible, function of genes that set up the molecular anatomy of the neuroectoderm and specify neuron types that emerge from distinct molecular coordinates. Complex similarities are detected that reveal aspects of neurodevelopment that most likely already occurred in a similar manner in the last common ancestor of the bilaterians, Urbilateria. Using this approach, different aspects of the molecular architecture of the urbilaterian nervous system are being reconstructed and are yielding insight into the degree of centralization that was in place in the bilaterian ancestors.

7.1 Introduction

Surprisingly little is known about the evolutionary origin of the CNS. It is not known when CNSs first appeared in animal evolution nor what their initial structure and function was. It is also unclear whether the CNSs of vertebrates and invertebrates trace back to a common CNS precursor (Arendt and Nübler-Jung, 1999) or whether they have independent evolutionary origins (Holland, 2003; Lowe *et al.*, 2003). This chapter addresses the questions

of when and in what form the CNS first came into existence, and how it further evolved in different animal phyla. To track the evolutionary transition from 'diffuse' to 'centralized' in the evolution of the bilaterian nervous system (Figure 7.1) we first define these terms. We then explain what the study of bilaterian neurodevelopment can reveal about this transition. Specifically, we focus on the role of *Decapentaplegic* (*Dpp*) signalling in triggering neurogenesis in a polarized manner along the dorsoventral body axis. We then outline the conserved mediolateral molecular anatomy of the bilaterian neuroectoderm (Figure 7.2 and Plate 5) and pinpoint a set of conserved neuron types that develop from corresponding regions (Figure 7.3 and Plate 6). We finally discuss the significance of these data for reconstructing the urbilaterian nervous system.

7.1.1 What is a CNS?

In physiological terms, a CNS integrates and processes sensory information coming from the periphery, and initiates body-wide responses via neurosecretion into the body fluid or via direct stimulation of the body musculature. Anatomically, a CNS is a delimited nervous tissue that comprises distinct agglomerations of functionally specialized neurons (nuclei) interconnected by axon tracts (neuropil). The CNS may be subdivided into separate parts (ganglia) and it connects to the periphery via nerves. A CNS thus defined is found in various shapes and degrees of complexity in different animal phyla, including vertebrates and many

(a)   (b)

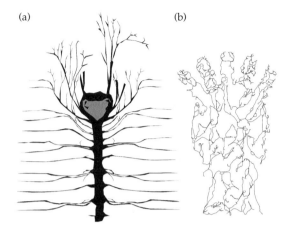

Figure 7.1 Different degrees of centralization in metazoan brains. (a) Centralized nervous system of an oligochaete worm. (b) Nerve net of a cnidarian polyp representing a typical non-centralized nervous system. Schematized drawings modified from Bullock and Horridge (1965).

invertebrates such as echinoderms, arthropods, nematodes, molluscs, and annelids (Figure 7.1a).

In contrast, a diffuse nervous system receives sensory input and processes locomotor or neurosecretory output only locally, without central integration. This is achieved by the direct interconnection of sensory neurons and effector neurons (Westfall *et al.*, 2002). A diffuse nervous system is present in the body wall epithelium of adult cnidarians, for example (Figure 7.1b).

Even though these definitions are straightforward, the categorization of some animal nervous systems remains ambiguous (Miljkovic-Licina *et al.*, 2004). For example, some cnidarian medusae possess an elaborate nerve ring around their central opening (manubrium) in addition to their diffuse nerve net (Mackie, 2004). This nerve ring reflects a considerable degree of centralization. Also, the nervous system of deuterostome enteropneusts exhibits aspects of both central and diffuse organization (reviewed and discussed in Holland, 2003). On the one hand, enteropneusts have axons tracts that run along the longitudinal body axis and show a strong concentration of neurons in the anterior part of the body, reflecting nervous integration. On the other hand, enteropneusts have a 'nerve net' interconnecting the cell bodies, dendrites, and

axons of sensory neurons, interneurons, and motor neurons, and neurons are embedded in the epidermis, as indicative of a diffuse system, rather than forming an anatomically distinct structure (Lowe *et al.*, 2003).

Given the vast differences in nervous system organization in Bilateria, what can we learn about the urbilaterian nervous system from comparative studies? So far, insight has been limited and proposals about the complexity and shape of the urbilaterian nervous system range from 'diffuse' (Mineta *et al.*, 2003; Lowe *et al.*, 2006) to 'centralized' (Denes *et al.*, 2007). Assuming a diffuse urbilaterian nervous system would imply independent centralization events at least in protostomes and deuterostomes (Holland, 2003; Lowe *et al.*, 2006). Assuming a centralized urbilaterian nervous system, on the other hand, would imply secondary simplification of the nervous system of enteropneusts and of many other invertebrate groups (Denes *et al.*, 2007). These two conflicting hypotheses can now be tested. If centralization occurred independently in protostomes and deuterostomes we would expect the neurodevelopment and molecular architecture of their CNSs to be generally divergent. If instead centralization pre-dated Bilateria, this should be reflected by similarities in neurodevelopment and CNS molecular architecture between the bilaterian superphyla.

7.2 Nervous system centralization—the evo-devo approach

A key strategy to unravelling the degree of centralization that was in place in the urbilaterian nervous system is the comparison of CNS development between protostome and deuterostome groups. However, depending on the amount of evolutionary change these groups have accumulated, their neurodevelopment will be more or less informative about ancestral characteristics of nervous system centralization in Bilateria. Ancestral features will be most apparent in the neurodevelopment of species that have changed relatively little during evolution and will be modified to a larger extent in faster-evolving species (Raible *et al.*, 2005). Distinct aspects of neurodevelopment are currently under study in a broad range of protostome and deuterostome model species:

1. *Polarized distribution of neuronal precursors with respect to the main body axes.* One important aspect of nervous system centralization is the early developmental segregation of the ectoderm into a 'non-neural' and a 'neural' portion, the neuroectoderm. In bilaterians, the neuroectoderm is located anteriorly where the brain and associated sensory organs develop, and on the 'neural' trunk side which is ventral in most invertebrates and dorsal in vertebrates due to dorsoventral axis inversion (Arendt and Nübler-Jung, 1994; De Robertis and Sasai, 1996; Lowe *et al.*, 2006). What are the signals that polarize the bilaterian ectoderm, and to what extent are they comparable between phyla?

2. *Subdivision of the neural anlage into regions ('molecular anatomy').* Another aspect of nervous system centralization amenable to comparative studies is how the developing nervous system relates to the 'molecular anatomy' of the body. Bilaterians have in common an early subdivision of the developing embryo (or larva) into regions of distinct molecular identities (St Johnston and Nusslein-Volhard, 1992; Arendt and Nübler-Jung, 1996; Lowe *et al.*, 2003; Schlosser and Ahrens, 2004; Yu *et al.*, 2007; Lowe, 2008). These are referred to as 'molecular anatomy' and can be used as a molecular map. A similar molecular anatomy of the CNS anlage at early developmental stages has been considered a good indication of CNS homology (Arendt and Nübler-Jung, 1996; Lichtneckert and Reichert, 2005). Note, however, that structures that develop from corresponding regions in two species are not necessarily homologous (Lowe *et al.*, 2003; Lowe, 2008). How similar is the molecular anatomy between species, of the whole body, and of the developing CNS in particular, and what is the significance of conserved expression regions for our understanding of CNS evolution?

3. *Spatial segregation of neuron types in the CNS.* Nervous system centralization not only implies local concentration of neurons but also their functional and spatial segregation and interrelation ('operational centralization'). This is exemplified by Herrick's longitudinal neuron columns in the vertebrate spinal cord, which comprise distinct sets of motor- and interneuron types. With the recent progress in the identification of conserved neuron

types by molecular fingerprint comparisons (Arendt and Nübler-Jung, 1999; Thor and Thomas, 2002; Arendt *et al.*, 2004), and using the conserved molecular anatomies as universal molecular maps, the localization and spatial segregation of neuron types can now be compared between remote bilaterians (Denes *et al.*, 2007; Sprecher *et al.*, 2007; Tessmar-Raible *et al.*, 2007). To what extent had neuron types already been spatially arranged in Urbilateria, and what does this tell about the ancestral state of nervous system centralization?

7.2.1 Central nervous systems develop from the non-*Dpp* body side

In all bilaterian animals investigated (with the exception of the nematodes) the bone morphogenetic protein (Bmp) signalling system sets up tissue polarity along the dorsoventral axis (Mizutani *et al.*, 2005; Lowe *et al.*, 2006; Levine and Brivanlou, 2007; Yu *et al.*, 2007). The Bmp system pre-dates the emergence of the bilaterian CNS (Matus *et al.*, 2006a; Rentzsch *et al.*, 2006) and was thus in place to be adapted for nervous system centralization, i.e. for the differential distribution of neuronal precursors along this axis. How similar is the role of Bmp signalling with respect to nervous system centralization in various bilaterians?

Whenever a CNS is present, it develops from the non-Bmp body side, in insects (Mizutani *et al.*, 2005, 2006), vertebrates (Sasai *et al.*, 1995; Levine and Brivanlou, 2007), amphioxus (Yu *et al.*, 2007), and also annelids (Denes *et al.*, 2007). Also, in early vertebrate (Harland and Gerhart, 1997) and fly development (Mizutani *et al.*, 2006) the antineurogenic activity of Bmps sets the limit of the neuroectoderm. These findings first suggested that Bmp signalling had an ancient role in the overall restriction of neurogenesis to the neural body side (e.g. Padgett *et al.*, 1993). Yet, this simple notion was not supported by recent additional comparative data: in enteropneusts (Lowe *et al.*, 2006) and in polychaetes (Denes *et al.*, 2007), the pan-neural marker *elav* is *not* downregulated by exogenously applied BMP4. How can we reconcile these findings?

The available data are consistent with a refined evolutionary scenario, which assumes that in early bilaterians the antineurogenic effect of Bmp signalling was on specific sets of motor neurons (and interneurons) only, restricting them to the neural body side, while there was a positive effect on the formation of sensory neurons that do not form part of the CNS proper (Rusten *et al.*, 2002). In line with this, Bmp signalling has been shown to trigger formation of the peripheral sensory neurons at later developmental stages, at the neural plate border and adjacent lateral placodes in the vertebrates (Schlosser and Ahrens, 2004), and in the lateral 'epidermal' ectoderm in *Drosophila* (Rusten *et al.*, 2002). In annelids, the types of sensory neurons characterized so far arise from the lateral and dorsal sides as opposed to motor- and interneurons that form from the ventral body side (Denes *et al.*, 2007); indeed, exogenous BMP4 strongly upregulates the sensory marker *atonal*, consistent with a conserved role of *Dpp*/BMP in the specification of peripheral sensory neurons (Denes *et al.*, 2007). Even in enteropneusts, where post-mitotic neurons are spread all around the circumference of the trunk (Lowe *et al.*, 2003), the distribution of motor neuron, interneuron, and sensory neuron precursors may not be uniform (Lowe *et al.*, 2006). For example, there is a small population of putative motorneurons in the ventral ectoderm (expressing conserved motor neuron markers) and motor neurons are reported to be enriched in the ventral axon tract. A more in-depth analysis of the role of Bmp signalling and of other signalling systems active along the dorsoventral axis will elucidate a possible conservation of neuron type segregation in annelid and enteropneust neurodevelopment.

Our revised scenario—that the ancestral role of Bmp signalling was to promote sensory neuron over motor neuron fates, rather than a general antineurogenic effect—fits well with the actual distribution of motor and sensory neurons in many invertebrates, where it appears to be the rule rather than the exception that sensory neurons emerge outside of the neuroectoderm on the 'non-neural' (='*Dpp*/Bmp') body side. If this were indeed an ancestral bilaterian trait this would imply that a certain degree of centralization was present in Urbilateria (i.e. the sorting out of motor versus sensory neurons along the secondary body axis).

7.2.2 A conserved pattern of mediolateral regions extending from head to trunk

To estimate the complexity of the urbilaterian CNS, we need to know the complexity of the underlying molecular anatomy that was in place in Urbilateria. Although comparative studies have addressed this for both the antero-posterior (Slack *et al.*, 1993; Schilling and Knight, 2001) as well as for the mediolateral (dorsoventral = neural/non-neural) axes (Cornell and Ohlen, 2000), our focus here is on mediolateral patterning. Previous comparisons of the molecular anatomy of the insect and vertebrate neuroectoderm had revealed a similar mediolateral sequence of *nk2.2+*, *gsx+*, and *msx+* neurogenic domains (reviewed in Arendt and Nübler-Jung, 1999, and Cheesman *et al.*, 2004) that also extends into the brain anlage (Urbach and Technau, 2003a,b; Sprecher *et al.*, 2007). Notably, in the developing forebrain, medial *nk2.2* expression is complemented by the medial expression of its sister gene, *nk2.1* (Zaffran *et al.*, 2000). *Nk6* genes also play a conserved role in mediolateral patterning because the neuroectodermal expression of the *Drosophila* orthologue shows medial restriction as observed in the vertebrates (Cheesman *et al.*, 2004).

Our recent work on the mediolateral anatomy of the developing annelid nerve cord has revealed an even higher degree of conservation in mediolateral patterning (Figure 7.2). In addition to the previously detected protostome–deuterostome similarities, we find that annelids and vertebrates share a *pax6+* column at similar mediolateral level that likewise extends up to the forebrain (violet in Figure 7.2; see Plate 5) (Denes *et al.*, 2007). In both groups the medial portion of the *pax6+* column overlaps the *nk6+* column (yellow in Figure 7.2). In addition to this, annelids and vertebrates share a lateral *pax3/7+* column (green in Figure 7.2; note that this gene is expressed strictly segmentally in the *Drosophila* neuroectoderm; Davis *et al.*, 2005). Our data also revealed that the positioning of the *gsx+* column is more variable than initially assumed and the vertebrate *dbx+* interneuron columns are probably vertebrate-specific evolutionary acquisitions (Denes *et al.*, 2007).

The conservation of mediolateral columns between vertebrates, annelids, and (to a lesser

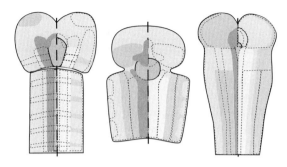

Figure 7.2 Comparison of mediolateral neurogenic columns across Bilateria. Expression of *nk2.2/nk2.1* (orange; Shimamura *et al.*, 1995), *Nk6* (yellow; Rubenstein *et al.*, 1998), *Pax6* (violet; Mastick *et al.*, 1997; Urbach and Technau, 2003a,b), *gooseberry/Pax3/7* (green; Matsunaga *et al.*, 2001; Puelles *et al.*, 2003), and *msh/Msx* (blue; Shimeld *et al.*, 1996) orthologues in the neuroectoderm of *Drosophila*, *Platynereis*, and mouse (left to right) at pre-differentiation stages. The *Drosophila* (left) and *Platynereis* (centre) schematics represent ventral views, the mouse (right) is a dorsal view with the neural tube unfolded into a neural plate for better comparison. Neurogenic columns are demarcated by expression boundaries and represent cells with a unique combination of transcription factors. All expression patterns are symmetrical but are shown on one side only for clarity. (See also Plate 5.)

extent) insects is in stark contrast to the situation in enteropneusts, where similar columns have not been observed, with the exception of the dorsal *dll*+ column and the ventral midline column (Lowe *et al.*, 2006; Lowe, 2008).

Two conclusions can be drawn. First, if the complex molecular mediolateral anatomy shared between annelids and vertebrates is indeed due to evolutionary conservation—and this notion seems inescapable given the overall complexity of this pattern (Figure 7.2)—it must have been present in Urbilateria. The immediate question then arises: what was the difference in developmental fate between these regions in Urbilateria? One plausible scenario is that these regions gave rise to distinct and segregated ancestral neuron types, as will be discussed in the next section. Second, these findings suggest that the mediolateral molecular anatomy in enteropneusts is secondarily simplified (Denes *et al.*, 2007), consistent with the notion of evolutionary loss in a slowly evolving species (see discussions in Lowe *et al.*, 2006, and Denes *et al.*, 2007).

7.2.3 Conserved neuron types develop from similar mediolateral progenitor domains

In insects and vertebrates, neuron types emerging from the medial *nk2.2*+ column pioneer the medial longitudinal fascicles as well as peripheral nerves (Arendt and Nübler-Jung, 1999, and references therein). Among these, the neuron populations that send out ascending and descending projections in the vertebrate hindbrain are serotonergic and they modulate spontaneous locomotor activity (Briscoe *et al.*, 1999; Schmidt and Jordan, 2000; Pattyn *et al.*, 2003). In *Platynereis*, serotonergic neurons likewise emerge from the medial *nk2.2* columns and pioneer the longitudinal tracts and segmental nerves (red in Figure 7.3; see Plate 6) (Denes *et al.*, 2007). One type of serotonergic neuron also emerges from the *nk2.1*+ brain regions, as evidenced for *Platynereis* and fish (Tessmar-Raible *et al.*, 2007) as well as sea urchin (Takacs *et al.*, 2004).

The *nk2.1*+ region in the developing forebrain of vertebrates and annelids gives rise to another conserved neuron type: early differentiating neurosecretory cells that synthesize the highly conserved neuropeptide arg-vasotocin/neurophysin (orange in Figure 7.3). These cells form in the vicinity of ciliated photoreceptor cells in the brain that share the expression of *rx* and of *c-opsin* orthologues in vertebrates and annelids (white in Figure 7.3) and of molecular clock cells positive for *bmal/cycle* (green in Figure 7.3) (Arendt *et al.*, 2004).

Somatic motor neurons exhibit the same transcription factor signature (*hb9*+, *lim3*+, *islet-1/2*+) in insects, nematodes, and vertebrates (Thor and Thomas, 2002). In the vertebrates, these neurons are cholinergic and emerge from the *pax*+, *nk6*+ progenitor domain (violet in Figure 7.3) (Ericson *et al.*, 1997). We found that the same is true for *Platynereis*, where the first cholinergic motor neurons that innervate the longitudinal musculature have the same transcription factor signature and emerge from the *pax6*+, *nk6*+ column (Denes *et al.*, 2007; AD, GJ and DA, unpublished).

Taken together, these data identify a considerable number of conserved neuron types that emerge from similar molecular coordinates in annelids and vertebrates. Obviously, this comparison is far from complete and awaits further characterization and localization of neuron types in both taxa.

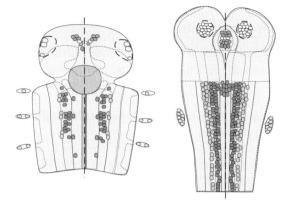

Figure 7.3 Conserved neural cell types in annelids and vertebrates. The neuron types emerging from homologous regions in the molecular coordinate systems in annelids and vertebrates and expressing orthologous effector genes are marked with the same colour. Homologous cell types include the molecular clock cells positive for *bmal* (dark green), ciliary photoreceptors positive for *c-opsin* and *rx* (white), rhabdomeric photoreceptors positive for *r-opsin*, *atonal*, and *pax6* (yellow), vasotocinergic cells positive for *nk2.1*, *rx*, and *otp* (orange), serotonergic cells positive for *nk2.1/nk2.2* (red), cholinergic motor neurons positive for *pax6*, *nk6*, and *hb9* (violet), interneurons positive for *dbx* (pink), as well as trunk sensory cells positive for *atonal* and *msh* (light blue). (See also Plate 6.)

As for the peripheral nervous system, we have so far identified and compared rhabdomeric photoreceptor cells in annelids and retinal ganglion cells in vertebrates (yellow in Figure 7.3) that form from the eye anlage in both species (dashed circles in Figure 7.3). In the trunk we found some conserved sensory neuron types that emerge from similar lateral molecular coordinates in annelids and vertebrates (blue in Figure 7.3) (*ath+* or *trpv+*) (Denes *et al.*, 2007); this comparison is ongoing.

7.3   Reconstructing the urbilaterian nervous system

In conclusion, the comparison of neurodevelopment in protostome and deuterostome animal models reveals a conserved molecular architecture of considerable complexity that was inherited from the Urbilateria. Beginning with a diffuse nerve net with homogeneously distributed neuron types, a first segregation of motor and sensory neurons occurred along the dorsoventral axis in the line of evolution leading to the bilaterians. This involved Bmp signalling and possibly other signalling cascades. These signals established a refined mediolateral molecular anatomy, involving at least four longitudinal neurogenic regions with distinct molecular identities (*nk2.2+/nk6+*, *pax6+/nk6+*, *pax6+/pax3/7+*, *msx+/pax3/7+*; Figure 7.2) that gave rise to spatially segregated neurons. Among these were medial serotonergic neurons, intermediate cholinergic motor neurons, some sort of interneurons and lateral sensory neurons (Figure 7.3) (Denes *et al.*, 2007). These neuron types presumably controlled ancestral locomotor patterns such as undulatory swimming and/or peristalsis. In the head region, specialized light-sensitive cell types evolved, integrating different kinds of photic input to set the molecular clock and to control neurosecretory and motor output (Tessmar-Raible *et al.*, 2007). While this already reflects a considerable degree of nervous system centralization that presumably was in place in Urbilateria, a renewed push in research combining developmental genetics with classical neuroethology in slowly evolving protostomes and deuterostomes will be needed to refine and complete this picture.

7.4   Acknowledgements

We thank an anonymous reviewer for very valuable comments. This work was supported by grants from the Marine Genomics Europe Network of Excellence [NoE-MGE (DA), GOCE-04–505403 (DA and FR)], fellowships of the Boehringer Ingelheim Foundation and of the Marie Curie RTN ZOONET [MRTN-CT-2004–005624 (KT-R)], and the Deutsche Forschungsgemeinschaft (Deep Metazoan Phylogeny; DA:Ar387/1–1 and HH: Ha4443/1–1). AD was supported by a Louis Jeantet Foundation fellowship.

CHAPTER 8

The origins and evolution of the Ecdysozoa

Maximilian J. Telford, Sarah J. Bourlat, Andrew Economou, Daniel Papillon, and Omar Rota-Stabelli

Ecdysozoa is a clade composed of eight phyla, three of which—arthropods, tardigrades, and onychophorans—share segmentation and have appendages, and the remaining five—nematodes, nematomorphs, priapulids, kinorhynchs, and loriciferans—are worms with an anterior proboscis or introvert. Ecdysozoa contains the vast majority of animal species and there is a great diversity of body plans among both living and fossil members. The monophyly of the clade has been called into question by some workers, based on analyses of whole-genome data sets, and we review the evidence that now conclusively supports the unique origin of these phyla. Relationships within Ecdysozoa are also controversial and we discuss the molecular and morphological evidence for several monophyletic groups within this superphylum.

8.1 Introduction

The Ecdysozoa is a widely accepted clade that encompasses the Euarthropoda (Insecta, Crustacea, Myriapoda, and Chelicerata), the arthropod-like Onychophora and Tardigrada, and five phyla of introvert bearing worms: the Nematoda, Nematomorpha, Priapulida, Kinorhyncha, and Loricifera. In terms of species numbers and niche diversity, the Ecdysozoa is far and away the most significant clade of animals ever to have existed, with over a million described species and an estimated total of more than 4.5 million living species (Chapman, 2005). The extraordinary number of insects is well known—there are estimated to be more than 10 times as many species of insects

than there are of all the deuterostomes and lophotrochozoans put together—yet even if the founder of the insect lineage had been eaten by a passing frog, the nematodes and the rest of the arthropods (myriapods, chelicerates, and crustaceans) would still easily outnumber all other living animals by close to a quarter of a million species (Chapman, 2005). Their characteristic tough cuticle also means that ecdysozoans are well represented in the fossil record, adding further wonderful forms to the diversity of the clade.

Despite the huge number of species and great niche diversity, the basic body plans of the Ecdysozoa are rather conservative, being either insect-like with a segmented body and jointed appendages or worm-like with an anterior circum-oesophageal nerve ring and a terminal mouth usually found on an introvert. All groups lack a primary larva as generally conceived and possess a moulted cuticle with concomitant lack of locomotory cilia. The periodic moulting or ecdysis of the cuticle gives the assemblage its name of Ecdysozoa. Although the morphological diversity of ecdysozoan phyla may be seen as fairly restrained when compared with the diversity of shapes seen among Lophotrochozoa, for example, these two ecdysozoan body plans happen to manifest themselves in the two most intensively studied invertebrates on the planet, the nematode *Caenorhabditis elegans* and the fruitfly *Drosophila melanogaster*.

Prior to 1997, the prevalent view of arthropod relationships linked them, via the onychophorans, to the annelid worms. This annelid–arthropod clade is called Articulata in recognition of the principal character uniting these phyla: a segmented body. Articulata was generally thought to be part of a larger assemblage of animal phyla linked by the possession of a coelomic cavity and called Coelomata. Although the concept of a relationship between arthropods and pseudocoelomate worms such as nematodes and priapulids existed much earlier (discussed in Schmidt-Rhaesa, 1998), the first support from molecular sequence data for such a relationship, and indeed the first reference to the Ecdysozoa, date to a paper by Aguinaldo *et al.*, (1997).

Our discussion is predicated on the assumption that the Ecdysozoa is a natural, monophyletic group; however, the existence of the Ecdysozoa is not yet universally accepted and so we will consider the evidence that has amassed in support of the monophyly of this group in the decade since the paper by Aguinaldo *et al.* The relationships among the introvertan worms, their position relative to Panarthropoda (Onychophora, Tardigrada, and Euarthropoda), and several aspects of the phylogeny within Panarthropoda and Euarthropoda themselves are all still controversial and we will consider recent arguments concerning each of these.

8.2   Ecdysozoa is a monophyletic group

The initial support for the Ecdysozoa came from a study of small-subunit (18S) ribosomal RNA (SSU rRNA) genes, that specifically addressed a common problem of phylogeny reconstruction—long branch attraction (LBA; Felsenstein, 1978). This systematic error is encountered when using molecular data derived from *C. elegans* and many other nematodes (Aguinaldo *et al.*, 1997) and stems from the fact that these genomes have evolved rapidly relative to those of most other animals. This instance of LBA would tend to cause the branch leading to the fast-evolving nematodes to be shifted towards the root of a tree. The use of short-branched nematodes in the analysis of Aguinaldo *et al.*, resulted in the nematodes moving from their position close to the root of the bilaterian tree (one also supported by consideration of their morphology, in particular

their lack of a coelomic cavity), to a close relationship with the arthropods and priapulid worms in a clade which the authors named the Ecdysozoa (Aguinaldo *et al.*, 1997).

Subsequent analyses of rRNA genes have confirmed this result and extended membership of the Ecdysozoa beyond Nematoda and Priapulida to include three further phyla of worms— Nematomorpha, Kinorhyncha, and Loricifera. The contribution of pseudocoelomate worms to the ecdysozoan clade had been anticipated by various authors who had already linked these five worm phyla in a group called the Cycloneuralia (Ahlrichs, 1995) or Introverta (Nielsen, 2001).

The finding of monophyletic Ecdysozoa has been replicated by other taxonomically well-sampled data sets, including combined small- and large-subunit (LSU) rRNAs (Mallatt and Winchell, 2002; Mallatt and Giribet, 2006) and myosin heavy-chain sequences (Ruiz-Trillo *et al.*, 2002), as well as *Hox* gene signature peptides (de Rosa *et al.*, 1999) and the (somewhat puzzling) shared presence in the nervous system of all studied ecdysozoans of an unidentified antigen recognized by the anti-horseradish peroxidase (anti-HRP) antibody (Haase *et al.*, 2001). The multimeric beta-thymosin gene found in flies and nematode worms (Manuel *et al.*, 2000) has been shown not to be a reliable synapomorphy of the Ecdysozoa (Telford, 2004).

Despite these congruent results, there exists a powerful series of papers arguing against the close relationship of nematodes and arthropods and supporting instead the traditional view of the monophyletic Coelomata linking arthropods such as *D. melanogaster* to humans rather than to nematode worms (Blair *et al.*, 2002; Wolf *et al.*, 2004; Philip *et al.*, 2005; Ciccarelli *et al.*, 2006; Rogozin *et al.*, 2007a,b, 2008). This specific phylogenetic question has the attraction of being approachable with the largest possible molecular data sets: the completely sequenced genomes of flies, worms, and humans. What almost all studies that have used this approach have found is that the evidence is strongly in favour of the Coelomata hypothesis and against Ecdysozoa.

The counter argument, naturally, is that these whole-genome studies suffer from precisely the problem that the Aguinaldo *et al.* paper addressed;

the systematic artefactual attraction of the nematode branch towards the root of the Bilateria due to LBA. This contention does seem to be borne out by a number of publications in the past few years. Copley *et al.,* (2004) compared the presence or absence of 1712 genes or distinct combinations of protein domains specific either to flies and humans or to flies and nematode worms. There were many more of the former, giving apparently strong support to the Coelomata hypothesis. However, they were able to show that this strong signal was an artefact resulting from a strong tendency towards secondary loss of genes in the nematode, a feature of its high rate of genomic evolution (Copley *et al.,* 2004). In parallel, Philippe *et al.,* (2005a) used large 'phylogenomic' data sets (whole genomes combined with data from expressed sequence tag projects and hence having much broader taxon sampling) and showed that experiments designed to reduce potential long-branch effects—using less distant outgroups, selecting slowly evolving nematodes, and discarding the more unevenly evolving genes—supported Ecdysozoa while Coelomata was supported without these efforts. Finally, a similar approach has been used (Irimia *et al.,* 2007; Roy and Irimia, 2008a,b) to show that claims of an excess of identical, rarely changing, amino acids and of specifically located introns shared by flies and humans and lacking in nematode worms (e.g. Rogozin *et al.,* 2007a,b, 2008) are biased by the use of distant outgroups and by the rapid evolution of *C. elegans.* They show that, when these biases are accounted for, there is significantly more support for Ecdysozoa than for Coelomata from this source of evidence (Irimia *et al.,* 2007; Roy and Irimia 2008a,b).

The phylogenomic approach has recently been extended to the slowly evolving Priapulida that are strongly supported as ecdysozoans (Webster *et al.,* 2006) and for Nematomorpha (T. Juliusdottir, R. Jenner, MJT, and R. Copley 2007, unpublished). This result was further strengthened by analyses of the very arthropod-like mitochondrial genome of *Priapulus caudatus.* Perhaps even more strikingly, the priapulid mitochondrial gene order can be reconciled with that of the arthropods by a single inversion (Webster *et al.,* 2006).

We would also like to highlight a further very convincing synapomorphy supporting the monophyly of Protostomia, and hence, we believe, definitively ruling out the Coelomata hypothesis. Papillon *et al.,* (2004) used the presence of a dozen rarely changing amino acids in the mitochondrial *nad5* gene of protostomes as a striking indication that the chaetognaths were protostomes and not deuterostomes as traditionally believed. The signature constitutes a very complex, conserved, derived character defining a monophyletic group of Protostomia. We have extended this analysis of the *nad5* gene, which appears to have undergone a significant burst of evolution within the lineage leading to the protostomes. Almost all of these signature amino acids are found in nematodes and priapulids as well as in other controversial protostomes, including rhabditophoran and catenulid flatworms and lophophorates. A monophyletic Protostomia, while not specifically proving the existence of Ecdysozoa, is clearly incompatible with a monophyletic group of coelomate animals and therefore contradicts the results from whole-genome studies supporting Coelomata.

We strongly support the notion of a monophyletic Ecdysozoa and feel that the only opposing evidence—the whole genome support for Coelomata—is flawed by systematic error, which has been addressed successfully by much better taxon sampling, in particular the use of a close outgroup (Philippe *et al.,* 2005a; Dunn *et al.,* 2008). In addition to the molecular systematic support, the monophyly of Ecdysozoa is supported by a number of morphological synapomorphies including ecdysis of a trilayered cuticle (consisting of epi-, exo-, and endocuticle), lack of locomotory cilia, lack of primary larva, terminal mouth, the HRP antigen in the nervous system, and conserved mitochondrial gene order that have been mentioned (see also Schmidt-Rhaesa, 1998).

8.3 Cycloneuralia, Introverta, Scalidophora, and Nematoida

While we do not have an equivalent of the *nad5* rare genomic change to support the monophyly of the Ecdysozoa within the Protostomia, as we have seen, we do have strong evidence from phylogenomic data sets of tens to hundreds of genes for the monophyly of Arthropoda plus Nematoda and Priapulida (Philippe *et al.*, 2005a; Webster *et al.*, 2006). More recently, data from Kinorhyncha and Nematomorpha have been added, and these phyla too were found to be part of the Ecdysozoan radiation (Dunn *et al.*, 2008). These worm phyla had previously been linked to one further phylum, the Loricifera, in a group collectively known as the Cycloneuralia (Ahlrichs, 1995). The name refers to their collar-shaped, circum-oral brain; something similar is seen in Gastrotricha which are, however, lophotrochozoans not ecdysozoans (Telford *et al.*, 2005; Todaro *et al.*, 2006). These phyla (but not Gastrotricha) also share an eversible anterior end, or introvert, which terminates in the mouth and gives the alternative name of Introverta (Nielsen 1995, 2001), although the introvert is only seen in the larvae of Nematomorpha and in isolated examples of Nematoda.

What we still do not have is much reliable information on the relationships between these phyla or their relationships to the Panarthropoda. This may be explained in part by the difficulty in working on the minute and hard to study Kinorhyncha and Loricifera. The two groupings that do seem credible are a close relationship between Nematoda and Nematomorpha and between Priapulida and Kinorhyncha. Nematodes and nematomorphs share a number of characters, including the reduced circular muscles in the body wall, the cloaca seen in both sexes, the aflagellate sperm, the cuticle (collagenous not chitinous), and the ectodermal ventral and dorsal nerve cords and were grouped by Nielsen (1995, 2001) and named Nematoida by Schmidt-Rhaesa (1998). This clade has weak support from SSU rRNA gene analyses (Peterson and Eernisse, 2001), combined analyses of LSU and SSU rRNA (Mallatt *et al.*, 2004), and from a recent phylogenomic analysis (Dunn *et al.*, 2008).

Morphologists have also united Priapulida, Kinorhyncha, and Loricifera in the Scalidophora (Schmidt-Rhaesa, 1998) or Cephalorhyncha (Nielsen, 1995, 2001), on the basis of an introvert with scalids and the presence of two rings of retractor muscles on the introvert. The close relationship between priapulids and kinorhynchs at least seems to hold up (Dunn *et al.*, 2008), but of the two existing phylogenetic studies that include data from the Loricifera, one showed their position to be

ambiguous (Park *et al.*, 2006) and the other linked them to the Nematomorpha (Sørensen *et al.*, 2008).

Early studies gave no strong indication that the Cycloneuralia is a monophyletic group, and in fact most evidence pointed to Priapulida (and therefore Kinorhyncha too) as being the earliest branch and the Nematoida as being the sister group of the Panarthropoda (e.g. Mallatt and Giribet 2006; Webster *et al.*, 2006). In the phylogenomic analysis of Dunn *et al.*, (2008), however, the Cycloneuralia are monophyletic. These alternative scenarios clearly have important consequences for the evolution of the arthropods, as the former implies either convergent evolution of complex structures such as segments and coeloms in the arthropods and annelids or parallel losses of such structures in independent clades of paraphyletic cyloneuralians. The latter result—monophyletic cycloneuralia—is a more parsimonious view of the evolution of morphology but is less informative regarding the origins of the Arthropoda.

8.4 Panarthropoda: Euarthropoda, Tardigrada, and Onychophora

The monophyly of the Euarthropoda plus Onychophora and Tardigrada seems, on the face of it, uncontentious. They are linked by a number of features, the most important of which are the segmentally repeated limbs with terminal claws; Onychophora translates as 'claw bearer'. The limbs in all three groups straddle parasegmental boundaries marked by the expression of the segment polarity gene *engrailed* (Patel *et al.*, 1989; Gabriel and Goldstein, 2007). Their segmental paired, saccate nephridia (reduced in number and functioning as excretory organs in euarthropods) and open circulatory system also seem to be valid synapomorphies, although both are missing in the miniaturized tardigrades (Hejnol and Schnabel, 2005). The circulatory system is characteristically formed as a fusion of both the coelomic cavities and the primary body cavity/embryonic blastocoel (i.e. a mixocoel), and there is a dorsal heart with characteristic openings (ostia) into the open circulatory system (Nielsen, 1995, 2001).

Despite the characters in common with Panarthropoda, not all molecular studies support their monophyly, some grouping Tardigrada with the nematodes rather than with Euarthropoda (Dunn *et al.*, 2008). This relationship between tardigrades and nematodes is biologically implausible and LBA is a strong contender for an explanation of this result. If we assume that Panarthropoda (including Tardigrada) is monophyletic, three trees could unite the euarthropods, tardigrades, and onychophorans: Euarthropoda with either (1) Tardigrada or (2) Onychophora, or (3) Tardigrada and Onychophora as sister groups. The branching order of these three taxa is not resolved by molecules or morphology. If we assume that the small size of tardigrades is derived and accounts for the absence of mixocoel, heart, and nephridia (Schmidt-Rhaesa, 2001), characteristics found in Onychophora and Euarthropoda), we suggest that the similarities of cuticle, ganglionated ventral nerve chord, and limbs in tardigrades and Euarthropoda may indicate a sister-group relationship.

8.5 Euarthropoda: Myriochelata versus Mandibulata

The relationships of the four euarthropod clades—Chelicerata, Myriapoda, Crustacea, and Hexapoda (Hexapoda = Insecta, plus the basally branching groups Diplura, Protura, and Collembola) have long been disputed. A decade or so ago there were even serious arguments over the single versus multiple origins of arthropodization, and therefore over the monophyly versus polyphyly of euarthropods (Fryer, 1997). Molecular analyses have emphatically supported the monophyly of euarthropods and a unique origin of their cuticularized body and jointed appendages, and in the past years attention has been focused more on the relationships between these four groups. One common feature of morphology-based interpretations of arthropod phylogeny was the close relationship between Myriapoda and Hexapoda in a clade called the Atelocerata (which means malformed horns and refers to their common lack of a pair of second antennae) defined additionally by unbranched (uniramous) appendages, Malpighian tubules, and tracheal breathing (Telford and Thomas, 1995). According to the polyphyleticists, the Atelocerata are grouped with the Onychophora in a clade

called the Uniramia. Arguably, the clearest result to date in arthropod phylogeny shows that the insects are not most closely related to the myriapods but to the crustaceans (Boore *et al.*, 1995) and, in all likelihood, constitute a subgroup within the Crustacea. This clade of crustaceans plus insects has been referred to as the Pancrustacea or as the Tetraconata due to the tetrapartite crystalline cones of the ommatidia (Dohle 1997, 2001; Harzsch 2002, 2004; Harzsch *et al.*, 2005).

More controversial, though, is the true position of Myriapoda which share numerous similarities of head organization not only with the insects (as discussed) but also with crustaceans, most notably the presence of a mandible on the third, appendage-bearing head segment and maxillae on the fourth and fifth head segments. The common head structure of myriapods, crustaceans, and insects with two pairs of antennae (at least primitively), paired gnathobasic mandibles, and two pairs of maxillae strongly supports their monophyly. This group is named the Mandibulata, reflecting the particular importance of detailed similarities seen between the mandibles of Pancrustacea and Myriapoda in terms of segmental identity, positioning relative to other body parts, gene expression, detailed similarities in terminal differentiation, and, of course, in function (Scholz *et al.*, 1998; Edgecombe *et al.*, 2003; Harzsch *et al.*, 2005). Surprisingly, a number of molecular studies using rRNA genes, nuclear protein-coding genes, and complete mitochondrial genome sequences do not support the mandibulate clade, instead linking the myriapods to the chelicerates in a group called the Paradoxopoda or the Myriochelata (Mallatt *et al.*, 2004; Negrisolo *et al.*, 2004; Pisani, 2004). An analysis of nuclear protein-coding genes (Regier *et al.*, 2005), however, did not find strong support for either Myriochelata or Mandibulata, suggesting that there is uncertainty over the affinity of myriapods. Recognizing that the distinction between the two possibilities comes down to the position of the root of the euarthropod tree, we have reanalysed the complete mitochondrial genome sequences of various arthropods using a priapulid as a short-branched, phylogenetically close relative of the arthropods (Rota-Stabelli and Telford, 2008). We find that, in contrast to previous studies that had used more distant outgroups (lophotrochozoans), our mitochondrial tree narrowly supports Mandibulata over Myriochelata (Figure 8.1; see also Pisani, 2004). We have also analysed a number of nuclear protein-coding genes and have reached the same conclusion (Bourlat et al., 2008; Figure 8.2). While our bias in support of a return to the Mandibulata is probably obvious, it is clear that this question remains to be resolved one way or the other. While there are specific characteristics shared by myriapods and chelicerates but absent from Pancrustacea (Dove and Stollewerk, 2003; Kadner and Stollewerk, 2004; Stollewerk and Simpson, 2005), it is difficult to demonstrate these as synapomorphies as we have insufficient data from an outgroup and the suspicion is that the Myriapoda/Chelicerata character state may be plesiomorphic and uninformative (Harzsch, 2004; Harzsch et al., 2005). While the same criticism may be made of some of the characters supporting Mandibulata, the chelicerate homologue of the mandible (the first walking leg; see Telford and Thomas, 1998) seems likely to represent the plesiomorphic condition as it strongly resembles adjacent, serially homologous, walking appendages. This implies that the mandible itself is a shared derived character uniting the mandibulates.

8.6 Pycnogonids are chelicerates

The pycnogonids, or sea spiders, have long been associated with the chelicerates due to the shared character of chelicerae (chelifores in pycnogonids) on the first limb-bearing segment. This phylogenetic link was questioned recently both by studies of their nervous systems and by molecular systematic analyses. The larval nervous system of a pycnogonid from the genus *Anoplodactylus* was studied and the chelifore appeared to be innervated from the frontmost portion of the brain (the protocerebrum), suggesting that this appendage was therefore not homologous to the chelicerae of other chelicerates which is innervated from the second portion of the brain (the deutocerebrum; Budd and Telford, 2005; Maxmen et al., 2005). This tied in with a molecular phylogenetic study placing the Pycnogonida at the base of Euarthropoda and not with Chelicerata (Regier et al., 2005). Subsequent analysis of *Hox*

(a)

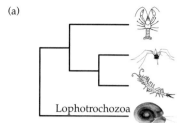

Phylogenetically distant outgroups: Paradoxopoda

(b)

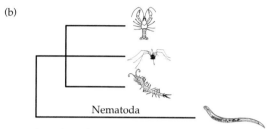

Long branch Ecdysozoa: unresolved Euarthropoda

(c)

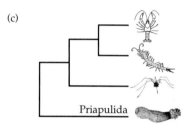

Short branch Ecdysozoa: Mandibulata

Figure 8.1 Different outgroups support different positions of the root of the Euarthropoda. (a) Support for Myriochelata (Myriapoda + Chelicerata) is strong using mitochondrial genome data when the Euarthropod tree is rooted using phylogenetically distant lophotrochozoans. (b) Support is equivocal using long branch but phylogenetically closer ecdysozoan nematodes (unresolved Euarthropoda). (c) The tree switches to supporting a monophyletic Mandibulata (Myriapoda with Crustacea + Hexapoda) when using the phylogenetically close and short-branched priapulid as an outgroup (Rota-Stabelli and Telford, 2008).

expression patterns have disproved the protocerebral position of the chelifores, showing that they are indeed in the same deutocerebral position as chelicerae (Jager et al., 2006) and most molecular data imply that the Chelicerata including Pycnogonida is a monophyletic group and that the contrary result was most likely derived from the rapid evolutionary rate of the Pycnogonida (Mallatt and Giribet, 2006).

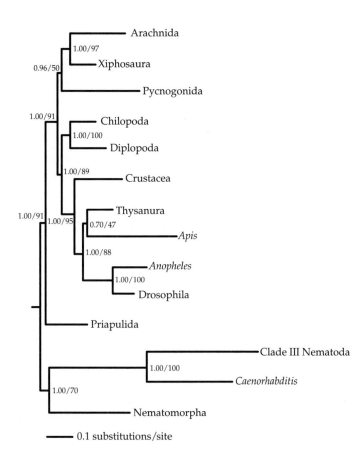

Figure 8.2 Phylogeny of the Ecdysozoa. Bayesian analysis using small-subunit (SSU) and large-subunit (LSU) ribosomal RNA sequences, complete mitochondrial genomes, and eight nuclear protein-coding genes (vacuolar ATP-synthase, enolase, glyceraldehyde 3-phosphate dehydrogenase, carnitine palmitoyl transferase, Na/K ATPase, RNA Pol II, dyskerin, and EF1-alpha). Some taxa with missing data have been merged into composite sequences. Support values are shown as Bayesian posterior probabilities and non-parametric bootstrap. Data are from Bourlat *et al.*, (2008).

8.7 The position of the Hexapoda within the Pancrustacea

The support for the monophyly of Crustacea + Hexapoda, which came most emphatically from the evidence of a shared mitochondrial genome rearrangement, has been bolstered by numerous subsequent molecular phylogenetic analyses (Friedrich and Tautz, 1995; Hwang *et al.*, 2001; Delsuc *et al.*, 2003; Nardi *et al.*, 2003; Regier *et al.*, 2005). Consideration of various aspects of morphology, in particular of nervous system ontogeny and structure, gives further weight to the integrity of this clade. Harzsch (2004) and Harzsch *et al.*, (2005) list neuroblasts, two pairs of serotonergic neurons per hemineuromere, a fixed number of excitatory motor neurons per limb muscle, and aspects of lateral eye ultrastructure in support of the Crustacea + Hexapoda clade, which they

term Tetraconata. More controversial has been the placement of Hexapoda within Crustacea and the monophyly versus polyphyly of Hexapoda, with a number of studies separating the Collembola from the Insecta. While it is generally agreed that Crustacea is paraphyletic rather than being the sister group of the Hexapoda (and that Hexapoda is in effect a terrestrial group of crustaceans), the closest crustacean sister group of the hexapods has been debated. Ignoring for the moment the little-studied Cephalocarida and Remipedia, there are two contenders among the main crustacean classes: the Malacostraca, that includes familiar species such as crabs and mantis shrimps, and Branchiopoda such as *Artemia* the brine shrimp and *Daphnia* the water flea. The Hexapoda–Malacostraca clade is supported by various features of brain anatomy; specifically, members of these two groups share the presence of three brain neuropils joined by chiasmata where other crustaceans have two neuropils linked by parallel fibres (Harzsch, 2002). Analyses of complete mitochondrial genome sequences on the other hand support a monophyletic Malacostraca and Branchiopoda clade as a sister group to Hexapoda (Cook *et al.*, 2005). Most other molecular analyses, however, support a sister-group relationship between Hexapoda and Branchiopoda (Regier *et al.*, 2005; Mallatt and Giribet, 2006).

We have recently gathered all available data from rRNAs, mitochondrial genomes, and various nuclear protein-coding genes, and our analyses support the close relationship between Hexapoda and Branchiopoda (Economou, 2008). This relationship is of great interest to the many workers interested in the evolution of the insects as it shows that *Daphnia*, a crustacean with a completely sequenced nuclear genome, is a relatively close sister group of insects. Our analyses also include data from Cephalocarida and Remipedia, and the placement of these two groups is less certain. Both taxa are atypical in terms of numbers of substitutions. While the remipedes consistently groups close to the hexapods, the position of the cephalocarids is very unstable (Economou, 2008). Although the relationships within the Hexapoda are beyond the scope of this discussion, the controversy over the placement of the Collembola is worth mentioning. While rRNA and nuclear protein-coding gene phylogenies recover the expected monophyly of the hexapods (Insecta, Diplura, Protura, and Collembola), analyses using complete mitochondrial genomes recover a diphyletic Hexapoda with the Insecta separated from the Collembola (Nardi *et al.*, 2003); mitochondrial sequences for Diplura and Protura were not available. The basis of this result has been questioned by subsequent authors, and one must conclude that although monophyly of Hexapoda ultimately seems the most likely result, this needs to be tested with larger data sets.

8.8 Conclusion

In Figure 8.3 we summarize our best current estimate of ecdysozoan phylogeny. The first thing that is obvious from this tree and from the preceding discussion is that while it seems clear that Ecdysozoa is a monophyletic group, the relationships between phyla and major classes within the clade are often uncertain. While the pattern of relationships of the Ecdysozoa has its own great intrinsic interest, the phylogeny should also be viewed as the basis for a further understanding of the evolution of the Ecdysozoa. The mapping of character states onto a phylogeny allows us to go beyond the relationships of organisms to the evolution of characters and ultimately a fuller understanding of the process of evolution. The characteristics of the common ancestor of Ecdysozoa is of particular interest, and it can be safely assumed to have possessed the synapomorphies of the group; Budd (2001a) has tentatively reconstructed the common ancestor as a large worm-like form with a terminal mouth, and to these characteristics we can add the shared characters discussed previously. The monophyly versus paraphyly of the Cycloneuralia becomes important now as, if paraphyletic, then their common ancestor becomes synonymous with the ecdysozoan ancestor, and suggests that it also possessed a cycloneuralian brain (not unreasonable considering the similar situation seen in onychophorans; Eriksson *et al.*, 2003), and an introvert.

More controversial is the possibility that the ecdysozoan ancestor was segmented. While the kinorhynch metameres are generally referred to as zonites rather than segments, this seems a rather pointless distinction and is one indication that segmentation may be primitive in the group (Müller and Schmidt-Rhaesa, 2003; Schmidt-Rhaesa and Rothe, 2006). The similar deployment of homologous genes ('segment polarity' or 'pair rule' genes) in arthropods and kinorhynchs would be a more direct indication of homology, and hence common ancestry of segmentation within the group, as would the demonstration that arthropod segmentation can be convincingly homologized with that of annelids (Prud'homme *et al.*, 2003) or even vertebrates (Damen, 2007).

Through comparison of the completely sequenced genomes of *D. melanogaster* and *C. elegans*, there is also the theoretical possibility of learning something about the genome of the ecdysozoan common ancestor, or perhaps something close to it depending on the position of the Arthropoda/Nematoida common ancestor. One significant con-

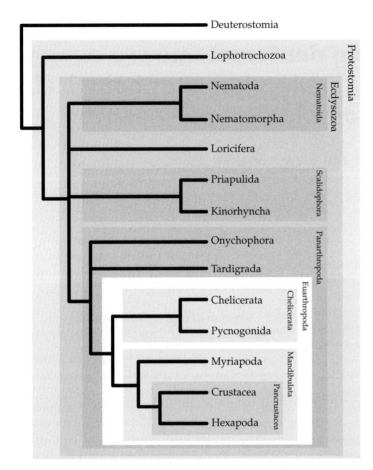

Figure 8.3 The phylogeny of the Ecdysozoa espoused in this chapter. Names of probable monophyletic groups are given for each box. Unresolved portions of the tree are shown as multifurcations. We have shown Cycloneuralia, Mandibulata (including Myriapoda), and Panarthropoda (including Tardigrada) as monophyletic groups, despite some uncertainty, as we feel the morphological evidence is particularly convincing for these clades.

clusion from comparative genomics to date has been the secondary loss of large numbers of genes in the two model ecdysozoans (Copley *et al.*, 2004; Putnam *et al.*, 2007). The problem, of course, is that the two model species appear to have very derived genomes making comparisons particularly difficult to interpret—are they different from other animals due to common ecdysozoan gene losses or through convergent gene losses in these two derived models? The ecdysozoan genome projects ongoing or recently announced, in particular that of the priapulid *Priapulus caudatus*, are very exciting for the purpose of reconstructing the ancestral ecdysozoan genome, and should also add further to our understanding of the evolutionary relationships of this huge, diverse, and fascinating group.

8.9 Acknowledgements

Research in the laboratory was supported by the BBSRC and by the Marie Curie RTN ZOONET (MRTN-CT-2004–005624). We are very grateful for careful reviews by Andreas Schmidt-Rhaesa and Davide Pisani.

CHAPTER 9

Deciphering deuterostome phylogeny: molecular, morphological, and palaeontological perspectives

Andrew B. Smith and Billie J. Swalla

9.1 Introduction

Deuterostomes form one of the three major divisions of the Bilateria and are sister group to the Lophotrochozoa plus Ecdysozoa (Eernisse and Peterson, 2004; Philippe *et al.*, 2005a; Telford *et al.*, 2005; Dunn *et al.*, 2008). Traditionally the group was recognized on the basis of a shared embryonic development pattern: gastrulation occurs at the vegetal pole and the blastopore becomes the anus, while the mouth forms secondarily (Chea *et al.*, 2005). Analysis of molecular data has consistently found the deuterostome grouping, with generally high levels of support (Turbeville *et al.*, 1994; Wada and Satoh, 1994; Halanych *et al.*, 1995; Cameron *et al.*, 2000). Five major clades make up the Deuterostomia: craniates, cephalochordates, echinoderms, hemichordates, and tunicates (see Figure 9.1). Because vertebrates, including ourselves, belong to the craniates, there has long been a fascination about their invertebrate origins. Theories of chordate evolution have abounded for over 100 years, but it is only in the last 10 to 15 years that deuterostome relationships have come into sharp focus, driven largely by new data from molecular genetics and the fossil record, and new analyses of traditional morphological and ontogenetic data. From this plethora of information, some complementary, others supporting contradictory conclusions, a more coherent picture of the phylogeny and early evolution of deuterostomes is starting to emerge.

Here we review four key areas where there has been the most heated debate in the last 5 years: phylogenetic relationships of the major deuterostome groups; the earliest fossil record and

divergence times of deuterostome groups; the evolution of body axes; and the characteristics of the ancestral deuterostome body plan.

9.2 Deuterostome phylogenetic relationships

Until 10 years ago there was little consensus about the relationships of the major deuterostome groups (see reviews by Gee, 1996, and Lambert, 2005). Depending upon whether emphasis was given to comparative adult morphology, embryology, or the fossil record, different sister-group relationships could be argued. Larval traits provided support for a grouping of echinoderms and hemichordates (Hara *et al.*, 2006; Swalla, 2006), adult traits provided support for a grouping of hemichordates and chordates (Cameron *et al.*, 2000) while palaeontological data were used to support an echinoderm–chordate pairing (Gee, 1996). Probably the most widely accepted view in the mid-1990s was that echinoderms were sister group to the rest and that chordates and hemichordates were sister taxa [i.e. (echinoderms {hemichordates [tunicates (cephalochordates + craniates)]})].

With the arrival of molecular data the problem of deuterostome relationships seemed to be solved. Early results were based on analyses of ribosomal gene sequences and pointed to echinoderms and hemichordates as sister groups (Turbeville *et al.*, 1994; Wada and Satoh, 1994) and to a monophyletic Chordata comprising tunicates, cephalochordates, and craniates (Turbeville *et al.*, 1994). Within the chordates, tunicates were identified as sister group

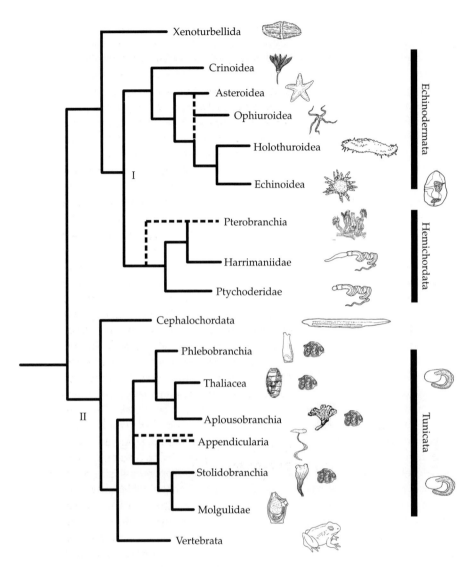

Figure 9.1 Current deuterostome phylogeny, according to available molecular and morphological data. Dotted lines show clades of uncertainty where conflicting data have been obtained; I and II mark clades where the evidence for a monophyletic group is very high. I, Ambulacraria is made up of hemichordates and echinoderms. Mitochondrial, ribosomal and genomic evidence are in agreement for this grouping. The Ambulacraria develop similarly through gastrulation and share larvae that feed by ciliated bands, strengthening their sister-group relationship. Genomic evidence suggests that Xenoturbellids may be a sister group to the Ambulacraria, but it is also possible that they are an outgroup to the rest of the deuterostomes. II, Chordates are a monophyletic group that share a specific body plan, but mitochondrial and genomic evidence are in conflict about the position of the tunicates. Mitochondrial and ribosomal evidence place cephalochordates as sister group to the vertebrates, whereas genomic evidence places tunicates as the sister group to the vertebrates. Figure modified from Zeng and Swalla (2005).

to Euchordates (= craniates plus cephalochordates) (Turbeville *et al.*, 1994; Wada and Satoh, 1994; Adoutte *et al.*, 2000; Cameron *et al.*, 2000; Winchell *et al.*, 2002; Bourlat *et al.*, 2003). Reassuringly these relationships appeared to be robust to tree-building

methods (Furlong and Holland, 2002) and morphological support for these groupings was forthcoming (Peterson and Eernisse, 2001). The pairing of echinoderms and hemichordates is further supported by shared *Hox* gene motifs (Peterson *et al.*,

2004), mitochondrial genes and gene arrangements (Bromham and Degnan, 1999; Lavrov and Lang, 2005), and amino acid sequences of selected nuclear genes, such as actins (Swalla, 2007).

However, as data accumulated and more careful probing of the molecular phylogenetic signal was carried out, some parts of the deuterostome tree came into question. Although the clades Ambulacraria and Chordata remained well supported (Figure 9.1), the position of tunicates was found to be unstable and varied according to which gene was used (Winchell *et al.*, 2002). A new group, Xenoturbellida, was also added to the deuterostomes (Bourlat *et al.*, 2003, 2006). Furthermore, in the last 2 years phylogenies have been increasingly constructed from concatenated amino acid sequences of large numbers of protein-coding genes (Blair and Hedges, 2005; Philippe *et al.*, 2005a; Bourlat *et al.*, 2006; Delsuc *et al.*, 2006; Dunn *et al.*, 2008). Most of these phylogenies place echinoderms and hemichordates as sister groups, and recognize Chordata as a clade, but place tunicates, not cephalochordates, as sister group to chordates (a grouping termed Olfactores). At the same time, an extensive cladistic reanalysis of morphological data found strong support for Olfactores (Ruppert, 2005). Figure 9.1 summarizes the current emerging view of deuterostome relationships.

While higher-level relationships have attracted most attention, extensive progress has also been made in constructing detailed phylogenetic hypotheses for taxa within each of these major groups. A summary of echinoderm relationships was given by Smith *et al.*, (2004) and, for the most part, morphological and molecular data point to the same branching order. Only the position of ophiuroids with respect to other eleutherozoan taxa remains problematic. The match between molecules, morphology, and the fossil record is particularly strong for echinoids (Smith *et al.*, 2006).

There is less certainty about the relationships within hemichordates. Traditionally the group has been divided into the worm-like enteropneusts and the tube-dwelling pterobranchs, with enteropneusts considered as derived from a pterobranch-like ancestor (e.g. Barrington, 1965). Morphological analysis places pterobranchs and enteropneusts as sister groups (Halanych *et al.*, 1995; Cameron *et al.*, 2000). However, molecular phylogenies based on ribosomal genes provide a mixed signal. The 18S rRNA gene data nest pterobranchs within a paraphyletic enteropneust grade while 28S rRNA gene sequence data place pterobranchs and enteropneusts as sister groups, but this may be due to the lack of informative sites in 28S rRNA for Hemichordata (Cameron *et al.*, 2000; Winchell *et al.*, 2002). As discussed below, whether pterobranchs are derived or basal with respect to other hemichordates is critical for reconstructing the ancestral body plan of the Ambulacraria.

Zeng *et al.*, (2006) provide a detailed 18S rDNA phylogeny of tunicates that mostly agrees with earlier molecular and morphological analyses (Swalla *et al.*, 2000). Within the tunicates, morphological and molecular data are largely congruent (Swalla *et al.*, 2000; Stach and Turbeville, 2002). Five separate clades of tunicates, corresponding to traditionally recognized groupings, are supported by molecular data: The ascidian clades Stolidibranchia, Phlebobranchia, and Aplousobranchiata, and the pelagic tunicates Appendicularia and Thaliacea (Figure 9.1). The Thaliacea are all colonial and are sister group to the Phlebobranchia plus Aplousobranchiata clade. In molecular phylogenies, the solitary, pelagic free-floating Appendicularia has a very long branch and falls either as outgroup to the rest of the tunicates (Swalla *et al.*, 2000; Stach and Turbeville, 2002) or as a sister group to the Stolidobranchia (Zeng *et al.*, 2006). If the tunicate ancestor resembles the Appendicularia plus Stolidobranchia clade, then it would be likely to be solitary. However, if the ancestor resembled the Phlebobranchia, Aplousobranchiata, and Thaliacea clade, then it would be more likely to be colonial (Zeng and Swalla, 2005). Whatever the tunicate ancestor, the tunicate adult body plan bears little resemblance to the rest of the chordates, suggesting that major changes in adult body plan happened early in tunicate evolution (Swalla 2006, 2007).

There are so few living cephalochordates that this group is represented by a single taxon in most molecular analyses, and the different extant genera are morphologically very similar (Zeng and Swalla, 2005). By contrast, extensive molecular and morphological phylogenies exist for craniates (Rowe, 2004). The basal relationships of hagfishes, lampreys, and other primitive craniates such as conodonts, has been detailed thoroughly by Donoghue *et al.*, (2000). The most recent molecular phylogeny based on protein-coding genes (Blair and Hedges, 2005) places hagfishes and lampreys as sister taxa (Cyclostomata) right at the base of the craniate tree, as in previous ribosomal-based analyses (Mallatt and Sullivan, 1998).

9.3 Insights from the fossil record

In the last 10 years a number of new and potentially important fossil deuterostomes have been recovered from Early and Middle Cambrian deposits. However, progress has been more difficult here than for molecular studies because of problems arising from incomplete preservation and ambiguity of interpretation. Fossils provide data on morphologies that once existed and, when included in phylogenetic analyses with extant taxa, can aid in recognition of ancestral character states and can alter relationships by displaying character combinations not seen in living forms. However, missing data can be a major problem for the interpretation of fossils, and even with the best soft-tissue preservation in the Cambrian we have only the outline of internal organs preserved without structural detail. Furthermore, fossil anatomy can only be interpreted through reference to extant organisms. For taxa without obvious modern counterparts (as for example in the case of vetulicolians discussed below), the choice of which modern analogue to select as a reference for interpretation can make a huge difference. The fact that the taxonomic placement of most Cambrian soft-bodied deuterostome taxa remains disputed attests to the difficulty of interpretation posed by these fossils.

An even more crippling problem for palaeontologists is that many of the high-ranking taxa are recognized on the basis of just a few, mostly embryological, biochemical, or genetic characters, or from molecular data alone, such as Xenambulacraria (Bourlat *et al.*, 2006). Even for classes and phyla, fossils can only be placed with certainty once one or more derived morphological

synapomorphies have evolved, so that early stem-group members are commonly much more difficult to identify, simply because there are no emergent features! This is a general problem of ancestral taxa in the fossil record—we know they probably exist, but we simply do not know how to identify them. Here we review the fossil evidence for the oldest members of each phylum or subphylum and then examine the case for more primitive deuterostome stem-group members.

9.3.1 Hemichordates

Of the two hemichordate morphologies, the colonial, tube-dwelling pterobranchs have by far the best fossil record. The tubes of fossil pterobranchs preserve well and show two excellent synapomorphies; a characteristic fusellar structure and an internal stolon. It has long been recognized that the extinct Graptolithina are a primarily planktonic group of colonial hemichordates closely related to the extant pterobranchs (Maletz *et al.*, 2005; Rickards and Durman, 2006). Indeed, as better knowledge of the Cambrian faunas has accrued, it has become increasingly difficult to draw a clear division between pterobranchs and graptolites (Maletz *et al.*, 2005). For example, the Ordovician *Cephalodiscus*-like genus *Melanostrophus* preserves the morphological ultrastructure of the tube in spectacular detail and displays a mixture of extant cephalodiscid and extinct graptolite features (Mierzejewski and Urbanek, 2004). It now seems likely that many of the primitive benthic dendroid graptolites are probably better classified as pterobranchs (Maletz *et al.*, 2005). Pterobranch tubes showing clear fusellar structure and an internal stolon are now recorded from the Middle Cambrian (Maletz *et al.*, 2005; Rickards and Durman, 2006), the oldest coming from the early Middle Cambrian (Bengtson and Urbanek, 1986). Both rhabdopleurids and cephalodiscids are present by the end of the Middle Cambrian (Rickards and Durman, 2006) showing that crown-group divergence had occurred by then.

Fossil enteropneust worms have proven to be much more elusive to identify. The most characteristic apomorphies likely to be seen in fossils are the proboscis and collar, but to date no such Cambrian worm-like organisms have been formally described from the Cambrian, although Conway Morris (1979) lists enteropneusts as present in the Burgess Shale. Shu *et al.,* (1996a) initially interpreted the fossil chordate *Yunannozoon* from the Lower Cambrian of China as a fossil enteropneust, but this was later corrected (Shu *et al.,* 2003b). Wignall and Twitchett (1996) recorded possible burrows of enteropneusts from the Lower Triassic of Poland, and the oldest body fossil of an enteropneust is from the Lower Jurassic of Italy (Arduini *et al.,* 1981).

9.3.2 Echinoderms

Echinoderms have a calcitic skeleton with a very distinctive structure termed 'stereom'. The genes responsible for stereom deposition are unique to echinoderms (Bottjer *et al.,* 2006). Thus although upstream regulator genes of skeletogenesis might be common to all deuterostomes (Ettensohn *et al.,* 2003), the specific skeletal construction of echinoderms is a reliable synapomorphy. Furthermore, all members of the crown group show radiate symmetry, so stereom-bearing fossils without radial symmetry are best considered as stem-group echinoderms. The morphology and relationships of stem-group echinoderms have been reviewed by Smith (2005) and some information on their soft-tissue anatomy can be deduced confidently because of the close correspondence between stereom microstructure and investing tissue (Clausen and Smith, 2005). Stem-group echinoderms are diverse and include a variety of asymmetrical forms. These reveal that the most basal echinoderms had external gill openings, a bilaterally symmetrical muscular stalk (solutes and stylophorans; Figure 9.1), and, less certainly, a pair of tentacles and a pharyngeal basket with atrium (cinctans) (Smith, 2005). These fossils thus provide an important window onto the early history of echinoderms, revealing key stages in the origins of the echinoderm body plan, one of the most divergent of any bilaterian. The oldest skeletal elements with stereom are mid Lower Cambrian, and both crown-group (eleutherozoan and pelmatozoan) and stem-group (solutes) members are present by the latest Lower Cambrian, showing

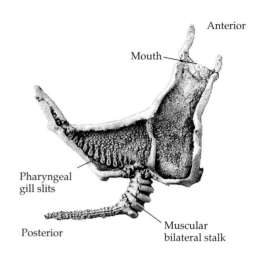

Figure 9.2 The stem group echinoderm *Cothurnocystis* (Ordovician, Scotland), showing the presence of pharyngeal gill slits.

that diversification of crown-group echinoderms had already occurred.

Vetulocystids, from the late Lower Cambrian of China, were described as stem-group echinoderms by Shu *et al.,* (2004). Three taxa were erected but, of these, two were badly preserved and not convincingly different from the better-known *Vetulocystis*. Furthermore, *Vetulocystis* shows close similarity to vetulicolans in having a bipartite body plan with a cuticular sheath, articulated posterior appendage, and at least one lateral pouch ('gill opening'—possibly paired but no specimen shows part and counterpart). These fossils do not share a single echinoderm synapomorphy so it is not clear why these were placed in the Ambulacraria.

9.3.3 Xenoturbellids

This small-bodied taxon has no known morphological synapomorphies and lacks a skeleton; not surprisingly, there is no known fossil record.

9.3.4 Tunicata

There are a number of synapomorphies that should allow firm identification of tunicates from the fossil record: their tunicine cuticle, pharyngeal basket, inhalent and exhalent openings to their atrium, and endostyle. However, they are soft-bodied and their fossil record is very sparse. Shu *et al.,* (2001a) reported a fossil tunicate (*Cheungkongella*) from the Lower Cambrian of China, but this was based on a unique and incomplete specimen that later, with the collection of additional material, turned out to be a junior synonym of *Phlogites*, a taxon with paired branched tentacles whose affinities are uncertain (Chen *et al.*, 2003; Xian-Guang *et al.*, 2006). However, another Lower Cambrian fossil from China, *Shanouclava*, is a tunicate and shows typical ascidian characters, including a pharyngeal basket and possible endostyle (Chen *et al.*, 2003). This is described as being a crown-group tunicate, related to aplousobranch ascidians. The fossil record thus clearly shows that, like echinoderms, the tunicate body plan was well established, and crown-group tunicate divergence had already occurred by the end of the Lower Cambrian.

It is also vaguely possible that the relatively diverse group of problematic Lower Cambrian fossils termed vetulicolians are appendicularian-like tunicates, but this is much less certain (Aldridge *et al.*, 2007). The problem associated with vetulicolians is discussed in more detail below.

9.3.5 Cephalochordata

This small group of fish-like invertebrates is readily identified from adult morphology, being jawless and having a continuous dorsal fin, myotomes, post-anal tail, and barred gill slits. *Pikaia*, from the Middle Cambrian Burgess Shale, has simple gill slits but is widely accepted to be a member of this clade. The Lower Cambrian *Cathaymyrus* is also attributed to this clade (Shu *et al.*, 1996b), though it is less well preserved and, consequently, the case here is less convincing.

9.3.6 Craniata

Hagfish, lampreys, and an important extinct group, the conodonts, branch close to the base of the craniate crown group (Donoghue *et al.*, 2000). The character combinations used to place fossil members within this clade are a well-developed brain with eyes, notochord, myotomes, and gill slits. Fossil agnathan fishes are present in the Lower Cambrian (Shu *et al.*, 1999; Zhang and Xian-Guang, 2004) and show evidence of a complex brain and well-developed eyes

(Shu *et al.*, 2003b). Two taxa were originally described (*Myllokunmingia* and *Haikouichthys*), but subsequently these have been shown to be synonymous (Xian-Guang *et al.*, 2002). There seems little dispute about their taxonomic affinities.

9.3.7 Potential representatives of more basal deuterostomes

Fossil representatives of the common stem of two or more phyla are difficult to recognize with confidence because adult morphological synapomorphies are lacking. Over the last 10 years various fossils have been championed as deuterostomes more basal than any of the five major groups, but few if any of these claims have stood up to detailed scrutiny. Vetulicolids are a clear case in point. First described as arthropods then later as basal deuterostomes, tunicates, or even possibly kinorhynchs, these animals have been the focus of much attention and debate, as recently reviewed (Aldridge *et al.*, 2007). If vetulicolids were indeed basal deuterostomes they would provide critical information on the ancestral body plan. However, the recent cladistic analysis of Aldridge *et al.* (2007) has highlighted just how tenuous the anatomical interpretations of these fossil remains are. Earlier reports of a vetulicolian mesodermal skeleton (Shu *et al.*, 2001a) are now considered unlikely. Indeed *Vetulicola* itself shows clear evidence of a jointed exoskeleton, and it is this that makes some (Caron, 2006) hesitant of rejecting arthropod affinities, despite their apparent lack of limbs. Most have a series of 'lateral pouches' which may or may not open externally and whose detailed structure is far from clear. These have been variously interpreted as gill slits, pouch-like arthropod gills, or midgut glands. Aldridge *et al.* (2007) accepted these as gills but could not determine whether they were internal or external openings. A third character, the presence of an endostyle, is unprovable, since the structure in question is no more than a dark line picked out by iron oxide—it exists, but its identification requires knowledge of taxonomic affinities and guesswork based on shape and position. Finally the nature of the thin tube running to the tip of the posterior is also critical. Shu *et al.*, (2001a) interpret this as the gut, making the anus terminal, whereas Aldridge *et al.,* (2007) raise the more likely possibility that it is another structure (possibly the notochord). In the face of such uncertainty the taxonomic placement of this group remains very much in doubt.

As Lacalli (2002) and Aldridge *et al.,* (2007) stress, vetulicolids share many similarities with tunicates, especially appendicularians, although there are still significant problems of interpretation. The jointed exoskeleton would have to be composed of tunicin and the lateral pouches would be internal gill slits, but both of these are potentially provable from the fossil record. Given that the interpretation of soft tissue imprints will always remain ambiguous, it seems likely that palaeontologists will continue to have to interpret fossils in the light of theories of deuterostome origins based on data from living organisms (*contra* Shu *et al.,* 2001b).

A potential stem group ambulacrarian, *Phlogites,* was recently redescribed by Xian-Guang *et al.,* (2006). These specimens have a pair of branched and erect tentacles, a large U-shaped gut and a muscular stalk. Xian-Guang *et al.,* (2006) compared *Phlogites* with various groups, noting its close similarity to uncalcified echinoderms, but eventually opting for it being a lophotrochozoan. It may, however represent a stem-group ambulacrarian, although the lack of obvious gill slits in *Phlogites* makes this somewhat less compelling.

The ancestral chordate was likely to have been fishlike with a post-anal tail, open pharyngeal slits, a notochord, and a single fin, with limited development of eyes and brain. The best contenders for representative stem-group chordates are the yunnanozoons. Yunannozoons were originally described as hemichordates (Shu *et al.,* 1996a) then later as stem deuterostomes (Shu *et al.,* 2003). However, one yunnanozoon, *Haikouella,* has been restudied by Mallatt and Chen (2003) and Mallatt *et al.,* (2003), and has been shown to be encephalized, with small eyes, a single fin, internal gill arches, and lateral and ventral myotomes. *Haikouella* was interpreted as a cephalochordate-like suspension feeder with an endostyle and tentacles forming a screen across the mouth. Detailed cladistic assessment (Mallatt and Chen, 2003) found that yunnanozoons lack any specific hagfish characters and fall closer to craniates than to cephalochordates. They are either stem-group craniates, stem-group Olfactores or

possibly late stem-group chordates. Again the disagreement over interpretation of structures (Mallatt and Chen, 2003; Shu *et al.,* 2003) highlights just how tenuous identification of soft-tissue anatomy in these forms can be. Basal deuterostomes are even less easy to diagnose, and no fossil has been convincingly identified.

In summary, members of all five major deuterostome groups can be distinguished on adult synapomorphies by the end of the Lower Cambrian, and crown-group divergence had taken place in at least three (echinoderms, hemichordates, tunicates) (Figure 9.3). By contrast, a lack of adult morphological criteria and the general difficulty of interpreting fossils that preserve only outline traces of soft-tissue organs has meant that no common stem-group member of two or more of these groups has yet been convincingly demonstrated, except possibly yunnanozoons, whose exact position within Chordata remains problematic.

9.4 When did deuterostome groups originate?

The palaeontological data discussed above, combined with our best molecular and morphological phylogenies, demonstrate that all five major clades of deuterostomes were already present in the Lower Cambrian, *c.* 520 million years ago (Ma) (Figure 9.3). Indeed, for some, crown-group divergence had already begun. There is no evidence from the rock record of an older fauna when just stem-group members existed, so unless one wishes to take a direct reading of the fossil record and argue for an almost instantaneous origin, the fossil record tells us only about the latest time by which these groups arose. Consequently, we must turn to molecular data to get estimates of their times of origin. This, however, is by no means without problems.

Most early attempts to use molecular similarity to estimate divergence times all pointed to a very deep divergence of deuterostomes (and other metazoans), pre-dating the first fossil evidence by several hundred million years (Wray *et al.,* 1996; Feng *et al.,* 1997; Gu and Li, 1998; Wang *et al.,* 1999; Nei *et al.,* 2001). These studies all calculate divergence times using a single molecular rate of change across the tree, and thus rely on all taxa having similar rates of molecular evolution. However, both Ayala *et al.*, (1998) and Peterson *et al.* (2004) noted that there was a significant decrease in the rate of molecular evolution in vertebrates that could mislead molecular clock estimates. To avoid this problem Ayala *et al.*, (1998) generated a linearized molecular tree after carefully checking for rate heterogeneity, and found that chordates and echinoderms split around 600 Ma. Peterson *et al.*, (2004) used a different approach, avoiding all vertebrates, and calculated divergence dates using a series of fixed calibration points based on the arthropod and echinoderm fossil records. This gave a minimum estimated divergence between hemichordates and echinoderms at 526–567 Ma.

Recently, methods of analysis that accommodate rate variation using a 'relaxed clock' model have been developed. These methods allow for rate variation across the tree to be modelled. A variety of methods are now available and can produce highly convincing results, both in simulation studies and with empirical data (Ho *et al.*, 2005; Near *et al.*, 2005; Smith *et al.*, 2006; Yang and Rannala, 2006), so long as the prior probabilities used are realistic. Early application of a Bayesian relaxed clock method to the problem of metazoan divergence times by Aris-Brosou and Yang (2003) found phylum-level splits in the Deuterostomia at around 520–530 Ma, matching Peterson *et al.*'s (2004) results. However, this used an unrealistic model of molecular evolution (Ho *et al.*, 2005), and other analyses that apply more appropriate models (Douzery *et al.*, 2004) have continued to find deeper divergence dates for the Bilateria. Using a much larger data set, Bayesian methodology, and multiple fossil calibration points, Blair and Hedges (2005) estimated the divergence of crown group craniates as 652 (605–742) Ma, the divergence of Olfactores at 794 (685–918) Ma and the divergence of Ambulacraria at 876 (725–1074) Ma. Based on the echinoderm divergence of echinoids from starfish, the more recent lower bound for these molecular estimates is in best accord with fossil evidence (Figure 9.3).

In summary, although there remains some debate about timing due to the methodological uncertainties associated with calculating divergence dates from molecular data, neither rapid diversification nor very deep diversification seem supported by current studies. Molecular data provide evidence that the earliest record of deuterostome evolution is missing from the fossil record, though whether this gap is only a few tens of millions of years or maybe as much as 200 million years remains unresolved.

If we cannot be precise about the timing of the event is it possible to determine from molecular data whether diversification occurred rapidly or slowly? Rokas *et al.*, (2005) raised an old argument in favour of metazoan diversification being compressed in time. They noted that whereas fungal relationships are clearly resolved by molecular data, those of some metazoans are not. As both groups evolved at around the same time, the early history of metazoans may have been a radiation compressed in time, in agreement with a direct reading of the palaeontological record. Baurain *et al.*, (2007), however, showed that their result was an artefact of inadequate taxon sampling and/or model of sequence evolution. By increasing the number of species, replacing fast-evolving species by slowly evolving ones, and using a better model of sequence evolution, Baurain *et al.*, (2007) found that resolution amongst metazoan clades was markedly improved. Thus with the right models and good taxon sampling, molecular data support the idea of a relatively gradual unfolding of deuterostome diversity over time.

9.5 Conserved gene networks pattern deuterostome axes and germ layers

The *Hox* complex is a duplicated set of genes that is frequently found in a single cluster on the chromosome, and is important for anterior to posterior patterning in bilaterian animals (Lemons and McGinnis, 2006). Within the deuterostomes, *Hox* clusters have been characterized from all of the major phylogenetic groups except xenoturbellids (Swalla, 2006). It is probable that the ancestral chordate had 14 genes linearly aligned on the chromosome, and that they were expressed collinearly in an anterior to posterior manner, as both cephalochordates and vertebrates have 14 genes, with the posterior six genes showing independent duplication from the protostome posterior genes

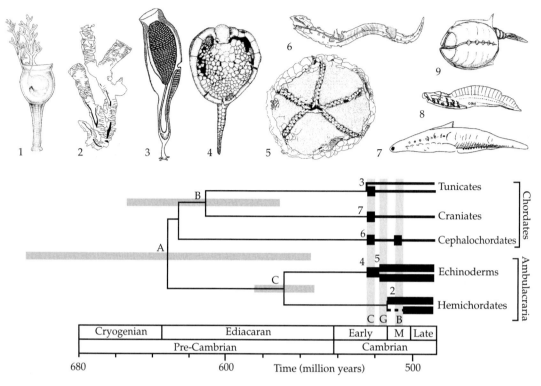

Figure 9.3 The fossil record of deuterostomes and molecular-clock estimates of divergence times. Thick black lines indicate known occurrence; thin lines indicate inferred range. Error bars show 95% confidence intervals on molecular estimates. Cambrian deuterostomes and possible deuterostomes as follows: 1, *Phlogites* (Lower Cambrian; possible stem-group ambulacrarian); 2, *Rhabdopleura* (Middle Cambrian pterobranch); 3, *Shanouclava* (Lower Cambrian ?aplousobranch ascidian tunicate); 4, *Trochocystites* (Middle Cambrian stem-group echinoderm); 5, *Stromatocystites* (Lower Cambrian crown-group echinoderm); 6, *Cathaymyrus* (Lower Cambrian cephalochordate); 7, *Haikouichthys* (Lower Cambrian craniate); 8, *Haikouella* (Lower Cambrian chordate; possible stem-group craniate); 9, *Vetulicolia* (Lower Cambrian problematica). Molecular dating for nodes A–C as follows. A, craniate–echinoderm split based on a linearized gene tree (from Ayala *et al.*, 1998); B, craniate–tunicate split based on the Bayesian relaxed clock model (from Douzery *et al.* 2004); C, echinoderm–hemichordate split based on a linearized gene tree (from Peterson *et al.*, 2004). Grey bands are deposits with soft bodied preservation: C, Chengjiang Formation; G, Guanshan Formation; B, Burgess Shale.

(Minguillón *et al.*, 2005; Lemons and McGinnis, 2006; Swalla, 2006).

A duplication of the posterior genes into three genes called *Hox 11/13a*, *Hox 11/13b*, and *Hox 11/13c* characterizes both echinoderm and hemichordate *Hox* gene clusters (Peterson, 2004; Morris and Byrne, 2005; Cameron *et al.*, 2006; Lowe *et al.*, 2006). The separate posterior genes share amino acid motifs, strongly suggesting that echinoderms and hemichordates share a common ancestor that differed from the chordate ancestor (Peterson, 2004). In addition, the sea urchin genome has been sequenced (Sodergren *et al.*, 2006) and the *Hox* cluster has an inversion, with *Hox 3* juxtaposed against *Hox 11/13c*, and the subsequent loss of *Hox 4*

(Cameron *et al.*, 2006). It will be very interesting to examine the arrangement of the *Hox* complex in other echinoderms to see if they have a similar inversion. If all echinoderms have this *Hox* inversion, then it will be interesting to examine enteropneust hemichordates, because they have collinear *Hox* expression in the ectoderm (Aronowicz and Lowe, 2006). The fact that hemichordates, cephalochordates, and vertebrates all show collinear expression of the *Hox* genes in an anterior to posterior manner, beginning directly after the first gill slit, suggests that the deuterostome ancestor had an antero-posterior axis determined by the *Hox* complex immediately posterior to the first gill slit formed (Swalla, 2006).

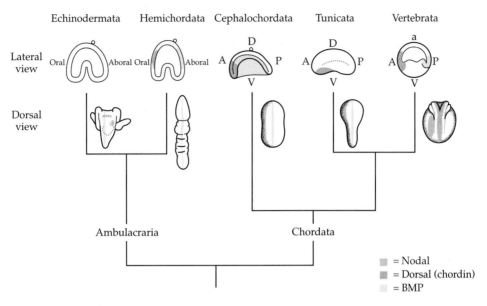

Figure 9.4 Summary of body axis determination in Deuterostomata: A, anterior; P, posterior; a, animal; v, vegetal. Expression of *nodal* is in red for all phyla. Dorsal in hemichordates is shown by *BMP* expression in yellow. In Echinodermata the aboral (dorsal) axis is shown by a yellow strip, with *nodal* expression marked red on the right side of the larvae, opposite where the adult rudiment will form. Dorsal in chordates is marked in blue (lower pictures) while expression of *chordin* is shown during gastrulation in Cephalochordata, Tunicata, and Vertebrata. The BMP–chordin axis is reversed in chordates from the Ambulacraria. Note that nodal expression is on the left side in chordates, and on the right side in echinoderms. Expression patterns taken from Duboc *et al.,* (2005) (urchin *nodal*), Lowe *et al.* (2006) (hemichordate *chordin*), Yu *et al.,* (2007) (cephalochordate *BMP* and *chordin*), Darras and Nishida (2001) (ascidian *chordin*), and Sasai and De Robertis (1997) (frog *BMP* and *chordin*). (See also Plate 7.)

In contrast, the chordate dorsoventral axis is inverted when compared to the hemichordate dorsoventral axis (Figure 9.4 and Plate 7; Ruppert, 2005; Lowe *et al.,* 2006; Swalla, 2007). The hemichordate gill slits and bars are located dorsally, opposite to the ventral mouth, while in cephalochordates and vertebrates the gill slits and bars are located on the same side of the mouth (Swalla, 2007). Gill openings are also dorsal in stem-group echinoderms (Smith, 2005). Recent results have shown that the dorsal side of a juvenile hemichordate is determined by a strip of bone morphogenetic protein (BMP) expressed dorsally, and a ventral expression stripe of BMP antagonists, including chordin, (Lowe *et al.,* 2006), as has been reported for arthropods (Sasai and De Robertis, 1997). In contrast, in vertebrates, BMP is expressed ventrally, and chordin is expressed dorsally (Sasai and De Robertis, 1997; Yu *et al.,* 2007), suggesting that the chordate dorsoventral axis is inverted compared with a hemichordate or arthropod dorsoventral axis (Lowe *et al.,* 2006). Considering these results,

evolution of the chordates would entail moving the mouth from the side opposite the gill slits to the same side as the gill slits, while also evolving a notochord and dorsal central nervous system from an enteropneust-like nerve net.

Nodal is so far one of the few described deuterostome-specific genes, evolving from a duplication of a *BMP*-like ancestor (Duboc *et al.,* 2004; Chea *et al.,* 2005). In deuterostomes, *nodal* signalling results in left–right asymmetry in bilateral embryos. In all chordates, *nodal* is expressed on the left side during development, specifying asymmetry (Chea *et al.,* 2005). However, in sea urchins, *nodal* is expressed on the right side (Duboc *et al.,* 2004, 2005). If *nodal* is expressed on the right side during hemichordate development, then it would give further weight to the evidence that the Ambulacraria have a dorsoventral axis similar to flies, and that chordates have an inverted dorsoventral axis (Figure 9.4).

Identifying the homologous body axes in echinoderms has long been problematic because of their lack of obvious anteroposterior or dorsoventral axes as adults. Indeed it has been traditional for echinoderm workers to avoid using such terms when referring to anatomical orientation, referring instead to oral–aboral and radial–interradial. This major innovation in body plan seems to have been triggered by a shift from posterior facultative to anterior obligate larval attachment (Smith, 2008). Anterior attachment necessitated the introduction of a phase of torsion in development to bring the mouth into a more appropriate orientation for filter feeding, which in turn rotated the axis of the developing adult 90° out of alignment with *Hox* and other body patterning genes. As a result the developing echinoderm rudiment came to receive a complex mosaic of antero-posterior signalling, and extensive co-option of signalling pathways was able to take place, allowing innovation. The fossil record provides important insights into both pre-attachment and post-attachment stages in this evolutionary process (Smith, 2008) .

9.6   What was the body plan of the earliest deuterostome?

By combining evidence from the most primitive fossil members of each clade with our best molecular phylogenetic hypotheses and current ideas of homology derived from both genetic network data and traditional comparative morphology we can start to build up a picture of the body plan of the earliest deuterostomes (e.g. Cameron *et al.*, 2000; Ruppert, 2005; Zeng and Swalla, 2005). Here we review the evidence relating to the latest common ancestors of Ambulacraria, Chordata, and Deuterostomia.

9.6.1   Chordata

The discovery that tunicates and craniates may be sister groups to the exclusion of cephalochordates implies that characters common to cephalochordates and craniates are basal to all Chordata. Tunicata, so distinct in many ways (Zeng and Swalla, 2005), have undergone considerable morphological and genomic simplification, including loss of metameres, nephridia, and some *Hox* genes (Swalla, 2006, 2007). As cephalochordates and the cartilaginous fishes both contain intact *Hox* clusters that are expressed collinearly along the antero-posterior

axis, the chordate ancestor presumably had a similar *Hox* cluster that determined the antero-posterior axis (Minguillón *et al.*, 2005). If tunicates and vertebrates are sister taxa, then striated heart muscle and recently discovered neural crest-like tissue were present in the latest common ancestor of tunicates and vertebrates. Based on shared features between cephalochordates and craniates, the latest common ancestor of all chordates might be expected to be a worm-like segmented organism with metamerism and myotomes, a dorsal notochord, dorsal hollow nerves, pharyngeal mucous net filtration with endostyle, undivided fins, locomotory post-anal tail, a pre-oral hood, and a hepatic portal system (Ruppert, 2005; Swalla, 2007). As neither Ambulacraria nor cephalochordates have a well-developed brain, encephalization may have been limited and sensory facilities poorly developed in the very earliest chordates (Brown *et al.*, 2008). However, the absence of some sensory systems in cephalochordates may be secondary if *Haikouella*, with its small eyes and weak encephalization, represents a late stem-group chordate. Reproduction was likely through direct development. Amongst fossils, yunnanozoons display the closest morphological similarity to this body plan.

9.6.2   Ambulacraria

Living echinoderms are so highly derived compared with other deuterostomes that meaningful comparisons of adult body plans have been more or less futile. Even determining the antero-posterior axis in adult echinoderms has proved to be very difficult until recently (Peterson *et al.*, 2000; Morris and Byrne, 2005; Swalla, 2006). However, with the inclusion of pre-radiate fossil stem-group echinoderms, comparison is greatly simplified and clarified. There is a clear antero-posterior axis, and structures that have been lost from all crown-group echinoderms, such as the pharyngeal openings and muscular stalk, are evident (Smith, 2005). There is another problem, however, and that is determining whether the enteropneust or pterobranch body plan is primitive for hemichordates. Because lophophorates were suggested as ancestral, the pterobranch model has traditionally been taken as primitive (Gee, 1996). Echinoderms and pterobranchs share a muscular stalk used for attachment and locomotion, a hollow, branched tentacular system derived from the same mesocoel and used for the same purpose, so these features would instead favour a stalked, tentaculate hemichordate form rather than a worm-like latest common ancestor (Smith, 2005). More recently, however, the enteropneust model has been suggested because the antero-posterior and dorsoventral axes in enteropneust hemichordates are determined by similar genetic pathways to the chordates (Cameron *et al.*, 2000; Zeng and Swalla, 2005).

Irrespective of whether an enteropneust or pterobranch body plan is ancestral for hemichordates, the latest common ancestor of echinoderms and hemichordates is expected to have a diffuse nerve network, no mesodermal skeleton, tricoelomic development, a planktotrophic dipleurula larva and abrupt metamorphosis, a nephridial system, and pharyngeal gill slits.

9.6.3 Deuterostomia

There are three competing models for how the latest common ancestor to the deuterostome body was constructed: an ambulacrarian model, a chordate model, or a xenoturbellid model. Given this uncertainty it is not yet clear whether the ancestral developmental mode would have been direct or indirect. Although both tunicates and ambulacrarians have larvae that are planktotrophic, they are very different in form and in the gene networks that are deployed during development (Swalla, 2006). Ambulacraria have dipleurula-type larvae that feed using ciliary bands, and there has been longstanding recognition that echinoderms and hemichordates larvae are homologous (Hara *et al.*, 2006). The tunicate larva is very different and clearly non-homologous, as well as non-feeding (Swalla, 2006). Ascidian tadpoles share a common plan with vertebrates including a notochord centred in the tail flanked dorsally by the ventral nerve chord, laterally by muscles, and ventrally by endoderm (Swalla, 2007). Both planktotrophic larval types presumably evolved independently in the late Precambrian at a time when the ocean surface waters were becoming rich in phytoplankton for the first time and benthic predation was dramatically increasing (Butterfield, 2007).

Ettensohn *et al.* (2003) have shown that there is a very similar homeodomain protein, Zlx1, in echinoderms and chordates that controls downstream genes required for biomineralization. This suggests that the ancestral deuterostome possessed a mesenchymal cell lineage that was able to engage in biomineralization and that a Zlx1-like protein was involved in the specification of these cells. However, the genes used for biomineralization in echinoderms and vertebrates are a completely different (Bottjer *et al.*, 2006), so the detailed biomineralization process evolved independently in echinoderms and vertebrates.

Given *Xenoturbella*'s phylogenetic position as basal to Ambulacraria (Bourlat *et al.*, 2003, 2006) it is possible that the latest common ancestor to deuterostomes was a small, delicate, ciliated marine worm with a simple body plan lacking a through gut, organized gonads, excretory structures, and coelomic cavities. Although this would accord with the long hidden history of deuterostomes suggested by molecular clocks (e.g. Davidson *et al.*, 1995) and may indeed pertain to the very earliest stem-group deuterostomes, it is definitely not parsimonious to have shared derived features, such as pharyngeal feeding and gill slits, a post-anal muscular tail, or stalk evolving independently in Chordates and Ambulacraria, and such features were surely present in the latest members of their stem group. There are therefore two realistic contenders:

The latest common ancestor might have been a benthic filter-feeding worm with gill slits similar to extant enteropneust worms (Cameron *et al.*, 2000). The common features are a simple nerve plexus without regionalization, a pharynx with gill slits used in filter feeding, well-developed circular and longitudinal muscles, and direct development. However, this would imply that tentacles were independently evolved in echinoderms and pterobranchs.

Alternatively a tentaculate pterobranch-like organism was long a popular model for the primitive deuterostome based on the idea that hemichordates were primitive and lophophorates were ancestral (Gee, 1996). With modern molecular phylogenies this has become less popular. However, it cannot be entirely dismissed if the latest common ancestor to Ambulacraria was pterobranch-like rather than enteropneust-like.

9.7 Summary

Deuterostomes are a monophyletic group of animals that include the vertebrates, invertebrate chordates, ambulacrarians, and xenoturbellids. Fossil representatives from most major deuterostome groups are found in the Lower Cambrian, suggesting that evolutionary divergence occurred in the late Precambrian, in agreement with molecular clocks. Molecular phylogenies, larval morphology, and the adult heart/kidney complex all support echinoderms and hemichordates as a sister grouping (Ambulacraria). Xenoturbellids represent a relatively newly identified deuterostome phylum that lacks a fossil record, but molecular evidence suggests that these animals are a sister group to the Ambulacraria. Within the chordates, lancelets share large stretches of chromosomal synteny with the vertebrates, have an intact *Hox* complex, and are sister group to the vertebrates according to ribosomal and mitochondrial gene evidence. In contrast, tunicates have a highly derived adult body plan and are sister group to the vertebrates by phylogenetic trees constructed from concatenated genomic sequences. Lancelets and hemichordates share gill slits and an acellular cartilage, suggesting that the ancestral deuterostome also shared these features. Gene network data suggest that the deuterostome ancestor had an anteroposterior axis specified by *Hox* and *Wnt* genes, a dorsoventral axis specified by a *BMP/chordin* gradient, and a left–right asymmetry determined by expression of *nodal*.

CHAPTER 10

Molecular genetic insights into deuterostome evolution from the direct-developing hemichordate *Saccoglossus kowalevskii*

Christopher J. Lowe

Progress in developmental biology, phylogenomics, and palaeontology over the past 5 years has made major contributions to a long-enduring problem in comparative biology: the early origins of the deuterostome phyla. A detailed characterization of the early development of the enteropneust hemichordate *Saccoglossus kowalevskii* has revealed close developmental genetic similarities between hemichordates and chordates during early body plan formation. The two phyla share close transcriptional and signalling ligand expression patterns during the early development of the anteroposterior and dorsoventral axes, despite large morphological disparity between the body plans. These genetic networks have been proposed to play conserved roles in patterning centralized nervous systems in metazoans, and probably play conserved roles in patterning the diffusely organized basiepithelial nerve net of the hemichordates. Developmental genetic data are providing a unique insight into early deuterostome evolution, revealing a complexity of genetic regulation previously attributed only to vertebrates. Although these data allow for key insights into the development of early deuterostomes, their utility for reconstructing ancestral morphologies is less certain; morphological, palaeontological and molecular data sets should all be considered carefully when speculating about ancestral deuterostome features.

10.1 Introduction

The deuterostome phyla form one of the major animal lineages of the bilaterians (Hyman, 1940; Brusca and Brusca, 1990). The evolutionary history of this group has been the subject of debate for over a century (Gee, 1996). The composition of deuterostomes has been in a state of flux since the advent of molecular systematics, making attempts to reconstruct the early history of the group very difficult. However, the bilaterian phyla belonging within the deuterostomes are now largely known (Turbeville *et al.*, 1994; Halanych, 1995; Halanych *et al.*, 1995; Bromham and Degnan, 1999; Cameron *et al.*, 2000; Bourlat *et al.*, 2003, 2006; Dunn *et al.*, 2008). The following four phyla make up the deuterostomes: chordates, hemichordates, echinoderms, and xenoturbellids.

Despite increased confidence in the relationships between the major deuterostome phyla, our understanding of early deuterostome body plan evolution remains quite murky. There are two main factors that contribute to this uncertainty: a poor fossil record (Swalla and Smith, 2008) and a large morphological disparity between the body plans of the four phyla. Both of these factors make the reconstruction of ancestral features of early deuterostomes particularly challenging. This chapter will focus on the molecular genetic data from hemichordates that facilitate more direct comparisons with the chordate body plan and provide novel insights into the genetic networks that must have been present in the common ancestor of all deuterostome phyla. First, I begin with a general introduction to the deuterostome phyla and the challenges associated with reconstructing early deuterostome evolution. Second, I summarize the molecular genetic information, most recently generated from enteropneusts, involved in the anteroposterior and dorsoventral patterning of hemichordates. Finally, I will discuss what sort of insights can be gained from molecular genetic data sets and their utility for testing both general axial or organizational homologies and more traditional morphological homologies.

10.2 Problems in the reconstruction of ancestral deuterostome characters

One of the most significant barriers to understanding the evolution of early deuterostome evolution has been the difficulty of making direct comparisons between the adult body plans of the four deuterostome phyla: chordates, echinoderms, hemichordates, and xenoturbellids. There are few uncontested deuterostome synapomorphies and a poor early deuterostome fossil record, making attempts to reconstruct an ancestral deuterostome body plan difficult. Echinoderms typify the difficulties of body plan comparison across the deuterostome phyla; the adult body is perhaps the most radical morphological departure of any of the bilaterian groups (Lowe and Wray, 1997). Extant species have lost ancestral bilateral symmetry and have become pentaradially symmetric as adults, while maintaining a bilaterally symmetric larva. There are two novel mesodermally derived structures that are key components of their unusual body plan: the calcitic endoskeleton and water vascular system (Hyman, 1940). Their nervous system is largely diffuse and organized as a basiepithelial nerve net, with some evidence of integrative abilities in the radial nerves (Bullock and Horridge, 1965). Even gross axial comparisons between extant echinoderms and other deuterostomes are problematic (Lowe and Wray, 1997) and it is not clear whether valid comparisons can be made to the anteroposterior and dorsoventral axes of the bilaterian groups. However, early stem-group fossil echinoderms give

some clues to their early bilaterian origins. These fossils show evidence of gill slits (Dominguez et al., 2002) and possibly even a muscular stalk (Swalla and Smith, 2008). There are a few molecular genetic studies that give insights into these questions (Peterson et al., 2000; Morris et al., 2004; Morris and Byrne, 2005; Hara et al., 2006), but many more data are required before any strong comparative conclusions can be drawn.

The most recent addition to the deuterostomes, xenoturbellids (Bourlat et al., 2003, 2006; Dunn et al., 2008) are morphologically rather unremarkable: they have a ventral mouth, a blind gut, and little in the way of external morphological features. They do share the general organizational features of the hemichordate nervous system (Pedersen and Pedersen, 1986), but little else currently described in their anatomy could be referred to as a deuterostome synapomorphy. There are still very few published studies of their biology, though preliminary developmental studies suggest that embryos are brooded (Israelsson and Budd, 2005). However, it is difficult to make any strong comparative conclusions based on current data and further study is needed, particularly in characterizing the development of this animal.

Hemichordates are perhaps the most promising of the non-chordate deuterostome groups for addressing issues of both early deuterostome evolution and the evolution of the chordate body plan (Cameron et al., 2000; Tagawa et al., 2001; Lowe et al., 2003). The phylum is divided into two classes: the enteropneust worms and the pterobranchs. Both groups possess a similar tripartite body organization, but are characterized by distinct feeding mechanisms. The pterobranchs are often small, colonial animals, and feed using a lophophore—a ciliated extension of the mesosome (Halanych, 1995). Enteropneusts are larger solitary animals, and use their highly muscular, ciliated proboscis (prosome) for direct particle ingestion and filter feeding (Cameron, 2002). I will focus exclusively on the body plan of enteropneusts, as there are currently few data on the body patterning of pterobranchs (Sato and Holland, 2008). The most recent molecular phylogenies describe two main groups of enteropneusts: the Harrimaniidae in one lineage and the Ptychoderidae and Spengelidae on the other (Cameron *et al.*, 2000). These two lineages have major life-history differences: harrimaniids are all direct developers, whereas the spengelids and ptychoderids are indirect developers with feeding larvae, which often spend many months in the plankton before metamorphosing into juveniles (Lowe *et al.*, 2004).

Phylogenetic relationships of the various hemichordate groups remain poorly resolved, and this area is in need of further research. Pterobranchs have traditionally been considered as basally branching hemichordates, based largely on the proposed homology of its lophophore with that in other lophophorate groups. However, reclassification of the lophophorates as protostomes reveals that the structural similarities of lophophores are due to convergence rather than homology (Halanych *et al.* 1995). Further molecular phylogenetic studies have proposed that pterobranchs are perhaps nested within the enteropneusts (Cameron *et al.*, 2000; Winchell *et al.*, 2002), but this is weakly supported by current data sets. Clearly this issue should be revisited with broader phylogenetic sampling.

Figure 10.1 outlines some of the main anatomical features of enteropneusts and shows a photomicrograph of a juvenile worm of the harrimaniid *Saccoglossus kowalevskii*. The tripartite body plan is divided into an anterior prosome or proboscis, a mesosome or collar, and a metasome or trunk. The proboscis is muscular, ciliated, and highly innervated with sensory neurons (Bullock, 1945; Knight-Jones, 1952), and its primary functions are digging and feeding. The mouth opens up on the ventral side and marks the boundary between the proboscis and the collar. In the most anterior region of the trunk, dorsolateral gill slits perforate the ectoderm. The gill slits can be very numerous and continue to be added as the animal grows (Bateson, 1885; Hyman, 1940). At the very far posterior end of the metasome, a ventral extension, sucker, or tail, extends ventrally from the anus and is used for locomotion by the post-hatching juvenile worm. A post-anal extension is present only in the juvenile of the harrimaniids, but not in other enteropneust groups, and is lost in adult worms. Given the uncertainty over the relationships of the major groups within the hemichordates, the possible homology

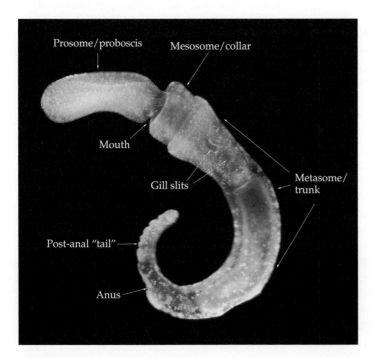

Figure 10.1 Organization of the adult body plan of enteropneusts. Light micrograph of a juvenile worm of the harrimaniid enteropneust *Saccoglossus kowalevskii* at day 13 of development. All major body regions (prosome, mesosome, and metasome) are well developed and several gill slits are perforated in the anterior metasome. The juvenile post-anal tail is still present, but is eventually lost in adult animals.

of this extension to the pterobranch stalk, and even more controversially, the chordate post-anal tail, remains unresolved (Cameron *et al.*, 2000; Swalla and Smith, 2008).

The earliest descriptions of hemichordate anatomy by Bateson (1884, 1885, 1886) and Morgan (1891, 1894) resulted in various hypotheses of morphological homologies between hemichordates and chordates (Bateson, 1886; Morgan, 1891; Nübler-Jung and Arendt, 1996). Most of these classical hypotheses are largely unsupported by both morphological and molecular data sets (Peterson *et al.*, 1999; Nieuwenhuys, 2002; Long *et al.*, 2003; Ruppert, 2005; Lowe *et al.*, 2006). However, only gill slits, as primary ciliated pouches, may represent an ancestral feature of the deuterostomes (Ogasawara *et al.*, 1999; Okai *et al.*, 2000; Tagawa *et al.*, 2001; Cameron, 2002; Rychel *et al.*, 2006; Rychel and Swalla, 2007) and possibly the post-anal tail (Lowe *et al.*, 2003). A more in-depth discussion of potential hemichordate and chordate morphological homologies is found in Ruppert (2005). Morphological homology aside, there has been little consensus over how to compare the body plans of hemichordates and chordates, even at a basic axial level.

10.3 The potential of molecular genetic data for providing insights into deuterostome evolution

Although establishing robust morphological homologies between deuterostome groups is problematic, progress in developmental biology over the past 20 years, mainly from studies of arthropods and chordates, has allowed unprecedented axial comparisons between distantly related groups (Gerhart and Kirschner, 1997; Carroll, 2005). However, representation of the deuterostome adult body plans in broad metazoan comparative studies has been dominated by chordate developmental biology. There is an impressive literature on early embryonic and larval patterning in echinoderms, but only a handful of studies on the development of the adult body plan (Lowe and Wray, 1997; Ferkowicz and Raff, 2001; Lowe *et al.*, 2002; Sly *et al.*, 2002; Morris *et al.*, 2004; Morris and Byrne, 2005; Hara *et al.*, 2006). Recent studies in hemichordates have revealed a novel way to compare the

adult body plans of chordates and hemichordates (Okai *et al.*, 2000; Harada *et al.*, 2002; Taguchi *et al.*, 2002; Lowe *et al.*, 2003, 2006). I will introduce these comparative data sets and their role in comparing anteroposterior and dorsoventral patterning of bilaterian groups. I will then review the current developmental genetic work from hemichordates and how this impacts upon our understanding of early deuterostome evolution.

Although it is now widely accepted that many of the developmental regulatory cascades controlling anteroposterior and dorsoventral axial patterning during nervous system development are probably homologous as regulatory modules (Gerhart and Kirschner, 1997; Carroll, 2005; Davidson, 2006), the extent to which the these data are effective for testing hypotheses of morphological homology remains controversial (Arendt and Nübler-Jung, 1996; De Robertis and Sasai, 1996; Lowe *et al.*, 2003, 2006; Lichtneckert and Reichert, 2005; Denes *et al.*, 2007). Most studies have focused on similarities in nervous system patterning along both dorsoventral and anteroposterior axes (Hirth *et al.*, 2003; Acampora *et al.*, 2005; Lichtneckert and Reichert, 2005). Classical morphological comparisons generally converged on the hypothesis that the central nervous systems of arthropods and chordates evolved independently and that early bilaterian nervous systems were generally quite simple (Holland, 2003). More recent interpretations of the molecular genetic data, based on model systems with central nervous systems, lead to quite different conclusions and propose a protostome–deuterostome ancestor with a complex, centralized nervous system with a regionalized brain. The homologous suites of genes involved in the patterning of the central nervous systems of model systems have very similar spatial domains during development (Arendt and Nübler-Jung, 1994, 1996; Finkelstein and Boncinelli, 1994; Sharman and Brand, 1998; Hirth *et al.*, 2003). Along the anteroposterior axis, the *Hox* genes are involved in patterning the nerve cords of both arthropods and chordates. The boundary of the trunk and the rest of the anterior nervous system is marked by the homeobox gene *gbx* or *unplugged* (Hirth *et al.*, 2003; Castro *et al.*, 2006). *Gbx* is expressed at the boundary between *Hox* genes and *otx* and marks a morphological transition in the organization of the nervous system. Other anteriorly localized homeobox genes such as *orthodenticle* (*otx*), *pax6*, *distalless* (*dlx*), *emx*, and *retinal homeobox* (*rx*) play conserved roles in brain patterning and exhibit similar relative spatial localization during the development of the central nervous system (Lowe *et al.*, 2003).

On the dorsoventral axis, early patterning events of the ectoderm are similarly conservative between arthropods, vertebrates, and annelids (Arendt and Nübler-Jung, 1994, 1996; De Robertis and Sasai, 1996; Holley and Ferguson, 1997; Cornell and Ohlen, 2000; Denes *et al.*, 2007). The secreted factor chordin/short gastrulation is released dorsally in vertebrates and ventrally in arthropods, protecting the ectoderm from the neural-inhibiting effects of the transforming growth factor (TGF)-β ligand Bmp, which is expressed ventrally in vertebrates and dorsally in flies. The interaction of these secreted ligands results in the formation of the central nervous system on the dorsal side in vertebrates and the ventral side in arthropods and annelids. These data have revived the venerable dorsoventral axis inversion hypothesis of Dohrn (1875), which proposed that the dorsoventral organization of chordates is best explained by a complete body axis inversion in the lineage leading to chordates. Further similarities have been revealed in later dorsoventral patterning of the neurectoderm of all three groups, but most closely between the annelid *Platynereis dummerilii* and vertebrates (Denes *et al.*, 2007). The similarities in relative expression domains and essential functions along both the dorsoventral and anteroposterior axes during central nervous system patterning have led to the proposal that a tripartite brain is ancestral for bilaterians, implying that the protostome/deuterostome ancestor and early deuterostomes were characterized by a complex, centralized nervous system. Within a purely phylogenetic framework, based on current molecular phylogenies, the outgroups to the bilaterians are acoel flatworms and cnidarians, both of which are characterized by a nerve net (Holland, 2003). Within the bilaterians, the basiepithelial nerve net is quite common. Proposing a protostome/deuterostome ancestor with a complex central nervous system and brain implies that this organization has been lost multiple times during the evolution of the bilaterian phyla.

Until recently, the vast majority of data generated in developmental biology have been from terrestrial model systems that are characterized by central nervous systems. This bias has begun to be addressed by a broad description of the genetic information involved in patterning the hemichordate enteropneust body plan, and its diffuse basiepithelial nerve net (Lowe *et al.*, 2003, 2006). The organization of the nerve net is based around a broad distribution of cell bodies throughout the ectoderm. Despite the general diffuse organization of the nervous system, there is a significant dorsoventral and anteroposterior polarity in its structure and organization, particularly in the dorsoventral dimension (Bullock, 1945; Knight-Jones, 1952). A mat of axons spreads out along the basement membrane, which is thickened in certain areas of the ectoderm—at the base of the proboscis, along the anterodorsal region of the body in the mesosome, and in both the dorsal and ventral midlines of the metasome. In the proboscis ectoderm, there is a dense concentration of nerve cells that have been proposed to be primarily sensory (Bullock, 1945; Knight-Jones, 1952) and are particularly thick at the base of the proboscis (Brambell and Cole, 1939).

Probably the most well known aspect of hemichordate anatomy is the mid-dorsal region of the dorsal cord, or collar cord, which is internalized into a hollow tube of epithelium in some species within the Ptychoderidae and in one species of the Spengelidae. However, in the Harrimaniids, there is no contiguous hollow tube, but scattered blind lacunae (Nieuwenhuys, 2002; Ruppert, 2005). This structure has been widely compared to the dorsal cord of chordates, not only due to the superficial similarities of the hollow nerve cord, but also because in some species, the collar cord forms by a process that resembles chordate neurulation (Morgan 1891). However, the similarities have generally been over-emphasized as it seems to be more of a conducting tract than a processing centre (Ruppert, 2005) as evidenced by both ultrastructural (Dilly *et al.*, 1970) and physiological data (Pickens, 1970; Cameron and Mackie, 1996). Another striking feature of the dorsal cord is the presence of giant axons in some species. The cell bodies project their axons across the midline and continue posteriorly within the collar cord. It is not known where the axons finally project; Bullock (1945) proposed that they innervate the ventrolateral muscles of the trunk and suspected that their primary function is to elicit a rapid contraction of these muscles. However, several groups do not possess giant axons and yet are still able to elicit a rapid retreat, so the role of the giant axons remains uncertain (Pickens, 1973).

In the metasome, the third body region, the most prominent features are the ventral and dorsal nerve cords, which are both thickenings of the nerve plexus. The dorsal cord is contiguous with the collar cord and projects down the entire length of the metasome. The ventral cord is comparatively much thicker, with more associated cell bodies, but both cords are largely described as through axon tracts. However, at least one study describes the ventral cord as having some integrative function (Pickens, 1970). It seems to play a role in rapid retreat of the animals following anterior stimulation (Knight-Jones, 1952; Bullock and Horridge, 1965; Pickens, 1973).

10.4   Anteroposterior patterning in hemichordates

Molecular genetic patterning information in hemichordates has the potential to address two major areas of comparative interest. First, these data could be another means to compare deuterostome body plans, giving insights into early deuterostome evolution. Second, hemichordates are representative of the first basiepithelial nervous system to be characterized molecularly and allow insights into whether the complex networks of regulatory genes involved in patterning complex central nervous systems play similar roles in less complex, more diffusely organized, nervous systems. These questions have been the focus of several papers over the past 10 years investigating the roles of body patterning genes in hemichordates. The first suite of papers focused on *Ptychodera flava*, an indirect-developing species with a ciliated feeding larva and an extended planktonic larval period: most of these initial studies focused on the establishment of the larval body plan (Dohrn, 1875; Peterson *et al.*, 1999; Harada *et al.*, 2000, 2002; Okai *et al.*, 2000; Tagawa *et al.*, 2001; Taguchi *et al.*, 2002). More recently, we

have developed a direct-developing species, the harrimaniid enteropneust *S. kowalevskii*, to investigate more directly the patterning of the adult rather than larval body plan of hemichordates.

In the first of two papers on body patterning, we investigated anteroposterior patterning (Lowe *et al.*, 2003) by examining the expression of orthologues of 22 transcription factors that have conserved roles in the patterning of the brain and spinal cord of vertebrates along the anteroposterior axis. At least 14 of these genes also play conserved roles in the patterning of the central nervous system of the fruit fly *Drosophila melanogaster*. The 22 transcription factors can be divided into three broad expression and functional domains during the development of the vertebrate brain and central nervous system: (1) expressed and involved in forebrain development, (2) expressed during midbrain development, and (3) expressed and functionally involved in the patterning of the hindbrain and spinal cord. In the first category a group of six transcription factors—*six3*, *brain factor 1(bf-1)*, *distalless (dlx)*, *nk2–1*, *ventral anterior homeobox (vax)*, and *retinal homeobox (rx)*—were all expressed in similar domains during the early development of the embryo and juvenile (Figure 10.2). Their expression was restricted, for the most part, to the developing proboscis ectoderm, the most anterior region of ectoderm. Unlike vertebrates and panarthropods, this expression is not restricted to the dorsal or ventral side, but rather forms rings encircling the entire dorsoventral aspect of the animal reflecting the inherent diffuse organization of the basiepithelial nerve net.

Vertebrate orthologues of the second group of genes are expressed with the posterior limit of expression marking the midbrain, and sometime hindbrain of vertebrates, including *pax6*, *tailless (tll)*, *barH*, *emx*, *orthopedia (otp)*, *dorsal brain homeobox (dbx)*, *lim1/5*, *iroquois (irx)*, *orthodenticle (otx)*, and *engrailed (en)*. Similar to vertebrates, this group of genes is expressed in a more posterior position along the anteroposterior axis of the developing hemichordate embryo, in the posterior proboscis, collar, and anterior trunk. Of these genes, *en* is particularly interesting as it forms a sharp single ring of expression in the ectoderm of the anterior metasome over the forming first gill slit. *En* is a

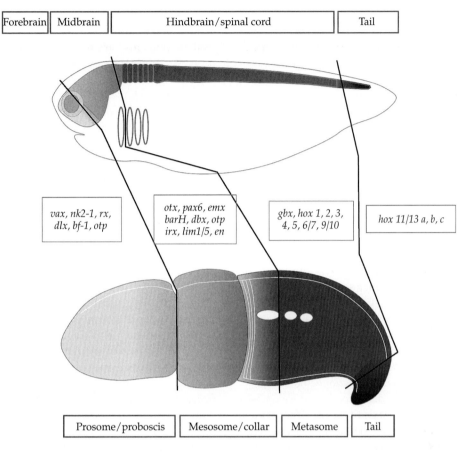

Figure 10.2 Summary of similarities between the enteropneust hemichordate *Saccoglossus kowalevskii* and vertebrates in the ectodermal expression of conserved transcriptional developmental regulatory genes. The upper panel represents an idealized vertebrate, and the bottom panel represents a juvenile hemichordate. The various shades of grey represent similarities in gene expression between the two groups.

critical gene in the formation of the vertebrate isthmus. This then makes a compelling case for investigating other genes involved in the formation of the midbrain/hindbrain division of the vertebrate brain and how much of this signalling regulatory cassette is conserved, as most studies of basal chordates have suggested that ectodermal signalling centres evolved in association with complex vertebrate neural anatomy (Canestro *et al.*, 2005).

The last group of genes includes *gbx* and *Hox* genes. In vertebrates, the regulatory interaction between *otx* and *gbx* is involved in positioning the isthmus along the anteroposterior axis, with *gbx* expressed posterior to *otx* (Rhinn *et al.*, 2005). In the cephalochordate amphioxus, *gbx* is also

expressed in a mutually exclusive domain to *otx* in the central nervous system, suggesting a conserved interaction between the two genes (Castro *et al.*, 2006). However, in ascidians *gbx* is absent from the genome. In *S. kowalevskii*, we observe a departure from chordates in that *otx* and *gbx* expression overlap extensively at all stages of development examined, suggesting that they do not share the same mutual antagonism as found in vertebrates.

A total of 11 *Hox* genes have now been cloned from *S. kowalevskii* (Aronowicz and Lowe, 2006). A study of the relative order of *Hox* genes in the genome has now been completed and will be reported elsewhere (Gerhart *et al.*, work in preparation).

Hox expression domains follow predictable nested domains with the most anterior *Hox* genes expressed in the most anterior regions of the metasome ectoderm, and the most posterior members in the most posterior domains. At the stages that were examined, expression of many of the genes was tightly grouped, with little evidence of difference in anterior expression limits. Expression has not been examined for all genes at late developmental stages when the trunk begins to elongate and become further regionalized. Perhaps the anterior expression limits of *Hox* genes become more markedly differentiated in later stages. Posterior *Hox* family members were examined at later stages when the ventral post-anal tail was developing, and the expression of these posterior members was restricted to the post-anal tail in juveniles, which is similar to the expression of their orthologues in the dorsal post-anal tail of vertebrates. These data would support the proposed homology of the chordate and enteropneust post-anal tails, although it is certainly also possible that independently evolved posterior extensions are likely to express posterior *Hox* genes already expressed in the posterior ectoderm.

A summary of the data from the development of the anteroposterior axis of the hemichordates is illustrated in Figure 10.2. These data clearly demonstrate similar relative expression of transcription factors with critical roles in anteroposterior patterning between chordates and hemichordates. Although most comparative studies and speculations on the nervous system of hemichordates have focused on the dorsal and ventral axon tracts as potential homologues of chordate central neural structures, the results from this study suggest, as was proposed by Bullock in 1945, that the appropriate comparison is with the entire net. The cords are probably local thickenings of the nerve plexus rather than integrative centres (Dilly *et al.*, 1970). The conclusions one can draw from these data are more complicated—particularly, whether these data can help to reconstruct early morphological evolution of deuterostomes. First, it is most parsimonious to conclude that the similarities in the relative expression domains of multiple genes between hemichordates and chordates are due to homology of a gene regulatory network. It is highly unlikely that all the similarities of gene expression along

the anteroposterior axes of both groups are a result of co-option of individual genes into convergently similar domains. However, the organizational difference in the nervous systems of both groups suggests that, despite this regulatory conservation, the evolutionary possibilities of the downstream morphologies have not been constrained. The nervous system, in particular, demonstrates this point effectively: the development of both the central nervous system of chordates and the basiepithelial nerve net of hemichordates is probably regulated by this conserved regulatory map (although it is important to note that this was not directly tested in Lowe *et al.*, 2003). Clearly, the forebrain of vertebrates and the proboscis of hemichordates are not homologous structures. This suite of genes is not a reliable marker of morphological homology between groups.

By considering these data alone, we can speculate that the deuterostome ancestor, and also the protostome/deuterostome ancestor, may have been characterized by a completely diffuse or fully centralized nervous system, and all possible intermediates. Reconstructing ancestral morphologies from gene expression data can be problematic, even with such large expression data sets. These data, however, do give a unique insight into the anteroposterior patterning of the deuterostome ancestor, revealing a degree of transcriptional complexity previously attributed to the complex nervous system of vertebrates. Finally, the nervous system of hemichordates has been described as barely more complex than the cnidarian nervous system (Bullock, 1945; Bullock and Horridge, 1965) and yet there is an exquisite level of transcriptional patterning in the ectoderm. This may suggest a level of neural diversity currently not recognized in this group. Perhaps the complexity of the basiepithelial net of the hemichordates has been underestimated and would benefit from a modern approach to describing the neural diversity? Detailed physiological and molecular studies would be required to address this hypothesis.

10.5 Dorsoventral patterning

Hemichordates have a distinctive and marked dorsoventral axis. The mouth opens on the ventral side by convention, and the most obvious dorsal markers are the paired dorsolateral gill slits. The stomochord is an anterodorsal projection from the gut supporting the axial complex or heart and kidney complex. As previously discussed, the nerve net also exhibits marked dorsoventral polarity in the distribution of dorsal and ventral cords, and the presence of giant axons in the dorsal cords. The TGF-β signalling ligand, Bmp, and one of its antagonists, chordin, are involved in establishing the dorsoventral developmental axis in arthropods and chordates. This molecular axis has recently been investigated in *S. kowalevskii* (Lowe *et al.*, 2006). Hemichordates occupy a key position for investigating the evolution of this developmental pathway in dorsoventral patterning of the bilaterians (Nübler-Jung and Arendt, 1996; Lowe *et al.*, 2006).

The most striking feature of *bmp* and *chordin* between vertebrates and arthropods is that their relative expression is inverted dorsoventrally with respect to each other (Arendt and Nübler-Jung, 1994, 1996; De Robertis and Sasai, 1996). In hemichordates, *bmp2/4* is expressed along the dorsal midline throughout all stages of development, along with all the members of the Bmp synexpression group (Niehrs and Pollet, 1999; Karaulanov *et al.*, 2004). At early developmental stages, *chordin* is expressed ventrally and very broadly on the opposite side to *bmp2/4*, almost up to the dorsal midline, but is increasingly restricted to the ventral side as development progresses. There are many genes that exhibit marked dorsoventrally restricted expression domains along either dorsal or ventral midlines in ectoderm (*tbx2/3, dlx, olig, netrin, pitx, poxN, lim3, admp, sim*), endoderm (*mnx, admp, sim, nk2.3/2.5*), and mesoderm (*mox/ gax*). These data reveal a molecular dorsoventral asymmetry that perhaps underlies the morphological asymmetry along this axis. Although in hemichordates the expression of *chordin* and *bmp*, in relation to the dorsoventral axis, is similar to protostomes, the early developmental action of Bmp and chordin does not result in segregation of a central nervous system from the general ectoderm: there is no central nervous system, but a diffuse and broadly distributed nerve net.

What is the early role of Bmp in an animal without a non-neural ectoderm? Over-expression and knockdown analyses addressed this question, and two main conclusions were presented; first, over-expression of Bmp did not result in the repression of neural fates; second, *bmp* plays a central and critical role in dorsoventral patterning (Lowe *et al.* 2006). In embryos incubated with recombinant vertebrate Bmp4 protein, endogenous hemichordate *bmp2/4* expression was activated throughout the ectoderm, rather than localized along the dorsal midline as in normal embryos. These treated embryos do not perforate a mouth, and with high levels of Bmp protein do not perforate gill slits. Additionally, in the endoderm, dorsolateral endodermal pouches, precursors to the gill slits, do not form, and the entire endoderm projects into the protocoel rather than a thin dorsal projection that would normally develop into the stomochord. Knockdown or diminished expression of *bmp2/4* by injection of short-interfering RNAs (siRNAs) resulted in a complementary phenotype, particularly in relation to the mouth, which normally perforates on the ventral side. In injected embryos, the mouth develops circumferentially, and eventually results in the detachment of the entire prosome.

The morphological interpretation of Bmp modulation experiments suggests that over-expression of Bmp 'dorsalizes' embryos, and knockdown of Bmp 'ventralizes' embryos. This was confirmed by further molecular analysis: markers of the dorsal midline, in both the ectoderm and endoderm, expanded into circumferential rings in Bmp ligand-treated embryos, suggesting that in normal embryos they are activated by Bmp signalling on the dorsal midline. Some of the same dorsal markers failed to activate expression following siRNA injection, adding further support for a role of Bmp in patterning dorsal cell fates. Further experimental evidence suggested that Bmp is involved in restricting the expression of ventrally expressed genes to the ventral midline, as Bmp ligand-treated embryos failed to express ventral markers, and these same markers expand to the dorsal side in siRNA-injected embryos.

The major differences between the hemichordates and vertebrates are summarized by two major criteria. First, in the disposition of the Bmp/chordin axis, which is inverted (Figure 10.3): hemichordates more closely resemble the protostomes with *chordin*

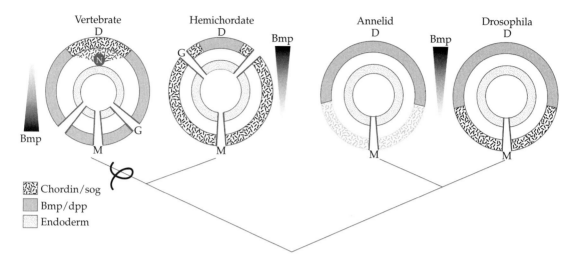

Figure 10.3 Expression of *bmp*/*chordin* in bilaterians. The diagrams represent idealized cross-sections through embryos, with dorsal (D) oriented up and ventral down. In vertebrates, the source of *chordin* is largely the notochord, here a grey circle marked with 'N', dorsal to the gut. The mouth in all panels is represented by M, and in hemichordates and vertebrates G represents the position of the gill slits. In annelids, the ventral, light grey colour represents the predicted domain of *chordin* expression, as this has yet to be published. The black symbol under the vertebrate model represents a potential dorsoventral axis inversion on the lineage leading to chordates.

expressed ventrally and *bmp* dorsally. Second, the mouth opens on the side of the embryo expressing *chordin* in hemichordates and other protostomes, but in the *bmp* domain in vertebrates. Functional experiments in hemichordates suggest that the Bmp/chordin axis is fundamental for the development of many components of the dorsoventral axis, and particularly important for the formation of the mouth. Lastly, although Bmp is directly involved in repressing neural fates in the developing epidermis of vertebrates, it plays no role in repressing neural cell fate in hemichordates, as neural markers are not down-regulated following Bmp4 treatment of embryos. This is also not surprising based on previous descriptions of the distribution of neural cell bodies, which are present along the dorsal midline where Bmp is normally expressed.

How can the differences between hemichordates and vertebrates be explained, and do these data give any critical insights into early deuterostome evolution? One way to explain the data is partially to accept the basic model of inversion as proposed by Dohrn (1875). However, the modification to the model is that a hypothetical ancestor was not necessarily characterized by a central

nervous system: the data from *S. kowalevskii* demonstrate that a dorsoventrally distributed Bmp/chordin axis, although fundamentally involved in dorsoventral patterning, is not always linked to the formation of a central nervous system. Therefore, issues of inversion and centralization can be uncoupled and considered separately. It is formally possible that inversion of an animal with a diffuse nervous system gave rise to the chordates, and centralization happened secondarily. Following inversion, the definitive chordate mouth must have either migrated from the dorsal side or a new mouth formed *de novo*. The new mouth of vertebrates seems to have a novel relationship to the Bmp/chordin axis as it forms in a region of Bmp expression, which in hemichordates inhibits the formation of the mouth.

10.6 Life-history considerations

There has been a diverse range of hypotheses to explain the early evolution of our phylum. Some of the most influential of these can be roughly divided into two kinds; ones that derive the chordate body plan from a larval life-history stage, and others from the adult life-history stage (Gee, 1996). The data I have presented from the direct-developing hemichordate exhibit extensive similarities with the adult body plan of vertebrates. I would argue that the extensive molecular similarities in patterning between adult body plans would argue strongly for adult life-history origins of chordates. However, there remain many supporters of larval origins of the chordate body plan (Nielsen, 1999; Tagawa *et al.*, 2000, 2001; Poustka *et al.*, 2004).

Walter Garstang proposed his Auricularian hypothesis over a century ago (Garstang, 1894, 1928), yet it remains as one of the most compelling of the plethora of hypotheses presented to explain the origins of the chordate body plan. The hypothesis derives the chordate dorsal nerve cord from a larval life-history stage by a dorsal migration and fusion of the ciliated bands, hypothesized to resemble that of many extant enteropneust and echinoderm species. Given the extensive similarities in the anteroposterior and dorsoventral patterning of many bilaterian groups, if chordates are derived from a larval life-history stage then we may expect to see some evidence of conserved regulatory networks during larval development, typical in the adult chordate body plan. There are some limited similarities, but so far the evidence is far from convincing. Here I review some of the molecular data for larval development of both echinoderms and hemichordates.

Hox genes, as discussed previously, are perhaps the best known of the body-patterning genes for their unique chromosomal cluster organization and how it relates to its collinear expression along the anteroposterior axis of bilaterian developing embryos. The expression of particular *Hox* genes conveys spatial information to cells along the developing anteroposterior axis. These homeobox genes have been used broadly as a comparative tool to investigate similarities in body plans between distantly related groups. The broad consensus from a wide range of bilaterians is that the *Hox* complex has played a central role in the evolution of the bilaterian anteroposterior axis (Krumlauf *et al.*, 1993; Lumsden and Krumlauf, 1996; Pearson *et al.*, 2005). If a larval life-history stage was involved in establishing the chordate

body plan, then we may expect to see some molecular evidence for a role of the *Hox* complex in ectodermal regionalization during the formation of the larval body plan. Within larval species of echinoderms and hemichordates, there is so far only expression information for *Hox* genes in echinoids (Arenas-Mena *et al.*, 2000). Interestingly there is only a small subset of the 13 *Hox* genes cloned from echinoderms expressed during the patterning of the larval body plan. This subset of *Hox* genes is not expressed in a coordinated and collinear fashion, but in a lineage-specific fashion. The first sign of colinear expression of the *Hox* cluster is later, during larval development as the adult radial echinoderm body plan is beginning to develop. This has also been found in lecithotrophic larvae of crinoids (Hara *et al.*, 2006) and direct-developing sea urchins (Morris and Byrne, 2005). Only the posterior cluster members have so far been examined, but their expression is detected in the posterior coeloms in larval species, much later in development.

Other homeobox genes with potentially conserved roles in anteroposterior patterning between *S. kowalevskii* and chordates similarly present little evidence of a conserved role in larval anteroposterior regionalization. For example, *otx* is a marker of the adult anterior nervous system during development of adult nervous systems (Finkelstein and Boncinelli, 1994; Acampora *et al.*, 2005). The expression of this gene has been examined quite extensively throughout different echinoderm groups and in *Ptychodera flava*, an indirect-developing species of hemichordate (Harada *et al.*, 2000). However, *otx* expression is not restricted to any particular region of the ciliated band in any of these larval species. *Distalless*, another conserved anterior neural marker, shows quite varied expression domains in different echinoderm larvae (Lowe *et al.*, 2002), but there is no evidence for a conserved role in larval anteroposterior patterning. Some authors have argued that there currently insufficient data to reject outright the paedomorphosis hypothesis proposed by Garstang (Poustka *et al.*, 2004) and show examples of gene expression domains that are consistent with chordate larval origins. Certainly broader comparisons should be carried out in the roles of body patterning genes during the early development of marine invertebrates with complex life histories. However, as Haag points out in his discussion of this point (Haag, 2005, 2006), whether or not an ancestral deuterostome was characterized by a feeding primary larva or not, it is important to consider carefully what exactly is being compared across groups, and the adult echinoderm body plan, however derived it may be, is probably the most relevant for body plan comparisons to the chordates. The data from Lowe *et al.* (2003, 2006) certainly support the hypothesis of adult origins of the chordate body plan. Echinoderm development has focused almost exclusively on the larva, and very few studies have been carried out on the development of the adult. Ultimately, a detailed comparison between direct- and indirect-developing echinoderms and hemichordates will be necessary to test between competing hypotheses of chordate origins.

10.7 Conclusions

The molecular genetic body patterning data presented in this chapter reveal some critical insights into the body plan of the deuterostome ancestor, and a unique way to compare the adult body plans of hemichordates and chordates. The detailed similarities in the transcriptional and signalling networks are not likely to be a result of recruitment of individual genes into convergently similar expression topologies. These exquisite similarities are almost certainly a result of homology. However, what we can most confidently reconstruct is an ancestral gene network rather than ancestral morphologies. Most of the gene networks discussed have been used comparatively to investigate the nature of ancestral nervous systems, and yet hemichordates are a good example of how homologous gene regulatory networks can be deployed to regulate the development of nervous systems with fundamental differences in their basic organization. While gene networks are conserved over large evolutionary timescales, the broad range of morphologies that they regulate has not been constrained by the higher-level regulatory control. Tight regulatory conservation is the foundation of both the highly complex vertebrate central nervous system and the basiepithelial nerve net of the hemichordates. Although these genetic networks have potential for testing hypotheses of morphological homology, their reliability as informative characters is questionable given the range of neural morphologies regulated by this network. Caution should be exercised when reconstructing ancestral neuroanatomies based on these data. Broader sampling and incorporation of fossil data sets will all be required for a more rigorous assessment of ancestral features of early deuterostomes.

PART III

Themes and perspectives

CHAPTER 11

Invertebrate Problematica: kinds, causes, and solutions

Ronald A. Jenner and D. Timothy J. Littlewood

11.1 Progress and remaining controversy

The field of high-level metazoan phylogenetics is moving extremely fast. Estimates of a consensus phylogeny for the Metazoa continue to change, particularly as ever-larger data sets begin to accumulate. Notable among the newer studies are phylogenomic analyses (Hausdorf *et al.*, 2007; Roeding *et al.*, 2007; Brinkmann and Philippe, 2008; Dunn *et al.*, 2008; Helmkampf *et al.*, 2008a,b; Lartillot and Philippe, 2008; Struck and Fisse, 2008), the results of which variously strengthen previous points of consensus (e.g. the dichotomy of Ecdysozoa and Lophotrochozoa), introduce new points of controversy (e.g. Ctenophora as sister group to all other metazoans), and leave other phylogenetic problems unresolved (e.g. the phylogenetic position of Ectoprocta). Through the application of increasingly sophisticated models of evolution to unparalleled quantities of data for larger numbers of taxa, these analyses underscore the value of the guidelines summarized in Jenner and Littlewood (2008) for continuing progress in our understanding of metazoan phylogeny. Nevertheless, as we discuss below, these increasingly comprehensive phylogenetic studies should not be uncritically accepted as being free from underlying flaws. Whereas phylogenetic analyses of relatively small data sets were chiefly marred by stochastic or sampling errors, analyses of larger data sets are subject to increasingly serious interpretational difficulties as systematic errors become visible.

Equally notable are new studies describing the detailed morphology or development of living and extinct taxa such as those by Maas *et al.* (2007) and Stach *et al.* (2008). Such studies shed light on steps involved in the evolution of body plans, and additionally provide new and independent evidence with which to evaluate molecular estimates of phylogenetic relationships. Stach *et al.*'s (2008) cell lineage analysis of the appendicularian *Oikopleura dioica*, for example, adds significantly to the debate about the phylogenetic position of appendicularians, which even with the addition of genomic information are labelled as an unstable rogue taxon (Brinkmann and Philippe, 2008).

Finally, synoptic perspectives, in which diverse sources of evidence have been compiled and synthesized, offer the most recent attempts to reconstruct the details of the evolution of animal body plans within the framework of the latest phylogenies (see, for example, Schmidt-Rhaesa, 2007, Sperling *et al.*, 2007, and Nielsen, 2008). The trees themselves are merely the first necessary step in our quest to understand metazoan evolution.

This chapter is modified from Jenner and Littlewood (2008), and although the general arguments of that paper are summarized here, we adopt a more taxon-focused perspective. We examine recent progress in high-level animal phylogenetics with specific attention to the invertebrate Problematica, i.e. those taxa that are particularly difficult to position in the animal tree of life.

In recent years great strides have been made in solving the phylogenetic positions of several classical Problematica, such as *Xenoturbella bocki* and *Buddenbrockia plumatellae* (Bourlat *et al.*, 2006; Jiménez-Guri *et al.*, 2007), principally by means of molecular phylogenetic analyses. However, new studies have also identified unexpected Problematica of a new kind, such as Acoela or Ctenophora (Brinkmann and Philippe, 2008; Dunn *et al.*, 2008). Classical Problematica were problematic chiefly as a result of the lack of data (Haszprunar *et al.*, 1991). In contrast, the new Problematica are problematic despite, or as a result of, the accumulation of large molecular data sets. Either phylogenetic methods are not able to deal with systematic errors inherent in large data sets, leading to rogue taxa that are very difficult to place (Acoela), or the large amounts of new data suggest a phylogenetic position that is unprecedented (Ctenophora), and which necessitates a fundamental rethinking of body plan evolution. Strikingly, as Figure 11.1 shows, roughly half of the 'phylum-level' taxa in the Metazoa can be labelled as Problematica on the basis of current evidence.

We review the methodological and biological causes of Problematica in the context of high-level metazoan phylogeny, and provide possible strategies for dealing with them. We discuss fossil and extant Problematica from the perspectives of morphological and molecular phylogenetics. A summary of attempts to grapple with Problematica provides insights into the relative abilities of different kinds of data and phylogenetic methods to deal with some of the most challenging problems in all of systematics.

11.2 Problematica—causes and recognition criteria

Problematica confront phylogeneticists with all the problems that can beset phylogenetic analysis. Problematica arise when we lack unambiguous phylogenetic signals that can relate them to other taxa. In many cases, such as the classical Problematica (Haszprunar *et al.*, 1991), this was simply the result of not (yet) having enough knowledge of a taxon. This is also the case for many fossil Problematica with unfavourable preservation. However, as large phylogenomic data sets become increasingly common, it has become clear

that even a large amount of data are no automatic solution to resolving interrelationships. In certain cases, the wealth of data can even be the cause of new problems, as phylogenetic methods fall victim to systematic errors that were undetectable in smaller data sets; here the source of the problem is in estimating interrelationships that leaves taxa in ambiguous positions.

We distinguish three main categories of reason for either the absence of sufficient phylogenetic signal or its obfuscation by other signals: (1) not enough phylogenetic signal has evolved; (2) the phylogenetic signal is lost through extinction; (3) the phylogenetic signal is lost or obscured by evolution of a non-phylogenetic signal.

In the first category, if lineage splitting events succeed each other rapidly, there may not be enough time for distinctive features to evolve that can be used to group descendant species. Although the length of the fuse of the Cambrian explosion is still debated, this has long been considered a distinct possibility for the divergence of the animal phyla.

In the second category, extinction may exacerbate the problem of inferring clades on the basis of homoplasy, or erase phylogenetic signal altogether if the organisms are not discovered. For example, reconstruction of the panarthropod stem group revealed that the subventral mouth shared by extant arthropods and onychophorans has evolved convergently (Eriksson and Budd, 2000). As is well known, fossils can contribute important phylogenetic signal (Cobbett *et al.*, 2007), and in view of the considerable differences between the body plans of extant phyla, extinction must have removed substantial amounts of morphological phylogenetic signal that can only be retrieved by the study of fossils.

The third category groups several causes related to evolutionary change that can erode or obscure phylogenetic signal with the same effects for phylogenetic analysis as extinction of taxa, even when all relevant taxa are included in the analysis. This is especially important when inferring phylogenies with short stems and long terminal branches (Rokas and Carroll, 2006), features common to estimates of metazoan phylogeny. Firstly, if newly evolved lineages have not yet evolved

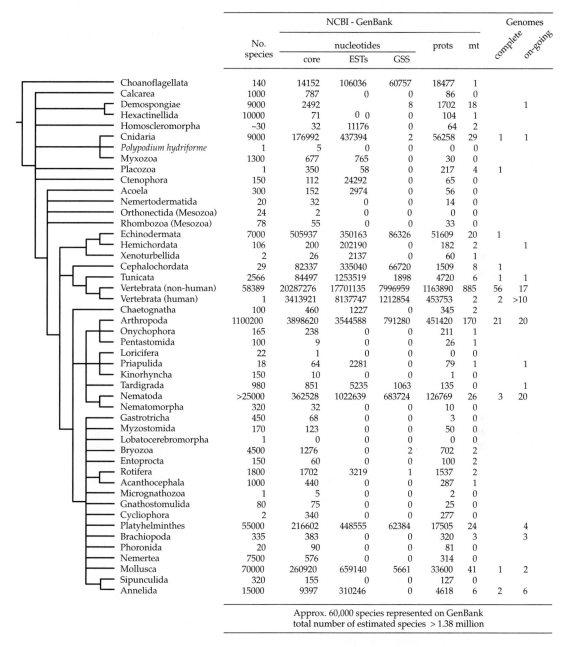

	No. species	NCBI - GenBank					Genomes	
		nucleotides			prots	mt	complete	on-going
		core	ESTs	GSS				
Choanoflagellata	140	14152	106036	60757	18477	1		
Calcarea	1000	787	0	0	86	0		
Demospongiae	9000	2492		8	1702	18		1
Hexactinellida	10000	71	0 0	0	104	1		
Homoscleromorpha	~30	32	11176	0	64	2		
Cnidaria	9000	176992	437394	2	56258	29	1	1
Polypodium hydriforme	1	5	0	0	0	0		
Myxozoa	1300	677	765	0	30	0		
Placozoa	1	350	58	0	217	4	1	
Ctenophora	150	112	24292	0	65	0		
Acoela	300	152	2974	0	56	0		
Nemertodermatida	20	32	0	0	14	0		
Orthonectida (Mesozoa)	24	2	0	0	0	0		
Rhombozoa (Mesozoa)	78	55	0	0	33	0		
Echinodermata	7000	505937	350163	86326	51609	20	1	
Hemichordata	106	200	202190	0	182	2		1
Xenoturbellida	2	26	2137	0	60	1		
Cephalochordata	29	82337	335040	66720	1509	8	1	
Tunicata	2566	84497	1253519	1898	4720	6	1	1
Vertebrata (non-human)	58389	20287276	17701135	7996959	1163890	885	56	17
Vertebrata (human)	1	3413921	8137747	1212854	453753	2	2	>10
Chaetognatha	100	460	1227	0	345	2		
Arthropoda	1100200	3898620	3544588	791280	451420	170	21	20
Onychophora	165	238	0	0	211	1		
Pentastomida	100	9	0	0	26	1		
Loricifera	22	1	0	0	0	0		
Priapulida	18	64	2281	0	79	1		1
Kinorhyncha	150	10	0	0	1	0		
Tardigrada	980	851	5235	1063	135	0		1
Nematoda	>25000	362528	1022639	683724	126769	26	3	20
Nematomorpha	320	32	0	0	10	0		
Gastrotricha	450	68	0	0	3	0		
Myzostomida	170	123	0	0	50	0		
Lobatocerebromorpha	1	0	0	0	0	0		
Bryozoa	4500	1276	0	2	702	2		
Entoprocta	150	60	0	0	100	2		
Rotifera	1800	1702	3219	1	1537	2		
Acanthocephala	1000	440	0	0	287	1		
Micrognathozoa	1	5	0	0	2	0		
Gnathostomulida	80	75	0	0	25	0		
Cycliophora	2	340	0	0	277	0		
Platyhelminthes	55000	216602	448555	62384	17505	24		4
Brachiopoda	335	383	0	0	320	3		3
Phoronida	20	90	0	0	81	0		
Nemertea	7500	576	0	0	314	0		
Mollusca	70000	260920	659140	5661	33600	41	1	2
Sipunculida	320	155	0	0	127	0		
Annelida	15000	9397	310246	0	4618	6	2	6

Approx. 60,000 species represented on GenBank
total number of estimated species > 1.38 million

Figure 11.1 A conservative consensus estimate of metazoan phylogeny based on the information in Table 11.1. It shows indications of estimated number of known species and, from the NCBI (GenBank) data bases, the number of nucleotide sequences (core), the number of nucleotides from large-scale expressed sequence tag (EST) or genome (GSS) projects, the number of protein (prots) sequences, the number of mitochondrial genomes (mt), and the number of completed and on-going genome projects (as of mid-2008).

complete intrinsic isolating mechanisms, extensive introgressive hybridization may occur, even of morphologically distinct species (Wiens *et al.*, 2006). Although extensive gene exchange between morphologically distinct species is considered rare (Coyne and Orr, 2004), this could scramble any original phylogenetic signal (Clarke *et al.*, 1996; Chan and Levin, 2005); it has recently been suggested to be a possible reason why even vast numbers of genome data may not be able to resolve high-level phylogenetic relationships (Hallström and Janke, 2008). Causes in this category also relate to the power of natural selection or shared internal constraints to produce extensive convergent evolution, and parallelisms (non-random non-phylogenetic signal) that may lead to the false inference of monophyletic taxa. This can be an important problem for both morphological and molecular phylogenetic analyses (Waegele and Mayer, 2007). Here we should distinguish between stochastic (sampling) error and systematic error. Small data sets can be prone to stochastic error as chance similarities (random noise) can incorrectly group unrelated taxa. Increasing the amount of data helps to avoid stochastic error, but can introduce the far more serious problem of systematic error.

Systematic errors are tree reconstruction artefacts that result from the inability of a method to deal with biases in a data set that can conflict with or obscure phylogenetic signal. Systematic error may result from, for example, base or amino acid compositional biases between taxa, differences in evolutionary rates between taxa or regions of the sequences, and shifts of position-specific evolutionary rates (heterotachy). As expertly discussed in a series of papers by Philippe and co-workers (Philippe *et al.*, 2005b; Brinkmann and Philippe, 2008; Lartillot and Philippe, 2008), these phenomena can cause strongly non-random, non-phylogenetic signals that can mislead phylogenetic analyses. The difficulty of trying to disentangle phylogenetic and non-phylogenetic signals is potently illustrated by the continuing debate about the validity of either Coelomata or Ecdysozoa using large data sets for a small sample of taxa (Rogozin *et al.*, 2007b; Roy and Irimia, 2008a). The different results reported by different authors reflect how well their methods are able to deal with systematic error.

The above causes can affect phylogenetic analyses of both fossil and extant taxa at any taxonomic level and independent of the type of evidence used. Difficulties generally become greater with increasing age of the divergence events we attempt to reconstruct, and all causes mentioned have probably confounded attempts to place particular Problematica in the tree of the Metazoa. In the following sections we pay more detailed attention to specific causes that are of relevance for certain Problematica.

Several criteria can be used to recognize Problematica: (1) the number of alternative sister-group hypotheses; (2) the phylogenetic spread and hierarchical range of alternative sister-group hypotheses; (3) controversial homology assessments; (4) absence of phylogenetically informative characters; and (5) assessment of molecular data quality.

The first two criteria are straightforward for recognizing Problematica when comparing different phylogenetic analyses, either by different workers or based on different treatments of the same data set. Classic Problematica, such as Chaetognatha, Ectoprocta, and Pogonophora, have long exhibited both a large number of alternative sister group hypotheses, and a large phylogenetic spread among these alternatives (covering both Protostomia and Deuterostomia). The phylogenetic spread of alternative hypotheses is positively related to the hierarchical depth across which the alternatives may be distributed. For example, the placement of Pentastomida is problematic only within the Panarthropoda, with a position either within Crustacea or in the arthropod stem group as the two main contending hypotheses (Waloszek *et al.*, 2005b). In contrast, the fossil vetulicolians are problematic on a much larger scale, across a wide phylogenetic spread (Bilateria), and a large hierarchical depth (ranging from being attributed to a separate 'phylum-level' clade, to belonging to a subtaxon of Tunicata) (Aldridge *et al.*, 2007). Vetulicolians also illustrate the challenges of homologizing imperfectly preserved and poorly understood features of fossils with key characters in extant taxa, with each decision strongly affecting the resulting phylogenetic hypothesis (Aldridge *et al.*, 2007; Swalla and Smith, 2008). Other taxa are problematic because of the lack of, or insufficient study of, informative characters. Myxozoa, for example, are very likely to be derived cnidarians (Jiménez-Guri *et al.*, 2007) that share so few characters with their closest non-parasitic relatives that most textbooks did not even include them in the Metazoa until very recently. Lacking detailed knowledge may also cause Problematica to be excluded from phylogenetic discussions. Species such as *Jennaria pulchra*, the lobatocerebrids, *Xenoturbella bocki* (until recently), *Buddenbrockia*, and myxozoans, but also myzostomids and pentastomids, are frequently excluded from morphological phylogenetic analyses. This is not because their phylogenetic position is so well understood. Finally, Problematica can be provisionally identified by the tell-tale signs of systematic errors in molecular data sets, such as mutational saturation of sequences, compositional biases in nucleotides or amino acids, and different evolutionary rates between taxa (Philippe *et al.*, 2005b; Waegele and Mayer, 2007; Brinkmann and Philippe, 2008). When such features are not properly dealt with they can cause tree reconstruction artefacts.

11.3   Fossil Problematica

11.3.1   The vagaries of preservation, typological thinking, and model choice

All the difficulties that beset phylogenetic analyses of extant taxa also play a role in the systematization of fossils. With fossils, however, several additional factors can cause problems, of which we think three are of particular importance. First, preservational artefacts can lead to formidable problems of interpretation. Although the majority of fossils can be related to extant body plans without much difficulty, 'unusual objects do occur in rocks' (Yochelson, 1991, p. 288). Problematica are particularly common from the fossil record of the late Neoproterozoic and earliest Phanerozoic (*c.* 575–500 million years ago) and it is especially these forms that may provide unique clues to the origin and diversification of early animal body plans. Yet many important taxa found in this time interval defy unambiguous interpretation because of the limits of preservation, and

taphonomic changes of the organism and surrounding sediment. This is clearly illustrated in recent debates over the putative Precambrian animal *Vernanimalcula* (a coelomate bilaterian?), the Cambrian animal *Odontogriphus* (segmented?), the oldest putative metazoan eggs and embryos (animals or bacteria?), and in the continuing debate about the Ediacaran biota (Dzik, 2003; Fedonkin, 2003; Bengtson and Budd, 2004; Chen *et al.*, 2004; Narbonne, 2005; Butterfield, 2006; Caron *et al.*, 2006; Bailey *et al.*, 2007a; Donoghue, 2007).

Budd and Jensen (2000) nominated typological thinking as another factor that may hinder the phylogenetic systematization of fossils, especially in the context of extant taxa. By a misguided emphasis on differences, fossils have automatically been labelled Problematica if their body plan did not exactly conform to that of a living phylum (see also Briggs *et al.*, 1992). Such reasoning is incompatible with established phylogenetic logic, but it is nevertheless prevalent (Jenner, 2006a).

A third factor that inescapably affects thinking about fossil Problematica is that fossils are predominantly interpreted in the light of our knowledge of living species. Consequently, disagreements about the phylogenetic placement of fossil Problematica frequently hinge upon the use of different living species as models for interpretation, as illustrated by the vetulicolians (Aldridge *et al.*, 2007). Related to this is that phylogenetic analyses of fossils may be strongly dependent upon a very small number of informative features that can be homologized between fossils and extant taxa. Consequently, the interpretation of these features can have a very strong effect on phylogenetic conclusions, whether that seems justified or not (for vetulicolians see Swalla and Smith, 2008).

11.3.2   Solving fossil Problematica: stem groups, new fossils, new techniques

Yochelson (1991, p. 289) remarked that he could only offer 'a few platitudes' about how 'to do' fossil Problematica. We hope the following suggestions are helpful. In essence, fossils should be treated like any other living taxon. Attempts to systematize fossils will lead to the establishment of stem groups (Conway Morris, 2000; Budd and

Jensen, 2000; Budd, 2001b, 2003). Although differences between fossils and extant taxa should not be ignored, they should not be interpreted typologically as evidence against affinities (Budd and Jensen, 2000; Jenner, 2006). Putative stem-group taxa are expected to exhibit some, but not all, of the diagnostic characters of crown groups, and by creating paraphyletic series of stem taxa we can illustrate the orderly sequential evolution of body plans. This may not be easy of course. If crucial information is not preserved, a fossil may not be reliably placed. Specifically, the lack of a diagnostic crown-group character state in a fossil, due to taphonomy, could bias a phylogenetic analysis by erroneously placing the fossil in the stem group, a problem that might be widespread (Donoghue, and Purnell, 2009). In such cases, unless new fossils are found or new techniques reveal new information, ambiguity will endure.

The main reason why fossil Problematica occur frequently in the late Neoproterozoic and early Phanerozoic is extinction. These fossils document the early evolution of animal body plans. The older fossils are, the more they are expected to fall outside the limits of extant body plans (Budd, 2003; Valentine, 2004). Unless body plan evolution takes large leaps, failure to systematize fossil Problematica is chiefly the result of not (yet) knowing related taxa that can bridge their morphology with those of the crown group. Hence, most progress is made with fossil Problematica when new specimens are found. Better-preserved fossils and forms with novel character combinations address the problems of taxon and character matrix completeness, allowing unknowns to be substituted with characters. Nevertheless, this approach relies on much fieldwork and a great deal of luck.

Palaeontological and analytical techniques are constantly being developed that present ways of discerning new characters, or of better resolving existing ones, and of handling existing data. For example, the three-dimensional reconstruction of fossil forms from thin serial sections has achieved remarkable levels of resolution, thanks to refinements in microscopy and computer rendering. This has provided valuable phylogenetic information for a diversity of taxa, ranging across the Bilateria (Sutton *et al.*, 2001a,b, 2005a,b,c; Thomson

et al., 2003). X-ray tomographic microscopy and Raman spectroscopy combined with confocal laser scanning microscopy have also yielded images and insights into the biomolecular nature of fossils with unrivalled resolution (Schopf and Kudryavtsev, 2005; Donoghue *et al.*, 2006a; Chen *et al.*, 2007).

Other advances will come from improvements in methods of phylogeny reconstruction. Model-based methods of analysis have proven their worth with molecular data, particularly in dealing with long-branch problems in phylogenetic reconstruction. Such methods, although still in their infancy, are now available for the analysis of morphological and fossil data as well (Lewis, 2001). This promises the chance to include incomplete taxa, such as fossil Problematica, with morphological and even molecular data from extant taxa using maximum likelihood or Bayesian techniques (Wiens, 2005), while at the same time parsimony-based methods are refined to be able to deal efficiently with large amounts of diverse phylogenetic evidence (Wheeler *et al.*, 2006).

11.4 Extant invertebrate Problematica

11.4.1 An apparent paradox: a weak molecular signal and large amounts of morphological evolution

It is not surprising that Problematica are encountered when metazoan phylogeny is analysed on the basis of extant taxa alone. First, any comparison between two extant species belonging to different phyla has to bridge in the order of 1 billion years of independent evolution. This is ample time to erase signs of ancestry, either through extensive modification or loss of characters, and for convergent evolution to obscure phylogenetic signal. It may thus be unsurprising that sessile taxa (ectoprocts, brachiopods, phoronids), very small (possibly miniaturized) taxa (tardigrades, placozoans, *Lobatocerebrum*), and parasitic taxa (pentastomids, myxozoans) have been particularly prominent Problematica. Another consequence is that molecular phylogenies of the Metazoa bear the typical signature of short stems and long terminal branches, providing ample opportunity for long branch

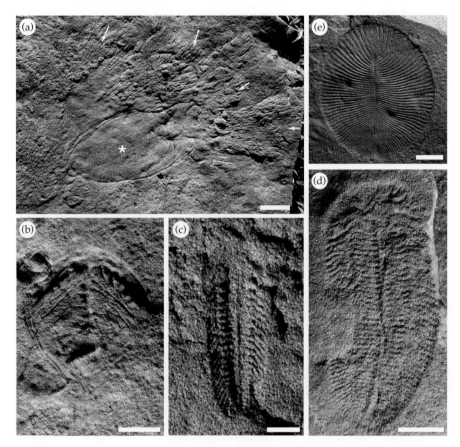

Plate 1: Putative ediacaran metazoans: (a) natural cast on bed base of *Kimberella* resting trace (asterisk) and *Radulichnus* radular feeding trace fans (arrows) (scale bar 1 cm); (b) *Dickinsonia costata* (scale bar 2 cm); (c) *Marywadea ovata* (scale bar 10 mm); (d) *Spriggina floundersi* (scale bar 10 mm); (e) *Parvancorina minchami* (scale bar 1 cm). See page 21.

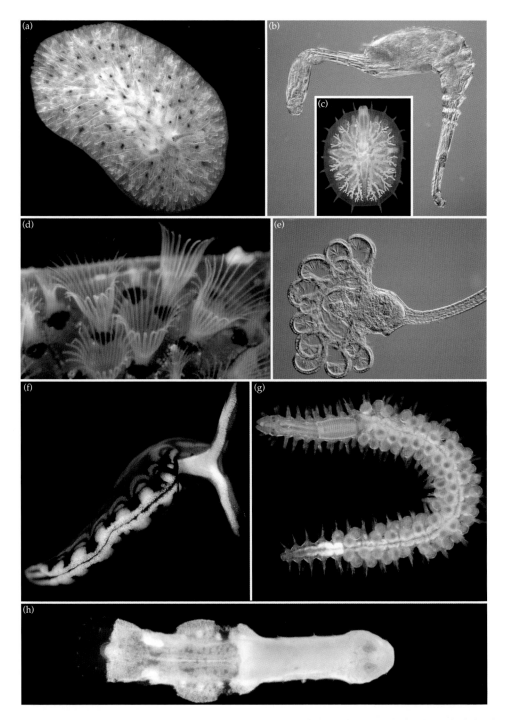

Plate 2: Examples of some spiralian taxa: (a), (b) Platyzoa; (c), (d), (h) uncertain; (e)–(g) Trochozoa. (a) The free-living platyhelminth *Hoploplana californica*. (b) An undescribed species of seisonid rotifer *Paraseison* taken from its crustacean host *Nebalia*. (c) A myzostomid *Myzostoma cirriferum* taken from its crinoid host. (d) Several zooids of a bryozoan colony. (e) Anterior end of entoproct *Pedicellina* sp. (f) Dorsal view of the sacoglossan mollusc *Thuridella picta*. (g) A syllid polychaete annelid brooding embryos on its dorsum. (h) A benthic spadellid chaetognath, *Spadella*, All photographs by G. W. Rouse. See page 56.

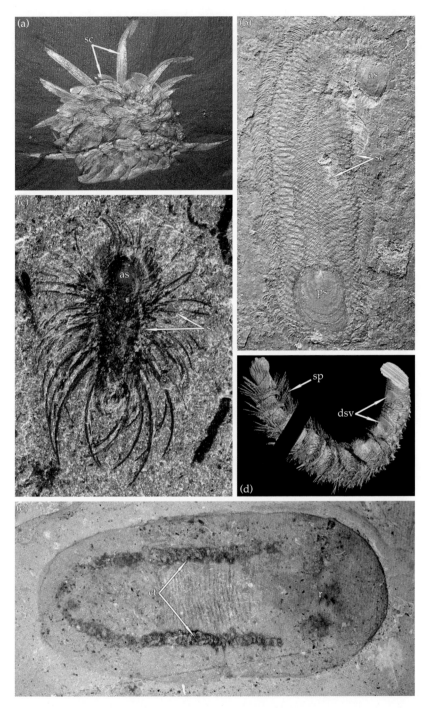

Plate 3: Exceptionally preserved Palaeozoic spiralian fossils. (a) *Wiwaxia corrugata* (Middle Cambrian, photo courtesy of Jean-Bernard Caron). (b) *Halkieria evangelista,* sclerites (sc) anterior shell (as) and posterior shell (ps) (Lower Cambrian, photo courtesy of Jakob Vinther). (c) *Orthrozanclus reburrus,* anterior shell (as), sclerites (sc) (Middle Cambrian, photo courtesy of Jean-Bernard Caron). (d) *Acaenoplax hayae,* dorsal shell plates (dsv), spines (sp) (Silurian, digital reconstruction courtesy of Mark Sutton). (e) *Odontogriphus omalus,* radula (r) and ctenidia (ct) (Middle Cambrian, photo courtesy of Jean-Bernard Caron). See page 59.

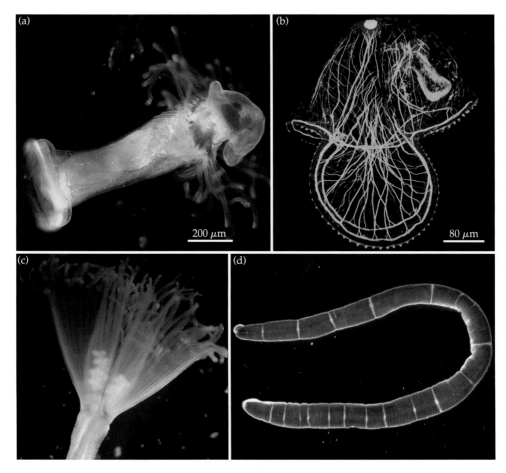

Plate 4: Examples of taxa and larval forms in Kryptotrochozoa, a new subtaxon of Trochozoa. (a) An actinotroch larva of an unidentified phoronid species. (b) Fluorescently labelled pilidium larva of the nemertean *Cerebatulus lacteus* (photograph by Patricia Lee and Dave Matus). (c) Anterior end of phoronid brachiopod *Phoronis hippocrepia* (photograph by G. W. Rouse). (d) Dorsal view of the nemertean *Micrura* sp. (photograph by G. W. Rouse). See page 63 .

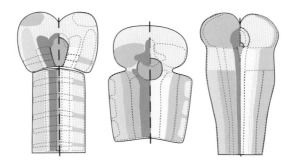

Plate 5: Comparison of mediolateral neurogenic columns across Bilateria. Expression of *nk2.2/nk2.1*) (orange; Shimamura *et al.*, 1995), *Nk6* (yellow; Rubenstein *et al.*, 1998), *Pax6* (violet; Mastick *et al.*, 1997; Urbach and Technau, 2003a,b), *gooseberry/Pax3/7* (green; Matsunaga *et al.*, 2001; Puelles *et al.*, 2003), and *msh/Msx* (blue; Shimeld *et al.*, 1996) orthologues in the neuroectoderm of *Drosophila*, *Platynereis*, and mouse (left to right) at pre-differentiation stages. The *Drosophila* (left) and *Platynereis* (centre) schematics represent ventral views, the mouse (right) is a dorsal view with the neural tube unfolded into a neural plate for better comparison. Neurogenic columns are demarcated by expression boundaries and represent cells with a unique combination of transcription factors. All expression patterns are symmetrical but are shown on one side only for clarity. See page 69.

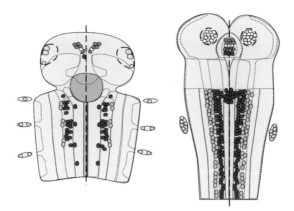

Plate 6: Conserved neural cell types in annelids and vertebrates. The neuron types emerging from homologous regions in the molecular coordinate systems in annelids and vertebrates and expressing orthologous effector genes are marked with the same colour. Homologous cell types include the molecular clock cells positive for *bmal* (dark green), ciliary photoreceptors positive for *c-opsin* and *rx* (white), rhabdomeric photoreceptors positive for *r-opsin*, *atonal*, and *pax6* (yellow), vasotocinergic cells positive for *nk2.1*, *rx*, and *otp* (orange), serotonergic cells positive for *nk2.1/nk2.2* (red), cholinergic motor neurons positive for *pax6*, *nk6*, and *hb9* (violet), interneurons positive for *dbx* (pink), as well as trunk sensory cells positive for *atonal* and *msh* (light blue). See page 70.

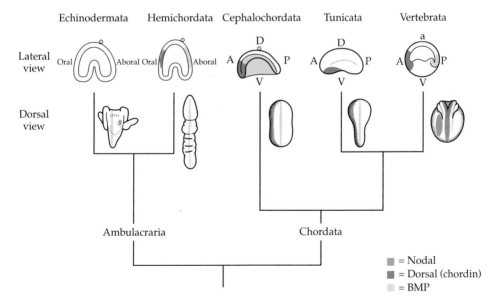

Plate 7: Summary of body axis determination in Deuterostomata: A, anterior; P, posterior; a, animal; v, vegetal. Expression of *nodal* is in red for all phyla. Dorsal in hemichordates is shown by *BMP* expression in yellow. In Echinodermata the aboral (dorsal) axis is shown by a yellow strip, with *nodal* expression marked red on the right side of the larvae, opposite where the adult rudiment will form. Dorsal in chordates is marked in blue (lower pictures) while expression of *chordin* is shown during gastrulation in Cephalochordata, Tunicata, and Vertebrata. The BMP–chordin axis is reversed in chordates from the Ambulacraria. Note that nodal expression is on the left side in chordates, and on the right side in echinoderms. Expression patterns taken from Duboc *et al.,* (2005) (urchin *nodal*), Lowe *et al.* (2006) (hemichordate *chordin*), Yu *et al.,* (2007) (cephalochordate *BMP* and *chordin*), Darras and Nishida (2001) (ascidian *chordin*), and Sasai and De Robertis (1997) (frog *BMP* and *chordin*). See page 89.

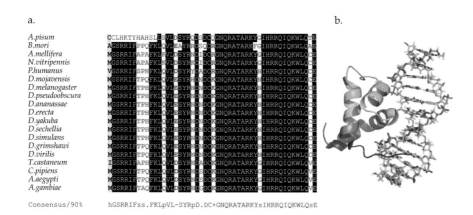

No.subs	WAG	GTR	CAT
0			
1			
2			
3			
4			
5			
6			
7			

Plate 8: Posterior predictive tests to analyse the behaviour of the WAG, GTR, and CAT models under substitutional saturation. A column of the alignment displaying only aspartic acid (D) and glutamic acid (E) was chosen at random, and for the three models, the probability of observing each of the 20 amino acids after n substitutions ($n = 0-7$), and starting from an aspartic acid, was estimated and visualized graphically. The height of each letter is proportional to the probability of the corresponding amino acid. The parameters of the substitution process were taken at random from the posterior distribution under each model. See page 130.

a.

A.pisum	CCLHKTYHAHSLESVLDSYRQESDCQGNQRATARKYCIHRRQIQKWLQTE
B.mori	AGSRRIEPPQEKLQVLEAYRRDSQRGNQRATARKFGIHRRQIQKWLQAE
A.mellifera	MGSRRIEAPAEKLKVLDSYRNLIDGRGNQRATARKYCIHRRQIQKWLQCE
N.vitripennis	MGSRRIEAPAEKLKVLDSYRKDIDGRGNQRATARKYCIHRRQIQKWLQCE
P.humanus	VGSRRIESPHEKLQVLDSYRYDADGRGNQRATARKYNIHRRQIQKWLQCE
D.mojavensis	MGSRRIETPQEKLQVLESYRHDNDGKGNQRATARKYNIHRRQIQKWLQCE
D.melanogaster	MGSRRIETPHEKLQVLESYRNDNDGKGNQRATARKYNIHRRQIQKWLQCE
D.pseudoobscura	MGSRRIETPHEKLQVLESYRNDNDGKGNQRATARKYNIHRRQIQKWLQCE
D.ananassae	MGSRRIETPHEKLQVLESYRNDNDGKGNQRATARKYNIHRRQIQKWLQCE
D.erecta	MGSRRIETPHEKLQVLESYRNDNDGKGNQRATARKYNIHRRQIQKWLQCE
D.yakuba	MGSRRIFTPHEKLQVLESYRNDNDGKGNQRATARKYNIHRRQIQKWLQCE
D.sechellia	MGSRRIETPHEKLQVLESYRNDNDGKGNQRATARKYNIHRRQIQKWLQCE
D.simulans	MGSRRIETPHEKLQVLESYRNDNDGKGNQRATARKYNIHRRQIQKWLQCE
D.grimshawi	MGSRRIETPQEKLQVLESYRNDNDGKGNQRATARKYNIHRRQIQKWLQCE
D.virilis	MGSRRIETPQEKLQVLESYRNDNDGKGNQRATARKYNIHRRQIQKWLQCE
T.castaneum	IGSRRIEAPHEKLQVLDSYRNDADGKGNQRATARKYGIHRRQIQKWLQVE
C.pipiens	MGSRRIETPQEKLQVLDSYRNDGDGKGNQRATARKYGIHRRQIQKWLQVE
A.aegypti	MGSRRIETPQEKLQVLDSYRNDSDGKGNQRATARKYGIHRRQIQKWLQVE
A.gambiae	MGSRRIETAQEKLQVLDSYRNDGDGKGNQRATARKYGIHRRQIQKWLQVE

Consensus/90%   hGSRRIFss.FKLpVL-SYRpD.DC+GNQRATARKYsIHRRQIQKWLQsE

b.

Plate 9: The DNA-binding domain of brinker is conserved within insects, but has no significantly similar sequences in other taxa. (a) The alignment shows the conserved core from a selection of insect species. Sequences of *Drosophila* species were taken from the UCSC web browser (http://genome.ucsc.edu/), *Anopheles* and *Aedes* from ENSEMBL (http://www.ensembl.org/), other predictions were made from sequences at the NCBI. GI accession numbers: *N. vitripennis* 146253130; *T. castaneum* 73486274; *C. pipiens* 145464888; *P. humanus* 145365328; *A. mellifera* 63051942; *B. mori* 91842977; *A. pisum* 47522326. (b) The three-dimensional structure of the aligned region when binding DNA. The structure was taken from the PDB file 2glo. See page 151.

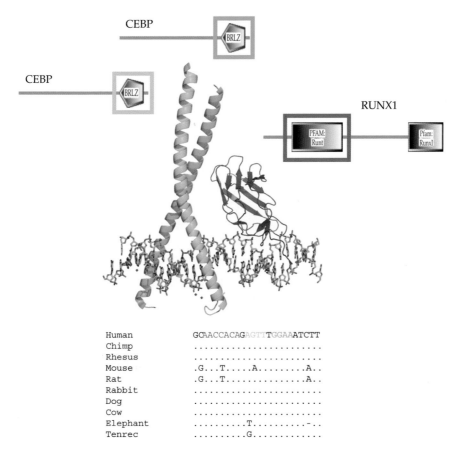

Human	GCAACCACAGAGTTTGGAAATCTT
Chimp	....
Rhesus	....
Mouse	.G...T.....A.........A..
Rat	.G...T..............A..
Rabbit	....
Dog	....
Cow	....
Elephant	.........T.........-..
Tenrec	.........G............

Plate 10: Adjacent transcription factor binding sites cause extended regions of DNA sequence conservation. Structure of CEBPβ homodimer and RUNX1 (Tahirov *et al.*, 2001). Three transcription factors (2× CEBPβ and RUNX1) bind in a region of 25 nucleotides conserved throughout placental mammals. The DNA-binding domains represented as three-dimensional structures are boxed and colour-coded in the schematic representation of the proteins. In each case, the majority of the protein is not represented in the structure; these regions could interact with other transcription factors, activators, and repressors. The human sequence coordinates are chromosome 5, bases 149,446,373–149,446,396 of the NCBI build 36. The alignment is taken from the UCSC web browser (http://genome.ucsc.edu/). See page 154.

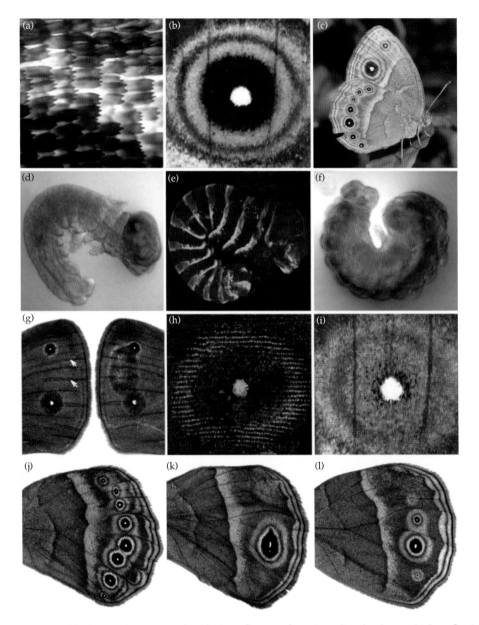

Plate 11: Conserved developmental processes implicated in butterfly eyespot formation. Coloured scales covering butterfly wings (a), and eyespot patterns formed by these scales (b), are key innovations in the lepidopteran lineage and are represented in the laboratory-tractable system, *Bicyclus anynana* (c, ventral view of female at rest). The formation of eyespots in *B. anynana* shares genetic commonalities with different conserved developmental processes such as embryonic development (d–f), wound healing (g), and wing vein patterning (j–l). Embryonic development in *B. anynana* has been characterized in wild-type and pleiotropic eyespot mutants (Saenko *et al.*, 2008): wild-type embryo after completion of blastokinesis (d), the characteristic expression of the segment polarity gene *engrailed* at that stage (e), and a homozygous *Goldeneye* embryo of the same age that has failed to undergo blastokinesis (f). Expression pattern of *engrailed* in pupal wings in association with the gold ring of the presumptive adult eyespot (h). This expression is altered in *Goldeneye* eyespots (Brunetti *et al.*, 2001; Saenko *et al.*, 2008) in a manner that matches the change in adult eyespot colour-composition (i). Damage with a fine needle applied to the distal part of the developing pupal wing (arrows in left panel) results in formation of ectopic eyespots around the wound site (right panel) (g). Wing venation mutants often affect eyespot patterns: the additional vein in *extra veins* mutants can lead to the formation of an extra eyespot (j), while partial vein loss in *Cyclops* (k) and vestigial venation in *veinless* (l) mutants typically result in changes in eyespot size, number, and/or shape. All photos show the ventral surface of the hindwing. See page 185.

attraction (Waegele and Mayer, 2007). This has been a problem for the placement of several taxa, ranging from myxozoans to acoels (Philippe *et al.*, 2007). Second, the major metazoan lineages may have radiated very rapidly, potentially allowing for very little phylogenetic signal to evolve. Although it remains disputed whether lack of resolution is a convincing signature of closely spaced cladogenetic events (Giribet, 2002; Rokas *et al.*, 2005; Rokas and Carroll, 2006; Baurain *et al.*, 2007; Whitfield and Lockhart, 2007), if current molecular clock estimates of metazoan divergence times are approximately accurate (Peterson *et al.*, 2004, 2005, 2008), the fact remains that the major metazoan lineages diverged over a time span that is significantly shorter than the subsequent independent history of modern phyla (including their stem groups). The appearance in the fossil record of a variety of crown phyla with their distinctive body plans as early as the Cambrian (Budd, 2003; Valentine, 2004) implies that important morphological traces of ancestry were probably already erased early in metazoan history.

Intriguingly, the relative branch lengths of morphological metazoan phylogenies seemingly contradict the absence of sufficient phylogenetic signal. These typically show a much smaller discrepancy between the length of stems and terminal branches, or even the opposite pattern of relatively longer stems and shorter tips (Zrzavý *et al.*, 1998, 2001; Nielsen, 2001; Peterson and Eernisse, 2001; Brusca and Brusca, 2003; Zrzavý, 2003). Large amounts of body plan evolution are commonly inferred along almost all stems. This raises interesting issues about the relationship between genetic and phenotypic evolution that we cannot address here. What is pertinent though is the large amount of body plan evolution inferred across a relatively small number of speciation events. For example, depending on the precise topology of the tree, possibly just six or seven nodes separate the body plan of the last common ancestor shared by (at least some) sponges and the remaining animals, and the last common ancestor of the chordates! Unless half a dozen speciation events are really all that is required to evolve from a sponge-grade organization to that of a protochordate, we must be missing something. That something is fossils.

Recent studies of the fossil record have yielded important insights that may help explain why extant Problematica are to be expected. First, Wagner (P. J. Wagner, 2000, 2001; Wagner *et al.*, 2006) drew the important conclusion that during evolutionary history taxa tend to exhaust their character state spaces. This means that, as clades age, homoplasies increase in frequency. Not surprisingly, homoplasies are common between the major lineages of animals (Valentine, 2004). Our estimates of homoplasy based on morphological phylogenetic studies are likely underestimates, giving a widespread problem of character coding (Jenner, 2004b).

Distressingly, P. J. Wagner (2001) noted that the inclusion of fossils into a phylogenetic analysis of extant species could reveal a significant amount of previously hidden character change along branches subtending extant taxa. This positive correlation between the amount of character change that is discovered and the number of taxa included is well known by molecular systematists, and is known as the node density effect. However, its effect for morphological phylogenetics and inference of body plan evolution has barely been acknowledged (Jenner and Wills, 2007). Hence, the inclusion of even incomplete fossil taxa has the potential to reveal that synapomorphies of extant taxa may in fact be homoplasies or symplesiomorphies, and their inclusion can improve accuracy of the phylogenetic relationships inferred between living taxa (Wiens, 2005). The reconstruction of stem groups is crucial for a complete picture of body plan evolution, and there is ample evidence that phylogenetic inferences based on extant taxa can be misled; for arthropod examples see Budd (2001b) and Eriksson and Budd (2000). The amount of character evolution that is missed by a focus on extant taxa is increasingly illustrated by studies showing that rates of morphological character change may be highest early in the history of a clade, which may go hand in hand both with the general early establishment of morphological disparity in the history of large clades and indications that morphological transformations had larger step sizes early in a clade's history (Valentine, 2004; Ruta *et al.*, 2006; Erwin, 2007). In combination, these insights suggest that by focusing on living taxa only we are missing a lot of character evolution, the recognition of which is crucial to prevent clades being based on homoplasies or symplesiomorphies.

11.4.2   From the unequal eye to morphological cladistics

To see all things with equal eye is not within our power: humans, and especially human narrators, always look upon the world with an unequal eye.

O'Hara (1992, p. 140)

Before computers came to the assistance of phylogenetic analysis, Problematica were an inescapable by-product of phylogenetic inference. Without the help of a computer it is impossible to achieve a balanced and unbiased evaluation of large numbers of comparative data for more than a few taxa. Emphasis on different aspects of available evidence as well as the lack of a uniform phylogenetic methodology fostered disagreement between workers. Consequently, from the beginning of our discipline one researcher's central insights were not uncommonly labelled another's 'fata morgana' [mirage] (Hubrecht, 1887, p. 641), and the coordinating theme of one school of zoological thought would deserve to be 'dead and buried' in the opinion of proponents of another (Hyman, 1959, p. 750).

The widespread adoption of cladistic reasoning in the second half of the 20th century increased the promise of reaching a general consensus on metazoan phylogeny. Yet, without the help of computers, progress was slow as the amount of conflicting evidence allowed many mutually exclusive conclusions. The computer-assisted morphological cladistic analyses of metazoan phylogeny published over the last decade greatly advanced the objectivity, explicitness, and testability of phylogenetic hypotheses. In this period the field progressed significantly beyond the traditional textbook trees (Adoutte et al., 2000), but perhaps the most important insight of this era of fruitful debate was discovering exactly how problematic many taxa and clades actually were. As reviewed elsewhere (Jenner, 2004a,b), differences in the construction of data matrices, including different strategies of character selection, character coding and scoring, and taxon selection,

resulted in many incompatible phylogenies. Taxa such as Chaetognatha and Ectoprocta behave like phylogenetic renegades, residing in as many different clades as there are studies, and although other aspects of the phylogenetic backbone seemed more secure (monophyly of Protostomia, Spiralia), total agreement between analyses is absent. Evidently, the phylogenetic signal residing in morphology needs to be supplemented with molecular evidence.

11.4.3   Old Problematica solved and new Problematica revealed

A new phylogenetic synthesis for the Metazoa (Halanych, 2004) (Figure 11.1) has emerged largely on the basis of molecular evidence. The backbone of this phylogeny is based on nuclear ribosomal sequences (18S and 28S rDNA), and despite challenges (Rogozin et al., 2007b) its major aspects are confirmed by increasingly sophisticated phylogenomic analyses based on larger amounts of data, and employing improved model-based analytical methods (Philippe et al., 2005b; Baurain et al., 2007; Brinkmann and Philippe 2008; Irimia et al., 2007; Hausdorf et al., 2007; Roeding et al., 2007; Dunn et al., 2008; Helmkampf et al., 2008a,b). These studies have done a fine job in solving some of the classical Problematica. For example, the enigmatic myxozoans and Polypodium have now been firmly placed within Cnidaria, the pogonophorans and vestimentiferans are now placed within Annelida, Xenoturbella is firmly placed as the sister group to Ambulacraria, and chaetognaths and the lophophorates are now definitely excluded from the Deuterostomia. The reliable placement of these taxa reveals why they were problematic before. They are all highly modified taxa and they have either lost complexity, or evolved an otherwise unique body plan.

However, in the case of Chaetognatha, for example, the 'solution' is not yet complete (Table 11.1), as their exact phylogenetic position remains uncertain. Our finding that over half of the major metazoan lineages listed in Table 11.1 can be classified as Problematica is quite remarkable. We classify as category I Problematica those taxa for which there is still no consensus about either their broad phylogenetic neighbourhood, let alone their precise position, or for which a precise understanding of their phylogenetic position is of particular importance for understanding major transitions in the evolution of animal body plans. Category II Problematica are those for which we have some idea about their general phylogenetic neighbourhood, but we are still far removed from reliably placing them. Knowing the precise position of these taxa will aid our understanding of body plan evolution mostly within the confines of relatively smaller clades, principally within the Lophotrochozoa. Only 21 out of 45 lineages in Table 11.1 can reasonably be labelled as non-Problematica, and six of these fall within the three 'phyla' Porifera, Cnidaria, and Annelida.

Probably the most important reason for the continued existence of Problematica is systematic error, despite the fact that most of them have now been included in at least one phylogenomic analysis. Even though the limited overlap in genes between published phylogenomic analyses (see supplementary information in Dunn *et al.*, 2008) may lead one to suspect sampling artefacts, systematic error is the inescapable explanation of several discrepancies noted between different analyses. In studies such as those of Lartillot and Philippe (2008), Dunn *et al.* (2008), Helmkampf *et al.* (2008b), and Struck and Fisse (2008), conspicuous differences between analyses of the same data set with different methods indicate sensitivity to systematic errors. The fit between the chosen method/evolutionary model and the data set is crucial, but one shoe does not necessarily fit all. For example, even though the CAT mixture model has been promoted as a superior model (Brinkmann and Philippe, 2008; Lartillot and Phillippe, 2008) for phylogenomics, particularly in the fight against long branch attraction, application to the data set of Struck and Fisse (2008) generated some likely nonsense results, such as the position of Syndermata within the Ecdysozoa (similar to the results of Helmkampf *et al.*, 2008b, and Marlétaz *et al.*, 2008, when they applied the CAT model). Dunn *et al.* (2008) noted that the position of Tardigrada is strongly model dependent for their data set. Clearly, data quality and model fit need to be assessed for each data set individually. Examples of taxa for which the lack of current consensus

is likely to be in part due to systematic error are Tardigrada, Acoela, Myzostomida, Bryozoa, Syndermata, Gastrotricha, Platyhelminthes, and Gnathostomulida. These taxa are unstable in phylogenomic analyses, and are often fast evolving for the sampled markers. We think that the clade uniting Myzostomida, Acoela, and Gnathostomulida in Dunn *et al.* (2008) is emblematic for this problem. If this clade goes, anything goes.

Systematic error may also be the reason why phylogenomic analyses may or may not support monophyly of Deuterostomia depending on the choice of evolutionary model (Lartillot and Philippe, 2008; Marlétaz *et al.* 2008). This illustrates that the overall relationships between larger clades may also be very uncertain, which holds true in particular for the topology within Lophotrochozoa.

Finally, new Problematica can be revealed by new discoveries or reinterpretations of established taxa on the basis of both molecular and morphological evidence. For example, detailed morphological study and preliminary molecular phylogenetic analysis of the interstitial worm-like genus *Diurodrilus* strongly suggests that it does not fall within the polychaetes, as previously assumed, but may instead represent an independent lineage of animals potentially related to gnathiferans (Worsaae and Kristensen, 2003; Worsaae and Rouse, 2008). This shows both the importance of the continued surveying of under-explored habitats for new groups of organisms, and the importance of properly integrating new discoveries in a phylogenetic framework so as to illuminate the evolution of animal body plans.

11.4.4   Guidelines for future progress in metazoan phylogeny

A large literature exists on troubleshooting molecular systematics. Some excellent recent reviews include: Gribaldo and Philippe (2002), Sanderson and Shaffer (2002), Delsuc *et al.* (2005), Philippe *et al.* (2005b), Boore (2006), Philippe and Telford (2006), Rokas and Carroll (2006), Wiens (2006), and Whitfield and Lockhart (2007). We extract a number of guidelines that we feel need to be kept in mind to ensure continued progress in understanding.

Table 11.1 A list of all major 'phylum-level' metazoan taxa with notes on Problematica. Problematic taxa for which either very little is known (e.g. *Lobatocerebrum* sp., *Planctosphaera pelagica*), or which are likely to be *incertae cedis* on lower taxonomic levels (Aeolosomatidae within annelids) are not included. Taxa are classified into categories I–IV, based on a consideration of the degree of controversy surrounding their phylogenetic placement, and their importance for understanding body plan evolution. Category I includes true modern Problematica for which there exists greatest uncertainty about their phylogenetic position and/or those for which an understanding of their true position is crucial for understanding major transitions in the evolution of body plans. Category II groups Problematica for which there is still serious uncertainty about their phylogenetic position, but for which the alternative hypotheses are more restricted in either number or phylogenetic depth. Categories III and IV group non-Problematica. Category III groups taxa for which the phylogenetic neighbourhood seems secure, and for which future efforts should primarily focus on positioning the taxa either close to one or a few, or within other 'phyla'. Category IV houses taxa for which their sister-group relationships now seem established. Please note that we did not try to achieve a comprehensive listing of alternative sister-group hypotheses. We restricted ourselves principally to recent phylogenomic and molecular phylogenetic analyses. Including a full consideration of available morphological and combined evidence analyses would have collapsed our classification into one bucket of Problematica that included all listed taxa. The table should be taken as a tool to facilitate discussion about the focus of future work, and as a framework for comparison with morphology-based studies. Note that cases of congruence between different phylogenomic analyses, in particular those of Dunn *et al.* (2008) and those of H. Philippe and colleagues, can be interpreted as providing largely independent support for phylogenetic hypotheses given the limited overlap between the genes upon which these analyses are based.

Cat.	Taxon	Alternative sister groups	Remarks	Recent references
I	Demospongiae	Hexactinellida, Calcarea (Homoscleromorpha) Eumetazoa	Borchiellini *et al.* (2004) reported the intriguing finding that the demosponges are only monophyletic (Demospongiae *sensu stricto*) when the homoscleromorphs are excluded. Newer analyses have upheld the separate status of the homoscleromorphs, but the first molecular phylogenetic analysis of hexactinellid sponges indicates that these fall within a paraphyletic Demospongiae (Dohrmann *et al.*, 2008). The position of Demospongiae *sensu stricto* + Hexactinellida with respect to homoscleromorphs and calcareans remains unresolved, as is the question of the monophyly of sponges (see below)	Borchiellini *et al.* (2004), Erpenbeck and Wörheide (2007), Sperling *et al.* (2007), Dohrmann *et al.* (2008)
I	Calcarea	Eumetazoa, Homoscleromorpha, Homoscleromorpha + Eumetazoa	Poriferan paraphyly, with calcareans and non-poriferan metazoans most closely related to each other, was reported repeatedly in papers from the late 1990s. However, newer and more comprehensive phylogenetic analyses now paint a different picture. Either Calcarea are most closely related to homoscleromorphs (previously considered to be demosponges, see below), a hypothesis supported by nuclear and mitochondrial ribosomal sequences (Dohrmann *et al.* 2008) as well as an unpublished phylogenomic analysis (R. Derelle, unpublished doctoral thesis, 2007), or calcareans are the sister group to Homoscleromorpha + Eumetazoa (Sperling *et al.*, 2007)	Erpenbeck and Wörheide (2007), Sperling *et al.* (2007), Dohrmann *et al.* (2008)
I	Homoscleromorpha	Eumetazoa, Calcarea	Paraphyly of sponges, in particular with homoscleromorphs as the sister group to the remaining metazoans, has recently been used as an interpretative framework for understanding the earliest steps of metazoan body plan evolution (Sperling *et al.* 2007; Nielsen 2008). However, although sponge paraphyly is supported by some studies (Sperling *et al.*, 2007), it is not supported by others (Dohrmann *et al.*, 2008). Interestingly, an unpublished phylogenomic analysis that includes representatives of calcareans, demosponges, homoscleromorphs, and hexactinellids supports poriferan monophyly (R. Derelle, unpublished doctoral thesis, 2007; M. Manuel, personal communication). If confirmed, this largely removes the rationale for using sponge body plans for understanding the origin of eumetazoan body plans	Borchiellini *et al.* (2004), Derelle (2007), Erpenbeck and Wörheide (2007), Sperling *et al.* (2007); Dohrmann *et al.* (2008)

I	Placozoa	The phylogenetic position of Placozoa remains profoundly puzzling, with morphology, mitochondrial genomes, nuclear ribosomal sequences, or combined morphological and molecular evidence providing no consensus whatsoever. We eagerly await their first inclusion in a phylogenomic analysis. Considering the fact that placozoans represent the morphologically simplest animal 'phylum' it is of great interest to see if they are primitively simple or secondarily simplified	Eernisse and Peterson (2004), Glenner et al. (2004), Wallberg et al. (2004), Dellaporta et al. (2006), Cartwright and Collins (2007), da Silva et al. (2007), Ruiz-Trillo et al. (2008)
	Cnidaria, Cnidaria + Bilateria, Myxozoa + Bilateria, Bilateria, Porifera + Cnidaria		
I	Ctenophora	The phylogenetic position of the Ctenophora is one of the biggest problems at the base of the Metazoa. Specifically, molecular sequence data consistently place the ctenophores outside a clade including cnidarians and bilaterians, whereas interpretations of morphological and embryological data instead suggest a closer relationship between ctenophores and bilaterians. With the exception of the unpublished PhD thesis of R. Derelle (2007) ctenophores were first included in the phylogenomic analysis of Dunn et al. (2008). Surprisingly, this placed them as a sister group to all remaining metazoans, in agreement with the analyses in Derelle's thesis. If confirmed, this either implies that sponges and placozoans have become greatly simplified, or that comb jellies have convergently evolved an astonishing amount of developmental and morphological complexity, shared with eumetazoans	Eernisse and Peterson (2004), Cartwright and Collins (2007), Derelle (2007), Dunn et al. (2008), Nielsen (2008)
	All other metazoans, Planulozoa (Cnidaria, Placozoa, Myxozoa, Bilateria), Porifera		
I	Acoela	Although acoels (and nemertodermatids) are placed at the base of the Bilateria when nuclear ribosomal and protein-coding genes are considered, they are unstable in more data-rich phylogenomic analyses, placing them either in (or sister to) Deuterostomia, or in (or sister to) Protostomia. Hox cluster data are ambiguous at the moment. Interestingly, data on the presence of miRNAs seems to support a placement of acoels at the base of the Bilateria. Given that so far miRNA data seem to be remarkably free of homoplasy, and the investigated acoels possess only a subset of miRNAs shared between protostomes and deuterostomes, this is strong support for their exclusion from Nephrozoa, the clade of bilaterians characterized by possession of complex organs such as nephridia. However, this result is apparently contradicted by phylogenomic analyses that place acoels in various positions higher in the tree, which would imply they are secondarily simplified. Thus, on the balance of current evidence it is impossible to place them with any confidence. Considering their morphological simplicity and the lack of a biphasic life cycle, proper placement of the acoels and nemertodermatids will have important consequences for character optimization, and thus our understanding of the evolution of complex morphology and life cycles	Sempere et al. (2006, 2007), Brinkmann and Philippe (2008), Philippe et al. (2007), Wallberg et al. (2007), Baguñà et al. (2008), Deutsch (2008), Dunn et al. (2008)
	Nemertodermatida + Nephrozoa, Nemertodermatida, various clades of deuterostomes, Gnathostomulida, Protostomia		
I	Nemertodermatida	The relationship of nemertodermatids and acoels on the basis of ribosomal sequence data remained uncertain. Although both were positioned basal to the remaining bilaterians (Nephrozoa), the monophyly of Acoelomorpha remained uncertain. The most recent molecular phylogenetic analyses (Wallberg et al., 2007; Baguñà et al., 2008) support the status of acoels and nemertodermatids as independent lineages. However, in view of the fact that nemertodermatids have not yet been included in phylogenomic analyses, and the unstable position of the acoels in such analyses, we cannot yet ascertain the precise positions for these two taxa	Wallberg et al. (2007), Baguñà et al. (2008)
	Nephrozoa, Acoela		

Table 11.1 (Continued.)

Cat.	Taxon	Alternative sister groups	Remarks	Recent references
I	Tardigrada	Nematoida, Onychophora, Onychophora + Arthropoda	Ribosomal sequence data have suggested the possibility that tardigrades and onychophorans are sister taxa (Mallatt and Giribet, 2006). However, the phylogenetic position of tardigrades in more recent phylogenomic studies has been more difficult to determine due to differences between the studies in taxon sampling. In analyses that exclude Nematomorpha and or Onychophora, such as Brinkmann and Philippe (2008), Roeding et al. (2007), Helmkampf et al. (2008b), and Lartillot and Philippe (2008), tardigrades are sister group to Nematoda. The study of Dunn et al. (2008) shows that the phylogenetic position of tardigrades is very sensitive to the choice of molecular evolutionary model, so that currently a choice is not possible. However, when both tardigrades and onychophorans are included, phylogenomic evidence suggests unequivocally that onychophorans are more closely related to arthropods than are tardigrades. Placement of tardigrades separate from onychophorans and arthropods could imply their independent evolution of limbs, which would be an astonishing case of convergent evolution	Mallatt and Giribet (2006), Brinkmann and Philippe (2008), Roeding et al. (2007), Dunn et al. (2008), Helmkampf et al. (2008b), Lartillot and Philippe (2008)
I	Ectoprocta	Entoprocta, Platyzoa = (Platyhelminthes, Acoela, Gastrotricha, Myzostomida, Gnathifera), Platyhelminthes, all other lophotrochozoans, Myzostomida	Bryozoans remain a true Problematicum, as neither morphological evidence nor available molecular analyses can agree on their monophyly, or their phylogenetic position. Dunn et al. (2008) identified them as an unstable taxon, for which increased species sampling is necessary for fully resolving their position. It is remarkable that the phylogeny of Dunn et al. (2008) has two main clades within Lophotrochozoa, to which the sessile entoprocts and ectoprocts, respectively, are sister taxa. Interestingly, the coelomate ectoprocts are the sister group to the clade of acoelomate groups, whereas the acoelomate entoprocts are sister to the clade of coelomate lophotrochozoans. Outgroup comparison with the coelomate chaetognaths would indicate that the acoelomate condition has evolved independently in entoprocts and the other acoelomate lophotrochozoans	Eernisse and Peterson (2004), Halanych (2004), Passamaneck and Halanych (2006), Hausdorf et al. (2007), Dunn et al. (2008)
I	Gastrotricha	Platyhelminthes, all other nephrozoans, Micrognathozoa, Cycliophora, Rotifera	Gastrotricha is another problematic taxon that has been labelled as unstable in phylogenomic analyses (Dunn et al., 2008). Dunn et al. (2008) support a sister-group hypothesis with Platyhelminthes. However, other analyses based on a smaller number of data (but with better taxon sampling or including morphology) suggest other possibilities. Morphology and molecules appear to conflict with each other, as cuticle characters group gastrotrichs with the ecdysozoans, and sequence data place them in Lophotrochozoa	Eernisse and Peterson (2004), Halanych (2004), Todaro et al. (2006), Dunn et al. (2008)
I	Chaetognatha	Lophotrochozoa, Protostomia, Onychophora, Priapulida	Phylogenomic analyses of this classic Problematicum have not yet reached a consensus on whether arrow worms are a sister group to Lophotrochozoa (Matus et al., 2006b; Dunn et al., 2008), or Protostomia (Marlétaz et al., 2006, 2008; Brinkmann and Philippe, 2008; Lartillot and Philippe, 2008), and their position can be sensitive to method of analysis (Helmkampf et al. 2008b). Note that Helmkampf et al. (2008a) united chaetognaths with priapulids within the Ecdysozoa. A position of the chaetognaths as sister group to a major clade(s) of bilaterians will have major consequences for how we reconstruct the evolution of a host of organ systems, given the chaetognaths' unique mix of what are traditionally perceived to be characters of distinct clades	Eernisse and Peterson (2004), Halanych (2004), Marlétaz et al. (2006, 2008), Matus et al. (2006b), Hausdorf et al. (2007), Dunn et al. (2008), Helmkampf et al. (2008a,b), Lartillot and Philippe (2008)

	Taxon	Hypotheses	Discussion	References
I	Myzostomida	Acoela + Gnathostomulida, within Ectoprocta, within Annelida	A genuine Problematicum, myzostomids possess morphological and embryological characters that seem to unite them to various phyla, from annelids to rotifers. However, uncritical treatment of this evidence has compromised morphological and combined evidence phylogenetic analyses (Jenner, 2003). Taxon sampling is a crucial parameter for resolving their position using molecular data. In Dunn et al. (2008) myzostomids are grouped with acoels and gnathostomulids in the Lophotrochozoa. Although this position far removed from the annelids finds apparent support in some previous molecular phylogenetic analyses as well (Giribet et al., 2004; Passamaneck and Halanych, 2006), Bleidorn et al. (2007) found that long branch attraction probably affected the results of these studies. It is notable that molecular phylogenetic studies that include a greater sample of annelids (Hall et al., 2004; Colgan et al., 2006; Rousset et al., 2007) consistently unite myzostomids with annelids. Hence, on the balance of current evidence, their position remains uncertain. If myzostomids are not annelids, the amount of convergent evolution of morphological and developmental details shared with particular annelid taxa will be astonishing	Jenner (2003), Giribet et al. (2004), Hall et al. (2004), Colgan et al. (2006), Passamaneck and Halanych (2006), Bleidorn et al. (2007), Roussett et al. (2007), Dunn et al. (2008),
II	Cnidaria	Porifera, Placozoa (Myxozoa Bilateria), Placozoa, Bilateria	The two chief alternative hypotheses that are based on phylogenomic analyses cannot decide whether cnidarians are sister group to bilaterians or poriferans. However, it should be kept in mind that phylogenomic analyses do not yet include placozoans. Nevertheless, irrespective of which of these alternatives will turn out to be correct, the shared morphological and developmental complexity of cnidarians and bilaterians is unlikely to be convergent	Halanych (2004), da Silva et al. (2007), Dunn et al. (2008), Ruiz-Trillo et al. (2008)
II	Rhombozoa (Dicyemida and Heterocyemida)	Triploblasts or lophotrochozoans	Although they are likely to be triploblasts, the phylogenetic positions of both rhombozoans and orthonectids remain essentially unknown. Nevertheless, being highly specialized parasites, we expect their body plans to be highly modified. A recent phylogenetic analysis of dicyemid Pax6 and Zic genes supported a bilaterian affinity of dicyemids (Aruga et al., 2007)	Zrzavý (2001), Halanych (2004), Aruga et al. (2007)
II	Orthonectida	Triploblasts or lophotrochozoans	As with the rhombozoans, the phylogenetic position of the orthonectids remains entirely unresolved on the basis of both scanty molecular and morphological evidence (Slyusarev and Kristensen, 2003; Halanych, 2004)	Slyusarev and Kristensen (2003), Halanych (2004)
II	Mollusca	Annelida, Annelida + Platyhelminthes, Annelida + Sipunculida + Phoronida + Brachiopoda + Nemertea, Annelida + Sipunculida, Nemertea, Nemertea + Sipunculida + Annelida, a diverse set of lophotrochozoan phyla, Entoprocta	Although our table lists a larger number of alternative sister-group hypotheses for the Mollusca than for any other taxon, this is in part due to differences in taxon sampling between different analyses, which artificially inflates the number of alternative sister-group hypotheses to some extent. Focusing on just those phylogenomic analyses with the broadest sampling of taxa (Helmkampf et al., 2008a,b; Dunn et al., 2008), we can conclude that although the exact sister group of the Mollusca remains elusive, it is at least part of a lophotrochozoan clade including Annelida, Sipunculida, Nemertea, Phoronida, and Brachiopoda. It is noteworthy that on the basis of new morphological evidence Haszprunar and Wanninger (2008) recently proposed that a sister-group relationship between Mollusca and Entoprocta 'is currently among the best documented interrelationships of two metazoan phyla'. Strikingly, no molecular support for this hypothesis exists	Eernisse and Peterson (2004), Passamaneck and Halanych (2006), Hausdorf et al. (2007), Wanninger et al. (2007), Dunn et al. (2008), Haszprunar and Wanninger (2008), Helmkampf et al. (2008a,b), Lartillot and Philippe (2008)

Table 11.1 (Continued.)

Cat.	Taxon	Alternative sister groups	Remarks	Recent references
II	Annelida (including the former pogonophorans and vestimentiferans)	Mollusca, Phoronida + Brachiopoda + Nemertea, Platyhelminthes, Mollusca + Nemertea	The most broadly sampled phylogenomic analyses available (Dunn et al., 2008; Helmkampf et al., 2008a,b) have not convincingly resolved the position of annelids. Whereas Dunn et al. (2008) support a relationship of annelids with phoronids, brachiopods, and nemerteans, Helmkampf et al. (2008a,b) instead favour a relationship with phoronids and ectoprocts. The other phylogenomic analyses either suggest a close relationship to Mollusca and Nemertea, or Platyhelminthes, but these studies have more restrictive taxon sampling that does not allow all hypotheses to be tested	Brinkmann and Philippe (2008), Dunn et al. (2008), Helmkampf et al. (2008a,b), Lartillot and Philippe (2008), Struck and Fisse (2008)
II	Nemertea	Brachiopoda, Brachiopoda + Phoronida, Neotrochozoa, Mollusca, Sipunculida + Annelida	As discussed in Jenner (2004b), available morphological and combined evidence analyses support the Nemertea as part of a clade including the Neotrochozoa (Mollusca, Annelida, Sipunculida, Echiura). Molecular sequence data in isolation, however, provide a less clear picture, partly as a result of differences in taxon sampling between studies. Intriguingly, the most comprehensive phylogenomic analyses to date (Dunn et al. 2008; Helmkampf et al. 2008b) support a sister-group relationship between nemerteans and brachiopods + phoronids, a relationship foreshadowed in some analyses based on ribosomal gene sequences (Glenner et al., 2004; Todaro et al., 2006). The phylogenomic analyses of Helmkampf et al. (2008a) and Struck and Fisse (2008), however, group nemerteans with various neotrochozoans, to the exclusion of brachiopods and phoronids	Eernisse and Peterson (2004), Jenner (2004b), Dunn et al. (2008), Helmkampf et al. (2008a,b), Struck and Fisse (2008)
II	Platyhelminthes	Gastrotricha, Annelida, Neotrochozoa + Brachiopoda + Phoronida + Nemertea, Ectoprocta, Syndermata, Mollusca + Annelida + Sipunculida, Neotrochozoa + Brachiopoda + Phoronida + Nemertea + Ectoprocta, other lophotrochozoans	Morphological phylogenetic analyses have failed to identify a sister group of Platyhelminthes (Jenner, 2004b). Differences in taxon sampling and different results due to the application of different reconstruction methods on the same data set make it difficult to evaluate the merits of molecular and phylogenomic analyses. The sister-group relationship with annelids (Brinkmann and Philippe, 2008; Lartillot and Philippe, 2008) is probably an artefact of insufficient taxon sampling, although the analysis of Todaro et al. (2006) based on 18S sequences and a broad taxon sampling also supports this hypothesis. A sister-group relationship of platyhelminths to a clade of neotrochozoans (plus brachiopods, phoronids, and nemerteans if these taxa are included) is supported by Helmkampf et al. (2008a) (plus Ectoprocta), Hausdorf et al. (2007), Marlétaz et al. (2008) (plus Ectoprocta and Entoprocta), and Baguñà et al. (2008). Helmkampf et al. (2008b) provide uncertain support for platyhelminths as sister group to the remaining lophotrochozoans. However, a closer relationship to non-coelomate lophotrochozoans, especially syndermates and gastrotrichs (when included) is found in other analyses (Passamaneck and Halanych, 2006; Hausdorf et al., 2007; Dunn et al., 2008; Helmkampf et al., 2008b; Marlétaz et al., 2008; Struck and Fisse 2008). Consequently, on the basis of current evidence it is still impossible to nominate a reliable sister group to Platyhelminthes within Lophotrochozoa. Very provisionally one may conclude on the basis of the most comprehensive analyses that Platyhelminthes is a part of one of two main clades within the Lophotrochozoa, which in turn is sister to a clade containing Neotrochozoa, and Nemertea, Phoronida, and Brachiopoda when these are included	Eernisse and Peterson (2004), Jenner (2004b), Passamaneck and Halanych (2006), Todaro et al. (2006), Hausdorf et al. (2007), Baguñà et al. (2008), Brinkmann and Philippe (2008), Dunn et al. (2008), Helmkampf et al. (2008a,b), Marlétaz et al. (2008), Struck and Fisse (2008), Lartillot and Philippe (2008)

=	Entoprocta	Neotrochozoa + Brachiopoda + Phoronida + Nemertea, Cycliophora, Ectoprocta, Mollusca	The phylogenetic position of Entoprocta remains problematic. Morphological phylogenetic analyses suggest a variety of different sister taxa ranging from ectoprocts to lobatocerebromorphans. Notably, recent studies by Haszprunar and Wanninger (2008) and Wanninger et al. (2007) strengthen a nexus of similarities between entoproct creeping larvae and a variety of adult and larval molluscan features. These similarities have been meant to strongly imply a sister-group relationship between entoprocts and molluscs. However, no molecular phylogenetic support for this hypothesis is available. The most comprehensive phylogenomic analysis available (Dunn et al., 2008) supports a sister-group relationship of entoprocts to the coelomate lophotrochozoans (Neotrochozoa, Nemertea, Brachiopoda, Phoronida). In contrast, the phylogenomic studies of Hausdorf et al. (2007) and Helmkampf et al. (2008b) found support for a monophyletic Bryozoa, with entoprocts and ectoprocts as sister taxa. The phylogenomic analysis of Marlétaz et al. (2008) finds some support for this hypothesis as well, although the result is dependent on the model of sequence evolution used. Combined molecular and morphological analyses, such as Glenner et al. (2004) and Eernisse and Peterson (2004), show a closer relationship between entoprocts and Cycliophora (and possibly Syndermata)	Eernisse and Peterson (2004), Hausdorf et al. (2007), Wanninger et al. (2007), Dunn et al. (2008), Haszprunar and Wanninger (2008), Helmkampf et al. (2008b)
=	Syndermata (Rotifera and Acanthocephala)	Myzostomida + Acoela + Gnathostomulida, Gnathostomulida, Gnathostomulida + Micrognathozoa, Platyhelminthes, Lophotrochozoa, Micrognathozoa, within Ecdysozoa, Ectoprocta	Morphological evidence strongly favours a relationship of syndermates to gnathostomulids and Micrognathozoa. However, robust molecular evidence that unites these taxa (Gnathifera) to the exclusion of others is not currently available. Previous molecular or combined evidence analyses suggest a variety of possible sister-group relationships. In recent phylogenomic studies, such as Dunn et al. (2008), Helmkampf et al. (2008b), and Marlétaz et al. (2008), Rotifera are very unstable (grouping alternatively within Lophotrochozoa or Ecdysozoa), and different molecular phylogenetic analyses support different sister-group hypotheses	Eernisse and Peterson (2004), Halanych (2004), Hausdorf et al. (2007), Passamaneck and Halanych (2006), Todaro et al. (2006), Baguñà et al. (2008), Dunn et al. (2008), Helmkampf et al. (2008b), Marlétaz et al. (2008), Struck and Fisse (2008)
=	Micrognathozoa	Rotifera, within Gnathifera, Cycliophora, Cycliophora + Gnathostomulida, Entoprocta	Morphological evidence firmly unites Limnognathia maerski with syndermates and gnathostomulids. However, molecular phylogenetic analyses are at the moment less conclusive (Giribet et al., 2004; Todaro et al., 2006), supporting either a relationship with syndermates, gnathostomulids, cycliophorans, or entoprocts. They have not yet been included in phylogenomic studies	Giribet et al. (2004), Halanych (2004), Todaro et al. (2006)
=	Gnathostomulida	Acoela, Rotifera, within Gnathifera, Gastrotricha + Rotifera + Micrognathozoa + Cycliophora	Morphological evidence strongly unites gnathostomulids with syndermates and Micrognathozoa. Labelled as an unstable taxon in the phylogenomic analysis of Dunn et al. (2008), gnathostomulids have not yet been placed reliably in molecular analyses. We suspect that their placement as sister group to the acoels in Dunn et al. (2008) is a systematic error due to long branch attraction	Eernisse and Peterson (2004), Halanych (2004), Todaro et al. (2006), Dunn et al. (2008)

Table 11.1 (Continued.)

Cat.	Tax on	Alternative sister groups	Remarks	Recent references
II	Cycliophora	Entoprocta, Syndermata, Rotifera + Micrognathozoa, Micrognathozoa	The phylogenetic position of Cycliophora on the basis of both morphological and molecular evidence remains uncertain. They have not yet been included in a phylogenomic analysis	Eernisse and Peterson (2004), Giribet et al. (2004), Halanych (2004), Passamaneck and Halanych (2006), Todaro et al. (2006)
III	Hexactinellida	Demospongiae *sensu stricto*, Demospongiae (Calcarea Eumetazoa), within Demospongiae	Although there is some 18S rDNA support for Hexactinellida representing the sister group to all other metazoans, recent molecular phylogenetic analyses based on several loci (Dohrmann et al., 2008) instead support the nesting of glass sponges within demosponges. An unpublished phylogenomic analysis based on more data, but fewer taxa (R. Derelle, doctoral thesis, 2007; M. Manuel pers. comm.) supports a sister-group relationship of hexactinellids and Demospongiae *sensu stricto* (excluding homoscleromorphs)	Nichols (2005), Derelle (2007), Erpenbeck and Wörheide (2007), Dohrmann et al. (2008)
III	Echiura	Annelida	Available evidence now reliably places echiurans inside the annelids as possible sister group to Capitellidae	Rouse and Pleijel (2007)
III	Sipunculida	Annelida	Available molecular evidence suggests that sipunculids are sister group to, or part of the Annelida	Eernisse and Peterson (2004), Hausdorf et al. (2007), Rouse and Pleijel (2007), Dunn et al. (2008), Struck and Fisse (2008)
III	Phoronida	Inarticulate brachiopods, Brachiopoda, Brachiopoda + Nemertea	Although it is beyond doubt that phoronids are closely related to brachiopods, it is at the moment unclear whether phoronids fall within brachiopods as sister group to inarticulates (Cohen and Weydman 2005), or whether they are separate lineages. Yet, in contrast, Helmkampf et al. (2008a) found a sister-group relationship between the single phoronid species and a species of ectoproct in their analyses, with brachiopods being more distantly related. This is in turn contradicted by Dunn et al. (2008) and Helmkampf et al. (2008b) which found support for brachiopods to be the sister group of phoronids, but in Dunn et al. (2008) the position of phoronids is sensitive to method of phylogenetic analysis	Halanych (2004), Cohen and Weydmann (2005), Dunn et al. (2008), Helmkampf et al. (2008a)
III	Brachiopoda	Phoronida, Nemertea, Ectoprocta + Phoronida + Nemertea + Mollusca + Annelida	Intriguingly, besides support for a connection to phoronids, the phylogenomic study by Dunn et al. (2008) found some support for a close relationship between brachiopods and nemerteans, a relationship also suggested in some previous analyses of ribosomal gene sequences (Passamaneck and Halanych 2006; Todaro et al. 2006). It should be noted that the phylogenetic position of brachiopods may change depending on the method of analysis for a given data set. This is obvious, for example, in the studies of Passamaneck and Halanych (2006) and Dunn et al. (2008)	Passamaneck and Halanych (2006), Todaro et al. (2006), Dunn et al. (2008)

IV	Myxozoa	Falls within Cnidaria	The status of myxozoans as parasitic and highly modified cnidarians is now robustly supported by a phylogenomic analysis. Previous suggestions of myxozoans being the sister group to Bilateria were the result of long branch attraction, but more studies are needed to establish with confidence whether myxozoans fall within Cnidaria or are sister group to Cnidaria	Jiménez-Guri et al. (2007), Evans et al. (2008)
IV	*Polypodium hydriforme*	Sister to or part of Hydrozoa	Combined 18S and 28S rDNA support the position of *Polypodium* within Cnidaria. However, due to long branch attraction problems affecting the analyses, the precise relationship between myxozoans and *Polypodium* remains unclear	Evans et al. (2008)
IV	Vertebrata	Urochordata	In a remarkable reversal of received opinion a new interpretation of morphological evidence and a phylogenomic analysis yielded support for the sister-group relationship between Tunicata (Urochordata) and Vertebrata (Craniata) (Ruppert, 2005; Delsuc et al., 2006). This previously heterodox hypothesis is now based on largely independent phylogenomic support (Dunn et al. 2008; Lartillot and Philippe 2008; Putnam et al., 2008), and has rapidly gained general approval	Ruppert (2005), Delsuc et al. (2006), Dunn et al. (2008), Lartillot and Philippe (2008), Putnam et al. (2008), Swalla and Smith (2008)
IV	Urochordata	Vertebrata	See under 'Vertebrata'	
IV	Cephalochordata	Urochordata + Vertebrata	The monophyly of chordates is beyond doubt, and the sister-group hypothesis between cephalochordates and a clade of Urochordata and Vertebrata is now robustly supported	Ruppert (2005), Dunn et al. (2008), Lartillot and Philippe (2008), Putnam et al. (2008), Swalla and Smith (2008)
IV	Echinodermata	Hemichordata	The sister-group relationship between echinoderms and hemichordates is robustly supported	Ruppert (2005), Dunn et al. (2008), Lartillot and Philippe (2008), Swalla and Smith (2008)
IV	Hemichordata	Echinodermata	See under 'Echinodermata'	
IV	Xenoturbellida	Echinodermata + Hemichordata	In contrast to the phylogenomic analysis of Philippe et al. (2007), which suggested that Xenoturbella was possibly the sister group of Acoela, more recent analyses support Xenoturbella as the sister group to Ambulacraria (Echinodermata Hemichordata), together named Xenambulacraria (Bourlat et al., 2006)	Bourlat et al. (2006), Philippe et al. (2007), Dunn et al. (2008), Lartillot and Philippe (2008), Swalla and Smith (2008)
IV	Nematoda	Nematomorpha	The emerging consensus on the sister-group relationship between nematodes and nematomorphs is now also supported by phylogenomic analyses (Dunn et al., 2008; T. Juliusdottir, R. Jenner, M. Telford, R. Copley, unpublished data)	Eernisse and Peterson (2004), Halanych (2004), Mallatt and Giribet (2006), Dunn et al. (2008)

Table 11.1 *(Continued.)*

Cat.	Tax on	Alternative sister groups	Remarks	Recent references
IV	Nematomorpha	Nematoda	See under "Nematoda"	Eernisse and Peterson (2004), Halanych (2004), Dunn et al. (2008)
IV	Priapulida	Loricifera or Kinorhyncha	As long as loriciferans are not yet included in phylogenomic analyses, the sister-group relationship between Priapulida and Kinorhyncha in such studies should be interpreted with caution. Morphological evidence allows no conclusive resolution, with either Priapulida or Kinorhyncha as the sister group to Loricifera. The phylogenomic analysis of Dunn et al. (2008) included kinorhynchs, which are resolved as the sister group of priapulids	
IV	Kinorhyncha	Loricifera or Priapulida	See under 'Priapulida'	
IV	Loricifera	Kinorhyncha or Priapulida	See under 'Priapulida'	
IV	Onychophora	Arthropoda, Chelicerata	Although the close relationship between velvet worms and arthropods is uncontested, it is currently unclear exactly how they relate to each other on the basis of phylogenomic analyses. A sister-group relationship to either Arthropoda (Roeding et al. 2007; Dunn et al. 2008) or, surprisingly, Chelicerata (Roeding et al. 2007; Marlétaz et al. (2008) is supported. The latter hypothesis is also supported by a recent phylogenetic analysis of neuroanatomical characters (Strausfeld et al. 2006), although most other morphological evidence has traditionally been interpreted as evidence for onychophorans and arthropods as separate lineages	Strausfeld et al. (2006), Roeding et al. (2007), Dunn et al. (2008)
IV	Arthropoda	Onychophora	See under 'Onychophora'	
IV	Pogonophora and Vestimentifera	Within Annelida	The pogonophorans and vestimentiferans are now confidently placed within the polychaetes	Rouse and Pleijel (2007), Rousset et al. (2007)

Several factors need to be balanced to produce a good phylogenetic analysis: number of taxa, number of characters, quality of data, and quality of analytical models. The interaction between these variables determines whether the results of a phylogenetic analysis are informative and reliable, or suffer from stochastic or systematic error. Stochastic error arises as chance correspondences overwhelm true phylogenetic signal when there are not enough informative data. Systematic error results when reconstruction methods are inaccurate and are unable to deal with bias in the raw data, which can have several causes (Philippe and Telford, 2006). The common problem of long branch attraction (Anderson and Swofford, 2004; Waegele and Mayer, 2007) can be both a stochastic or a systematic error.

In trying to avoid stochastic error by increasing the number of characters in a data set, systematic errors may become increasingly prominent due to insufficient taxon sampling, uneven amounts of data across taxa, and a failure to detect non-phylogenetic signals in the data. So far the molecular data generated for different phyla is wildly uneven (Figure 11.1) because of the bias towards key taxa that are important as model organisms, or organisms of biomedical or economic importance, or simply because they are the easiest to collect. To avoid systematic error it is therefore important to strive for a better balance in the number of taxa and characters (Philippe and Telford, 2006):

1. Avoiding stochastic error:

• Increase the number of characters. In molecular systematics this is the main rationale for doing phylogenomics, based on large numbers of data generated through genome projects, EST projects, or large-scale projects targeting particular genes with degenerate primers (Delsuc *et al.*, 2005; Philippe *et al.*, 2005a; Philippe and Telford, 2006; Baurain *et al.*, 2007). However, workers should be aware that uncritically concatenating information from different genes may cause systematic errors (Bapteste *et al.*, 2008; Hartmann and Vision, 2008).

2. Avoiding systematic error:

• Develop better models of sequence evolution that can deal with problematic data to prevent systematic error (Delsuc *et al.*, 2005; Philippe *et al.*, 2005b; Rokas and Carroll, 2006; Baurain *et al.*, 2007).

• Move towards less homoplastic characters such as rare genomic changes (Boore, 2006; Rokas and Carroll, 2006).

• Sample more taxa, including at least several species representing a high-level metazoan taxon, which may do more to prevent systematic error than aiming to have whole-genome sequences for fewer taxa (Hillis *et al.*, 2003).

• Recognize and remove problematic data, such as fast-evolving taxa or characters, or characters the evolution of which violates phylogenetic model assumptions (Lecointre and Deleporte, 2004; Delsuc *et al.*, 2005; Philippe *et al.*, 2005b).

3. Other considerations:

• Care should be taken not to be misled by gene duplication (paralogy), causing gene trees to diverge from the species tree.

• Be aware that heuristic analyses can get caught in local optima, with different methods showing different degrees of sensitivity to this (Brinkmann and Philippe, 2008).

• To maximize the power to test the phylogenetic position of a particular taxon, try to include at least all the taxa that have previously been proposed to be its closest relatives.

• If practical, reconstruct a phylogenetic scaffold based on a restricted number of taxa scored for many characters. Additional taxa can then be added sequentially on the basis of smaller numbers of characters (Wiens, 2006). The addition of incompletely known taxa can boost accuracy and confidence. To prevent systematic error it may be better to add a smaller number of characters scored for many taxa, rather than many characters for fewer taxa.

• If there is not enough phylogenetic signal, focus on characters with higher rates of evolution.

• Assess data quality as a standard part of any phylogenetic analysis (Brinkmann and Philippe, 2008; Waegele and Mayer, 2007).

• Exploit combined evidence analyses, where possible including fossil data (Giribet, 2002;

Eernisse and Peterson, 2004), while recognizing the interpretational difficulties associated with combining molecular exemplar species and inferred morphological ground patterns.

• Sample different genes that evolve at different rates to be able to resolve different regions of the tree (Sanderson and Shaffer, 2002; Glenner *et al.*, 2004; Philippe and Telford, 2006).

• Boost the number of descriptive and comparative morphological studies to revise outdated received wisdom, and provide more data crucial for the inference of body plan evolution (Nielsen, 2001; Jenner, 2006b). Papers such as those by Wanninger *et al.* (2007) and Stach *et al.* (2008) are valuable in their contribution to phylogenetic debate and our understanding of body plan evolution.

• Adopt an experimental approach (sensitivity analysis) to phylogenetic analysis to see how results change depending on different assumptions.

• Re-evaluate contentious morphological evidence in the light of independent molecular phylogenies, especially to detect cases of unrecognized character loss (Jenner, 2004c).

• Carefully construct morphological data sets to maximize testing power (Jenner, 2004a).

• Adopt standardized methods for the presentation, annotation, and analysis of molecular data. To this end Leebens-Mack *et al.* (2006) have called for a standard for reporting on phylogenies, the MIAPA (minimum information about a phylogenetic analysis), in which each component of a phylogenetic analysis (alignment procedures, alignment, sequences, voucher specimens, methods and parameters used, etc.) is outlined using universally accepted criteria. This will facilitate better evaluation and comparison of results of different analyses.

In summary, the recognition of Problematica reveals more than the sum of their missing or ambiguous parts. In avoiding fragmentary fossils or extant organisms with combinations of chimaeric or autapomorphic features, and by excluding long branching taxa or heavily biased nucleotide and protein sequences from molecular analyses, we may bring near completeness to data matrices and greater stability to our phylogenetic analyses, but probably at the expense of accuracy and an understanding of the full evolutionary picture. Problematica reveal themselves as supremely important; for without their inclusion and accurate placement, other relationships are liable to change. In understanding how to deal with Problematica we understand the limits of systematics and our ability to have faith in our reconstructions of the tree of life.

11.5 Acknowledgements

RAJ gratefully acknowledges the UK Biotechnology and Biological Sciences Research Council for financial support. RAJ thanks Dr M. Obst for his stimulating thoughts on metazoan evolution shared during the scientific collecting expedition Pandalina V. We thank Drs Michaël Manuel and Katrine Worsaae for sharing unpublished information about their research.

CHAPTER 12

Improvement of molecular phylogenetic inference and the phylogeny of Bilateria

Nicolas Lartillot and Hervé Philippe

Inferring the relationships among Bilateria has been an active and controversial research area since the time of Haeckel. The lack of a sufficient number of phylogenetically reliable characters was the main limitation of traditional phylogenies based on morphology. With the advent of molecular data this problem has been replaced by another, statistical inconsistency, which stems from an erroneous interpretation of convergences induced by multiple changes. The analysis of alignments rich in both genes and species, combined with a probabilistic method (maximum likelihood or Bayesian) using sophisticated models of sequence evolution, should alleviate these two major limitations. We have applied this approach to a data set of 94 genes from 79 species using the CAT model, which accounts for site-specific amino acid replacement patterns. The resulting tree is in good agreement with current knowledge: the monophyly of most major groups (e.g. Chordata, Arthropoda, Lophotrochozoa, Ecdysozoa, Protostomia) was recovered with high support. Two results are surprising and are discussed in an evo-devo framework: the sister-group relationship of Platyhelminthes and Annelida to the exclusion of Mollusca, contradicting the Neotrochozoa hypothesis, and, with a lower statistical support, the paraphyly of Deuterostomia. These results, in particular the status of deuterostomes, need further confirmation, both through increased taxonomic sampling and future improvements of probabilistic models.

12.1 Introduction

12.1.1 The limits of morphology

The inference of animal phylogeny from morphological data has always been a difficult issue. Although a rapid consensus was obtained on the definition of phyla (with a few exceptions: vestimentiferans, pogonophores, or platyhelminths), the relationships among phyla has long remained unsolved (Brusca and Brusca, 1990; Nielsen, 2001). The dominant view, albeit far from being universally accepted, was traditionally biased in favour of the *Scala Naturae* concept of Aristotle, which postulates an evolution from simple to more complex organisms (Adoutte *et al.*, 1999). Briefly, acoelomates (platyhelminths and nemertines) were considered as emerging first, followed by pseudocoelomates (nematodes), and then coelomates, representing the 'crown group' of Bilateria. A similar gradist view was proposed for deuterostomes, with the successive emergence of Chaetognatha, Echinodermata, Hemichordata, Urochordata, and Cephalochordata, culminating in Vertebrata (e.g. Conway Morris, 1993a).

Irrespective of its underlying 'ideological' preconceptions, however, this traditional bilaterian phylogeny was based on very few morphological and developmental characters (position of the nerve cord, cleavage patterns, modes of gastrulation, etc.) whose phylogenetic reliability may sometimes be disputable (either because of the description, or the coding and analysis; see Jenner, 2001). This general lack of homologous characters is related to the wide disparity observed between body plans. For some phyla (such as echinoderms), the body plan is nearly exclusively characterized by idiosyncrasies, leaving few characters to compare with other bilaterian phyla. Traditional animal phylogenies based on morphological data were thus hampered by an insufficient amount of reliable primary signal.

12.1.2 The difficult beginnings of molecular phylogeny

Great hopes were placed in the use of molecular data (Zuckerkandl and Pauling, 1965). Unfortunately, the first phylogenies based on ribosomal RNA (rRNA) turned out to be quite controversial (Field *et al.*, 1988). They contained some results that were difficult to accept, such as the polyphyly of animals. We will not review in detail this turbulent early history, but rather note that these trees were based on a scarce taxon sampling and inferred using overly simple methods (e.g. the Jukes and Cantor distance). As a consequence, tree building artefacts were frequent. The problem was mainly addressed through improved taxon sampling: over a period of about 10 years, rRNAs were sequenced from several hundred species. In part because of the sheer improvement due to a denser taxonomic sampling, but also thanks to a systematic selection of the slowest-evolving representatives of the majority of animal phyla, a consensus rapidly emerged, reducing the diversity of Bilateria into three main clades: Deuterostomia, Lophotrochozoa (Halanych *et al.*, 1995), and Ecdysozoa (Aguinaldo *et al.*, 1997). The statistical support for most of the nodes was nevertheless non-significant (Philippe *et al.*, 1994; Abouheif *et al.*, 1998), thus preventing any firm conclusions.

This brief historical overview provides a clear illustration of the problems of phylogenetic inference. The resolution of the morphological and rRNA trees is limited because too few substitutions occurred during the evolution of this set of conserved characters, yielding too few synapomorphies. At the same time, unequal rates of evolution across characters imply that some characters concentrate numerous multiple substitutions (convergences and reversions). These multiple substitutions can be misinterpreted by tree reconstruction methods and lead to incorrect results. In particular, the well-known long branch attraction

(LBA) artefact (Felsenstein, 1978) leads to an erroneous grouping of fast-evolving taxa, often resulting in an apparent earlier emergence (Philippe and Laurent, 1998). For instance, this is the reason for the initial absence of monophyly of animals, with Bilateria evolving too fast and being attracted towards the outgroup. Similarly, the recognition of the LBA problem played a major role in the establishment of the Ecdysozoa hypothesis (Aguinaldo *et al.*, 1997); i.e. when the fast-evolving *Caenorhabditis* is considered, nematodes emerge at the base of Bilateria, but when the slowly-evolving *Trichinella* is included, nematodes cluster with arthropods. Compositional heterogeneity can also generate artefacts, especially for trees based on mitochondrial sequences (Foster and Hickey, 1999).

Twenty years ago, the expectations from molecular data were further boosted by the prospect of using genomic data. The underlying assumption is that the joint analysis of numerous genes potentially provides numerous synapomorphies, thus eliminating the problem of stochastic errors. Yet although it is true that stochastic errors will naturally vanish in such a phylogenomic context, systematic errors, which are due to the inconsistency of tree building methods, will not disappear. Indeed, they should even become more apparent (Philippe *et al.*, 2005b).

We recognized two main avenues to circumvent systematic errors (Philippe and Laurent, 1998): (1) the use of rare, and putatively slowly evolving, complex characters, such as gene order (Boore, 2006), which should be homoplasy-free; (2) the use of numerous genes combined with inference methods that deal efficiently with multiple substitutions, which should avoid artefacts due to homoplasy. We will briefly review the application of the second approach to the question of the monophyly of Ecdysozoa as a way of demonstrating the importance of using numerous species and models that handle the heterogeneity of the evolutionary process across positions.

12.1.3 Illustration of the misleading effect of multiple substitutions in the case of Ecdysozoa

The first phylogenies based on numerous genes (up to 500) significantly rejected the monophyly of

Ecdysozoa (e.g. Blair *et al.*, 2002; Dopazo *et al.*, 2004; Wolf *et al.*, 2004). To exclude the possibility that this was due to a LBA artefact, the use of putatively rarely changing amino acids was proposed (Rogozin *et al.*, 2007b), an approach that also supported Coelomata (i.e. arthropods as sister group of vertebrates rather than nematodes). At first, phylogenomics seemed to reject strongly the new animal phylogeny, which was mainly based on rRNA.

However, these phylogenomic analyses were characterized by a very sparse taxon sampling, and used only simple tree reconstruction methods, rendering them potentially sensitive to systematic errors. We will show that, as in the first rRNA phylogenies, the monophyly of Coelomata was an artefact due to the attraction of the fast evolving *Caenorhabditis* to the distant outgroup (e.g. Fungi). As detailed below, three different and independent approaches that reduce the misleading effect of multiple substitutions lead to a change the topology from Coelomata to Ecdysozoa.

12.1.4 Removal of the fast-evolving positions

An obvious way to reduce systematic errors is to remove the fastest-evolving characters from the alignment (Olsen, 1987). In principle, the phylogeny has to be known to compute the evolutionary rate, rendering simplistic circular approaches potentially hazardous (Rodríguez-Ezpeleta *et al.*, 2007). The SF method (Brinkmann and Philippe, 1999) partially circumvents this issue by computing rates within predefined monophyletic groups. Only the relationships among these groups can be studied and an equilibrated species sample should be available for each predefined group. When the SF method is applied to a large alignment of 146 genes with four representatives each from Fungi, Arthropoda, Nematoda, and Deuterostomia (Delsuc *et al.*, 2005), the removal of fast-evolving sites leads to an almost total disappearance of the support in favour of Coelomata. Interestingly, this does not correspond to a loss of phylogenetic signal, since support in favour of Ecdysozoa steadily increases (up to a bootstrap support value of 91%). The simplest interpretation of this experiment is that the misleading effect of multiple substitutions creates a LBA artefact that disappears when

fast-evolving positions are discarded. Note that this way of selecting slowly evolving characters (SF method) differs from the one used by Rogozin *et al.* (2007b) by the use of a rich taxon sampling that allows us to select positions that are more likely to reflect the ancestral state of the predefined monophyletic group, therefore reducing the risk of convergence along the long terminal branches.

12.1.5 Improvement of taxon sampling

Another obvious way of reducing the misleading effect of multiple substitutions is to incorporate more species, breaking long branches (Hendy and Penny, 1989) and thus allowing one to detect convergences and reversions more easily. In the case of Bilateria, simply adding a close outgroup (Cnidaria) to an alignment containing only a distant outgroup (Ascomyceta) is sufficient to change from strong support for Coelomata to strong support for Ecdysozoa. This is true for the analysis of both complete primary sequences (Delsuc *et al.*, 2005) and rare amino acid changes (Irimia *et al.*, 2007). Undetected convergences between the fast-evolving nematodes and the distant outgroup therefore create a strong but erroneous signal that biases tree building methods. Accordingly, none of the phylogenomic studies that have used dense taxon sampling found any support in favour of Coelomata (e.g. Philippe *et al.*, 2005a; Marlétaz *et al.*, 2006; Matus *et al.*, 2006b).

12.1.6 Improvement of the tree building method

Probabilistic methods are now widely recognized as the most accurate methods for phylogenetic reconstruction (Felsenstein, 2004). However, to avoid the problem of systematic errors, they require good models of sequence evolution. We recently developed a new model, named CAT, which partitions sites into categories, so as to take into account site-specific amino acid preferences (Lartillot and Philippe, 2004). When applied to the difficult case of the bilaterian tree rooted by a distant outgroup (Fungi), the CAT model provides strong support for Ecdysozoa, whereas the WAG model (Whelan and Goldman, 2001) strongly favours Coelomata (Lartillot *et al.*, 2007). Posterior predictive analyses demonstrate that the CAT model predicts homoplasies more accurately than the WAG model. In other words, the CAT model detects multiple substitutions more efficiently and is therefore less sensitive to systematic errors.

The greater robustness of CAT against the Coelomata artefact is related to the fact that it accounts better for site-specific restrictions of the amino acid alphabet. Amino acid replacements in most proteins tend to be biochemically conservative, with typical variable positions in a protein accepting substitutions among only two or three amino acids. This has important consequences for phylogenetic reconstruction using amino acid sequences, since it implies that convergences and reversions (homoplasies) are much more frequent than what would be expected if all amino acids were considered equally acceptable at any given position. In practice, the typical number of amino acids observed per position is indeed overestimated by classical site-homogeneous models, based on empirical matrices such as WAG, which in turn results in a poor anticipation of the risk of homoplasy, and thereby in a greater prevalence of artefacts. In contrast, site-specific models such as CAT will anticipate these problems better, and will be less prone to systematic errors (Lartillot *et al.*, 2007).

Figure 12.1 (see also Plate 8) illustrates the strikingly different behaviour of site-homogenous and site-heterogeneous models when a position is saturated. Under WAG and GTR (general time reversible) models, the substitution process rapidly converges to a nearly flat distribution over the 20 amino acids. Therefore, according to these two models, a position cannot be saturated and at the same time display a strong preference for a few amino acids. This is at odds with common intuition about strong site-specific effects of purifying selection related to the protein's conformational and functional constraints. In contrast, under the CAT model, the position underwent substitution almost solely between the two negatively charged amino acids aspartate and glutamate, and all other 18 amino acids are rarely encountered, even over long time periods (Figure 12.1). Thus, under CAT, the effective substitutional alphabet at the position under investigation is essentially of size 2, which automatically implies that convergence towards the

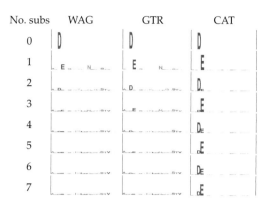

Figure 12.1 Posterior predictive tests to analyse the behaviour of the WAG, GTR, and CAT models under substitutional saturation. A column of the alignment displaying only aspartic acid (D) and glutamic acid (E) was chosen at random, and for the three models, the probability of observing each of the 20 amino acids after n substitutions ($n = 0–7$), and starting from an aspartic acid, was estimated and visualized graphically. The height of each letter is proportional to the probability of the corresponding amino acid. The parameters of the substitution process were taken at random from the posterior distribution under each model. (See also Plate 8.)

same amino acid is very likely, much more likely than if all 20 amino acids were allowed at this site. Thanks to this phenomenon, the CAT model more easily detects saturation in protein alignments, compared with standard models such as WAG or GTR, and is therefore less sensitive to long branch attraction artefacts (Lartillot *et al.*, 2007).

Of course, there are many other potential causes of error, all of which can in principle be traced back to model misspecification problems: in all cases, it is a matter of correctly modelling various features of the substitution process that may potentially lead to an increase in the level of homoplasy (e.g. compositional biases). Improving the models of sequence evolution is thus an essential requirement for phylogenetics, and is currently a very active area of research. In principle, it should be preferred to the two other approaches detailed above because (1) it avoids the risk of stochastic errors implied by the use of the rare slowly evolving positions and (2) it applies even when the taxon sampling is naturally sparse (e.g. coelacanth).

Finally, it should be noted that an incorrect handling of multiple substitutions does not necessarily lead to a robust incorrect tree (as in the case of Coelomata) but possibly to an unresolved tree. For instance, an analysis based on 50 genes using a sparse taxon sampling (21 species, with most of the animal phyla being represented by a single, often fast-evolving, species) and a simple model of sequence evolution (RtREV+I+G), resulted in a poorly resolved tree, in which even the monophyly of Bilateria was not supported (Rokas et al., 2005). Since the approach used does not allow efficient detection of multiple substitutions, we decided to do a comparable study in which we simultaneously improved the species sampling (from 21 to 57, including many slowly evolving species) and the model of sequence evolution (i.e. using the CAT model). Interestingly, the statistical support was high (e.g. bootstrap values > 95% for Bilateria, Ecdysozoa, and Lophotrochozoa) in the resulting tree (Baurain et al., 2007). This illustrates that incorrect handling of multiple substitutions can create an artefactual signal that is not sufficiently strong to overcome the genuine phylogenetic signal and to create a highly supported erroneous topology, but is sufficient to lead to a poorly resolved tree.

In summary, a combination of many positions, corresponding to multiple genes, and a dense taxonomic sampling are a necessary prerequisite to obtain reliable phylogenies. Ideally, these sequences should then be analysed with probabilistic models that correctly describe the true evolutionary patterns of the sequences under study. In practice, one may perform analyses with alternative models of evolution among those currently available, compare the fit of those models, check for possible model violations, and test the robustness of the analyses by site and taxon resampling. This is the method that we apply to the phylogeny of Bilateria. Full details of the materials and methods used can be found associated with the original article (Lartillot and Philippe, 2008).

12.2 Results

12.2.1 Comparison of phylogenies based on CAT and WAG models

We analysed our large data set (79 animal species and 19,993 positions) using two alternative models of amino acid replacement: the WAG empirical matrix (Whelan and Goldman, 2001), which is currently one of the standard models (Ronquist and Huelsenbeck, 2003; Jobb et al., 2004; Hordijk and Gascuel, 2005; Stamatakis et al., 2005); and the CAT mixture model (see above). The trees obtained under CAT (Figure 12.2) and WAG (Figure 12.3) models are very similar and in good agreement with current knowledge (Halanych, 2004). The following major aspects can be noted:

• Ecdysozoa and Lophotrochozoa receive a stronger bootstrap support under CAT (bootstrap proportion (BP) of 99% and 100%) than under WAG (53% and 71%). Under WAG, platyhelminths are slightly attracted by nematodes, as can be seen by the low bootstrap support values along the path between the two groups. The attraction is nevertheless less marked than with a poorer taxon sampling (Philippe et al., 2005a).

• Within Lophotrochozoa, many phyla are unsampled, but the three that are present (platyhelminths, molluscs, and annelids) are reasonably well represented. Interestingly, with the CAT model annelids and platyhelminths are sister groups (94% BP), while the analysis under WAG recovers a more traditional grouping of annelids and molluscs (Neotrochozoa, 97% BP). A sister-group relationship between annelids and platyhelminths had already been observed in a combined large subunit (LSU)–small subunit (SSU) rDNA analysis (Passamaneck and Halanych, 2006), and in an analysis based on mitochondrial gene order (Lavrov and Lang, 2005), but was not found in previous analyses based on expressed sequence tags (ESTs) (Philippe et al., 2005a).

• The relationships among Ecdysozoa are not well resolved. This is mainly due to the fluctuating position of tardigrades and priapulids, whose sequences are incomplete (39.9% and 75.8% of missing data, respectively). The most likely configuration displays priapulids at the base of all other Ecdysozoa, and tardigrades as sister group of nematodes, but two major alternatives are also proposed by the bootstrap analysis: tardigrades as sister group of priapulids, together at the base of nematodes and arthropods, or priapulids at the base of nematodes.

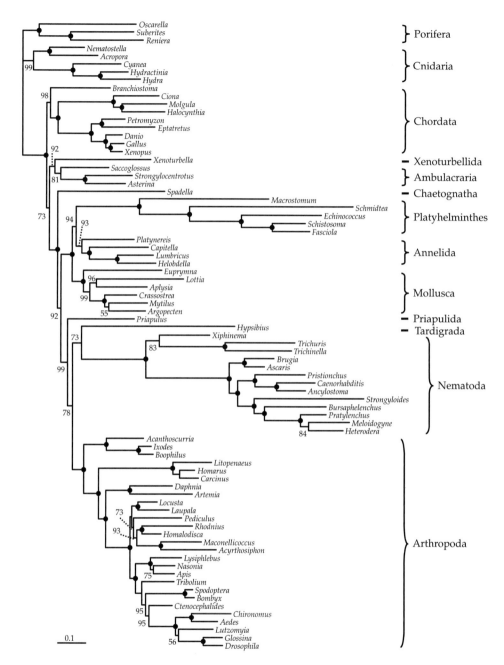

Figure 12.2 Phylogeny inferred using the CAT model. The alignment consists of 19,993 unambiguously aligned positions (94 genes and 79 species). The tree was rooted using sponges and cnidarians as outgroups. Nodes supported by 100% bootstrap values are denoted by black circles while lower values are given in plain style. The scale bar indicates the number of changes per site.

• Chaetognaths appear at the base of all protostomes (92% CAT BP, 52% WAG BP), which is in accordance with Marlétaz *et al.* (2006).

• The monophyly of deuterostomes is weakly supported under the WAG model (76% BP), whereas the CAT model favours a paraphyly, also weakly

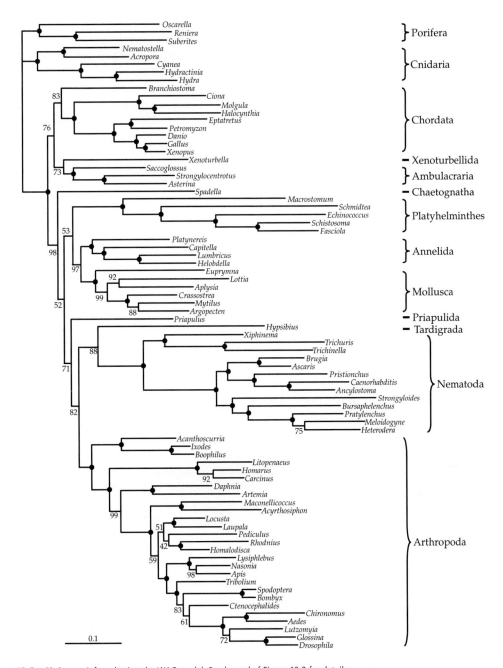

Figure 12.3   Phylogeny inferred using the WAG model. See legend of Figure 12.2 for details.

supported, with chordates emerging first (73% BP, only 19% for deuterostome monophyly).

• Chordates are monophyletic (98% CAT BP, 83% WAG BP), receiving a stronger support than in previous phylogenomic studies (Bourlat *et al.*,

2006; Delsuc *et al.*, 2006). In addition, under both analyses, urochordates are closer to vertebrates than cephalochordates with 100% BP confirming the monophyly of Olfactores (Delsuc *et al.*, 2006).

• The phylogenetic position of Xenoturbellida, as sister group of echinoderms + hemichordates (Bourlat *et al.*, 2006), is also recovered (92% CAT BP, 73% WAG BP).

In summary, the WAG and CAT models agree with each other on 73 nodes, and disagree for a minor change at the base of insects and two major points: the monophyly of deuterostomes, and the relative order of molluscs, annelids, and platyhelminths. In the case of lophotrochozoans, the two models strongly disagree, whereas concerning deuterostomes, the difference is not statistically significant.

12.2.2 Model comparison and evaluation

The discrepancy between the two models indicates the presence of artefacts due to systematic errors. A statistical comparison of the two models may help in deciding which of them offers the most reliable phylogenetic tree. In addition, the observed artefacts are the symptoms of model violations, which we will analyse using a standard statistical method, namely posterior predictive analysis. Note that the WAG empirical matrix is just one among the available empirical matrices, and the results obtained with WAG may not be representative of the general class of site-homogeneous models. To address this point, we also tested the GTR model along with WAG and CAT.

First, based on cross-validation tests (see electronic supplementary material in Lartillot and Philippe, 2008), the CAT model was found to have a much better statistical fit than either WAG (a score of 3939 ± 163 in favour of CAT) or GTR (2765 ± 128). The better score of GTR relative to WAG (a difference of 1174 in favour of GTR) indicates that the data set is big enough for the parameters of the amino acid replacement matrix to be directly inferred, rather than taken from an empirically derived empirical matrix. On the other hand, the improvement in doing so is less significant than that accomplished by using the site-heterogeneous CAT model (1174 versus 2765).

Second, we performed posterior predictive analyses, using two test statistics: one, vertical (i.e. computed along the columns of the alignment), is

the mean number of distinct amino acids per column (mean site-specific diversity) (Lartillot *et al.*, 2007); the other one, horizontal (i.e. computed along the rows of the alignment), is the chi-square compositional homogeneity test (Foster, 2004); see electronic supplementary material in Lartillot and Philippe (2008). Violation of the horizontal statistic indicates that the model does not handle compositional biases correctly, whereas violation of the vertical statistic means that the model does not correctly account for site-specific biochemical patterns.

Concerning the vertical test (Figure 12.4a), the mean number of distinct amino acids per column of the true alignment (mean observed diversity) is 4.45. Site-homogeneous models predict a much higher value (6.89 ± 0.04 for WAG, 6.53 ± 0.04 for GTR). Thus, the assumptions underlying the site-homogeneous models are strongly violated ($P <$ 0.001, z = 62.3 for WAG; $P <$ 0.001, z = 72.6 for GTR). In contrast, the CAT model performs much better (mean predicted diversity of 4.59 ± 0.03), although it is also weakly rejected ($P <$ 0.001, z = 4.6). As explained above, since overestimating the number of states per position lead to underestimating the probability of convergence (Lartillot *et al.*, 2007), one may expect a greater risk of LBA under WAG for the present analysis. On the other hand, all models—CAT, GTR, and WAG—fail the horizontal test to the same extent (compositional homogeneity, Figure 12.4b). This is not too surprising, given that all of them are time-homogeneous amino acid replacement processes. However, the violation is strong, as measured by the z-score ($z > 11$), which warns us that there is a risk, whichever model is used, of observing artefacts related to compositional biases.

12.3 Discussion

12.3.1 Towards better phylogenetic analyses

Phylogenetics is still a difficult and controversial field, because no foolproof method is yet available for avoiding systematic errors. In this study we have tried to combine two methods that have proven efficient at alleviating artefacts while obtaining sufficient statistical support. First,

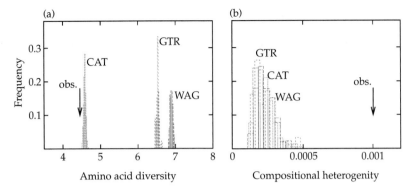

Figure 12.4 Posterior predictive tests. The observed value (arrow) of the test statistic is compared with the null distribucns under CAT and WAG models: (a) mean biochemical diversity per site; (b) maximum compositional deviation over taxa.

relying on EST projects, we have tried to combine an increase in the overall number of aligned sequence positions, so as to capture more of the primary phylogenetic signal, with an improved taxonomic sampling (Philippe and Telford, 2006). Second, we have also brought particular attention to the problem of probabilistic models underlying the phylogenetic reconstruction. As is obvious from our statistical evaluations, the standard model used in phylogenetics today, WAG, is not reliable, at least for deep-level phylogenies such as that of animals. Essentially, it is strongly rejected for its failure to explain either site-specific biochemical patterns or compositional differences between taxa. As indicated by our analysis of GTR, this failure is not specific to WAG and is likely to apply to all the site-homogeneous models. The alternative model used here, CAT, is significantly better, but may not be reliable enough, in particular against potential artefacts induced by compositional biases.

Interestingly, the weaknesses of the WAG model also result in an overall lack of support, which is probably due to the unstable position of some fast-evolving groups (in particular, platyhelminths). This confirms previous observations (Baurain *et al.*, 2007), and also illustrates that improving taxonomic sampling is not in itself a sufficient response to systematic errors but should be combined with an in-depth analysis of the probabilistic models used.

In the light of our model evaluation, the position of platyhelminths proposed by CAT as a sister group to annelids should be taken seriously. In this perspective, the Neotrochozoa (molluscs + annelids) found by WAG is interpreted as an artefact. This would not be too surprising, given the overall saturation of the platyhelminth sequences. Note that the vestige of artefactual attraction between platyhelminths and nematodes observed under WAG should in itself warn us that the position of platyhelminths within Lophotrochozoa may not be reliably inferred under WAG. The phylogenetic position of platyhelminths, relative to other lophotrochozoans, is a long-standing question, the potentially important implications of which have already been pointed out (Passamaneck and Halanych, 2006).

The other point of disagreement is about deuterostomes: monophyletic under WAG, they appear to be paraphyletic under CAT. This progressive emergence of deuterostome phyla is unusual. In fact, Deuterostomia *sensu stricto* (echinoderms, hemichordates, and chordates) have long been considered as one of the most reliable phylogenetic groupings in animal phylogeny (Adoutte *et al.*, 2000). A possible explanation of the monophyly of deuterostomes obtained with WAG in terms of LBA would be that the fast-evolving protostomes are attracted by the outgroup. Given its implications (see below), this potential artefact would certainly deserve further attention. On the other hand, caution is needed since the basal position of chordates observed under CAT does not receive a high bootstrap support (73%). It is also unstable upon small variations of the taxon sampling: for instance, deuterostome monophyly is recovered with CAT if either *Spadella* or *Xenoturbella* are removed from the analysis (data not shown). Similarly, it appears that the removal of the non-bilaterian outgroup leads to the non-monophyly of Xenambulacraria (Philippe *et al.*, 2007). In addition, the fast-evolving acoels probably emerge close to the base of deuterostomes, further shortening internal branches (Philippe *et al.*, 2007). In summary, this suggests that the signal for resolving this part of the tree is weak, all the more so when the outgroup is distantly related. Additional data should be analysed with improved methods before taking sides on this issue.

In a few respects our analysis is in contradiction with the results found by Dunn *et al.* (2008) (see also Chapter 6). First, the support in favour of deuterostomes is higher in Dunn *et al.*'s study than in our own investigation, although it still remains lower than 90%. Second, in their study, chaetognaths are found in a sister-group relationship with Lophotrochozoa (albeit without support), whereas we found them at the base of Lophotrochozoa + Ecdysozoa. Third, Dunn *et al.* found molluscs and annelids to be closely related, whereas platyhelminths fall within a larger group of 'Aschelminthes' (Platyzoa), including gastrotrichs, rotifers, acoels, and myzostomids. Compared with our analysis, Dunn *et al.*'s investigation relies on a richer taxon sampling, which may confer more robustness to their conclusions. On the other hand, the Platyzoa hypothesis is not totally convincing. First, the relatively long branches of most of these groups (in particular acoels, but also platyhelminths) raise some suspicion about a possible artefactual attraction between these fast-evolving phyla. Second, Platyzoa are not congruent with recent findings about the phylogenetic position of acoels (Philippe *et al.*, 2007). Third, missing data are much more frequent in the data set of Dunn *et al.* (2008) than in ours (44.5% versus 20%). In any case, the discrepancies observed between the two analyses suggest that both taxonomic sampling

and probabilistic models still need to be improved before a consensus about the details of the animal phylogeny can be reached.

12.3.2 Implications for the evolution of Bilateria

Converging towards a reliable picture of the animal phylogenetic tree is an interesting objective in itself. But more important are the implications of this phylogenetic picture for our vision of the morphological evolution of Bilateria (Telford and Budd, 2003). As mentioned in Section 12.1, morphological and developmental characters were traditionally the primary source of data used to infer phylogenetic trees. It has now become clear that many of those characters, such as cleavage patterns or the fate of the blastopore, are not reliable phylogenetic markers. It is nevertheless interesting to map their evolution on a tree that has been inferred from independent (molecular) data and use this to learn as much as possible about the history of morphological diversification of animal body plans. In this respect, comparative embryology, or evo-devo, is probably the primary customer of animal molecular phylogenetics.

Much has already been said about how the 'new animal phylogeny' changes our way of looking at the evolution of animals (Adoutte *et al.*, 1999; Halanych, 2004). One of the most important, and most frequent, messages has been that secondary simplifications of morphology and of developmental processes are common. This has been repeatedly implied by most of the successive changes brought to the animal phylogeny over the last 10 years, such as the repositioning of Nematoda and Platyhelminthes within coelomate protostomes, or of Tunicata as the sister group of Vertebrata.

In the context of the present chapter, the position of platyhelminths alongside two neotrochozoan phyla, as proposed by our CAT analysis, has similar implications, specifically concerning the evolution of development. Molluscs and annelids, together with sipunculans, have a canonical spiral development, characterized by a four-quartet spiral cleavage, an invariant and evolutionarily conserved cell lineage, including a single stem cell (4d, or mesen-toblast) giving rise to the mesodermal germbands, and a typical trochophore larva (Nielsen, 2001). In contrast, platyhelminths display atypical forms of spiral cleavage, and pass through a larval stage (Mueller's larva) that can only loosely be homolo-gized to a trochophore. In this context, a basal pos-ition of platyhelminths in the lophotrochozoan group, as previously often found, is compatible with the intuitively appealing idea that evolution proceeds from simple to complex forms. Namely, platyhelminths would be 'proto'-spiralians, outside a series of nested phyla, Trochozoa, Eutrochozoa, and Neotrochozoa (Peterson and Eernisse, 2001), corresponding to a graded series of increasingly complex forms of spiral development. Yet, the phylogeny favoured by CAT is at odds with the neotrochozoan hypothesis and implies that the development of platyhelminths is a secondarily modified (and ancestrally canonical) spiral devel-opment. Further taxonomic sampling within lophotrochozoans will be important, as it may not only allow a more robust inference of the position of platyhelminths, but also bring additional phyla that do not display a canonical spiral development among Eu- or Neotrochozoa, thereby leading to a completely different view of the evolution of spiral development.

The paraphyly of deuterostomes favoured by our CAT analysis, if confirmed, would also have deep implications concerning the way we inter-pret the evolution of Bilateria. First, it would result in a paraphyletic succession of three groups (Chordata, Xenambulacraria, and Chaetognatha), all of which display a radial cleavage, a deuter-ostomous gastrulation, and an enterocoelic mode of formation of the body cavity. Although these embryological characters are known to be evolu-tionarily labile (for instance, brachiopods have a deuterostomous gastrulation, and enterocoely is observed in nemerteans), this may be interpreted as phylogenetic evidence in favour of an ancestral deuterostomy. Similarly, the gill slits, found in chordates and hemichordates, would also have to be considered as ancestral to all Bilateria. In add-ition, with respect to all other Bilateria, chordates would then be of basal emergence, which turns

the traditional preconceptions radically upside down: in the perspective of this new phylogen-etic hypothesis, the chordate body plan is no longer the pinnacle of a progressive evolution through a succession of body plans of increas-ing complexity. Rather, chordates are one of the first bilaterian offshoots. This in turn would have consequences concerning the polarization of the morphological characters: thus far, most chordate-specific morphological and developmen-tal features (for instance their unique dorsoven-tral polarity, with dorsal nerve cord and ventral heart, have generally been assumed to be derived; Arendt and Nübler-Jung, 1994). In the context of the more traditional hypothesis of deuterostome monophyly, this assumption is justified, provided that the ancestral condition is clearly and jointly recognized in protostomes and in Ambulacraria (Arendt and Nübler-Jung, 1994). But the argument does not hold if chordates are the sister group of all other Bilateria: in that case, it is possible that some characters of chordates, such as the dorso-ventral polarity, may well have been ancestral to all bilaterally symmetrical animals.

12.4 Conclusion

Several phyla, in particular brachiopods and ony-chophorans, are still missing in phylogenomic analyses, and some others are poorly represented (aschelminths, chaetognaths, and hemichordates, among others), but the most species-rich phyla are now well sampled. Accordingly, one can be increasingly confident concerning the few robust aspects of the phylogeny of bilaterians that emerge from this and previous phylogenomic analyses. Essentially, the overall structure of protostomes (a split between lophotrochozoans and ecdyso-zoans, with chaetognaths at their base) seems stable, as well as the monophyly of Chordata and Ambulacraria. On the other hand, the monophyly of deuterostomes appears to be the most import-ant point yet to be settled in order to draw a complete picture of the scaffold of the bilaterian tree. Many aspects of the detailed relationships within each supergroup (in particular, Ecdysozoa and Lophotrochozoa) remain to be investigated.

Ongoing EST projects will soon bring many new species into this emerging picture, which will not only inform us about the phylogenetic position of those new species, but also result in an enriched taxonomic sampling, having a positive impact on the overall accuracy of phylogenetic inference. Yet, as suggested by the present chapter, this will not be sufficient, and will have to be combined with a significant improvement of the underlying probabilistic models. Much work is still needed, both concerning the acquisition of primary data and the methodological side, if one wants to converge towards a reliable, possibly final, picture of the bilaterian tree.

12.5 Acknowledgements

We thank Max Telford and Tim Littlewood for giving us the opportunity to write this chapter. The Réseau Québécois de Calcul de Haute Performance provided computational resources. This work was supported by the Canadian Institute for Advanced Research, the Canadian Research Chair Program, the Centre National de la Recherche Scientifique (through the ACI-IMPBIO Model-Phylo funding program), and the Robert Cedergren Centre for bioinformatics and genomics. This work was financially supported in part by the '60ème Commission Franco-Québécoise de Coopération Scientifique'.

CHAPTER 13

Beyond linear sequence comparisons: the use of genome-level characters for phylogenetic reconstruction

Jeffrey L. Boore and Susan I. Fuerstenberg

The first whole genomes to be compared for phylogenetic inference were those of mitochondria, which provided the first sets of genome-level characters for phylogenetic reconstruction. Most powerful among these characters have been comparisons of the relative arrangements of genes, which have convincingly resolved numerous branching points, including some that had remained recalcitrant even to very large molecular sequence comparisons. Now the world faces a tsunami of complete nuclear genome sequences. In addition to the tremendous amount of DNA sequence that is becoming available for comparison, there is also the potential for many more genome-level characters to be developed, including the relative positions of introns, the domain structures of proteins, gene family membership, the presence of particular biochemical pathways, aspects of DNA replication or transcription, and many others. These characters can be especially convincing because of their low likelihood of reverting to a primitive condition or occurring independently in separate lineages, so reducing the occurrence of homoplasy. The comparisons of organelle genomes pioneered the way for the use of such features for phylogenetic reconstructions, and it is almost certainly true, as ever more genomic sequence becomes available, that further use of genome-level characters will play a big role in outlining the relationships among major animal groups.

13.1 Why do we need anything other than molecular sequence comparisons?

Over the past few decades, the comparison of nucleotide and amino acid sequences has revolutionized our understanding of evolutionary relationships for many groups of organisms. The broader field of systematics has been reinvigorated and a generation of evolutionary biologists have come to accept that molecular sequence comparisons are an essential component for inferring the phylogeny of any group. These studies have led to extensive revision of animal systematics and to the overturning of previous reliance on features of the coelom and segmentation (Adoutte et al., 1999).

In the 1980s, when comparing molecular sequences for phylogenetic inference was first becoming common, some asserted with great confidence that all evolutionary relationships would soon be convincingly resolved solely with this type of data, leading to much consternation. However, some of the relationships that were equivocal in early molecular studies have remained highly recalcitrant even with many more DNA sequence data in hand. There are several potential explanations, including:

1. Multiple nucleotide or amino acid substitutions may have occurred at a single site, obscuring any accumulated signal.

2. Convergent or parallel substitutions may have occurred among different lineages due to having only four (for nucleotides) or 20 (for amino acids) possible character states, exacerbated by convergent biases in base composition (Naylor and Brown, 1998), which may even cause ever-increasing confidence measures for incorrect associations with ever larger data sets (Phillips *et al.*, 2004).

3. The analysis may show artefactual association of the more rapidly changing lineages (Felsenstein, 1978), including the attraction of long branches to the base of the ingroup in association with the outgroup (which is almost always a long branch; Philippe and Laurent, 1998).

4. In some cases, non-orthologous gene copies may be inadvertently compared among various lineages due to ancestral gene duplications followed by differential losses, or due to incomplete sampling.

5. Differing views of scientists on alignments, exclusion sets, and weighting schemes frequently cannot be arbitrated based on objective criteria and can lead to radically different phylogenetic reconstructions.

6. The most difficult problems are when the time of shared ancestry is short relative to the subsequent time of divergence, where there has been little opportunity to accumulate signal and ample time for it to have been erased.

Molecular sequence comparison is now a mature field that has influenced the culture of systematics. Many have come to expect that the future of systematics will be dominated by creating ever more sophisticated methods for teasing a weak signal from noisy data. This causes concern that differing preferences for various methods will ensure that no consensus on many evolutionary relationships will ever be reached.

However, an alternative is possible, that there may be other, less explored, types of characters that could be powerful for resolving these contentious relationships. There is no doubt that comparisons of some characters have identified certain robust synapomorphies (shared and derived character states) that have supported long-standing, little-contested evolutionary relationships, such as the monophyly of mammals, tetrapods, and echinoderms. These synapomorphies are subjectively judged to be of characters so unlikely to revert to an earlier condition or to occur multiple times in parallel that they could only have arisen once in the common ancestor of the group. Can new sets of characters be found that would meet these criteria to provide confident resolution of some problematic evolutionary relationships? Although there is a broad range of character types to explore, we will focus here specifically on comparison of features of genomes.

13.2   Comparisons of mitochondrial genomes have laid the foundation

Sequences from mitochondrial genes and genomes have been used extensively for phylogenetic inference, with complete mtDNA sequences being publicly available for more than 1000 animal species. (For a summary of the characteristics of animal mtDNAs, see Boore, 1999.) It has been long-argued (e.g. Boore and Brown, 1998) that the relative arrangement of the (normally) 37 genes in animal mitochondrial genomes constitutes an especially powerful type of character for phylogenetic inference, and so constitutes the first set of genome-level features to be used extensively for animal phylogeny. Briefly summarized, these genes are present in nearly all animal groups, are unambiguously homologous, and can potentially be rearranged into an enormous number of states such that convergent rearrangements are very unlikely (and demonstrated to be uncommon). All genes on each strand are transcribed together in cases where it has been studied (Clayton, 1992), so selection on gene arrangements is expected to be minimal. A summary of the evolutionary relationships convincingly demonstrated by these types of data (and in many cases left unresolved by all other studies) is found in Boore (2006), but here are a few of the more significant conclusions of deep-branch phylogenetic relationships: (1) the superphylum Eutrochozoa includes cestode platyhelminths (von Nickisch-Rosenegk *et al.*, 2001) and the phylum Phoronida (Helfenbein and Boore, 2004); (2) Sipuncula is closely related to Annelida rather than to Mollusca (Boore and Staton, 2002); (3) Annelida is more closely related to Mollusca than to Arthropoda (Boore and Brown, 2000);

Table 13.1 URLs for the largest public DNA sequencing centres

DNA sequencing centre	website
Wellcome Trust Sanger Institute	http://www.sanger.ac.uk/
DOE Joint Genome Institute	http://www.jgi.doe.gov/
Washington University Genome Sequencing Center	http://genome.wustl.edu/
Broad Institute	http://www.broad.mit.edu/
Baylor College of Medicine Genome Center	http://www.hgsc.bcm.tmc.edu/
Beijing Genomics Institute	http://www.genomics.org.cn/en/index.php
Riken Genomic Sciences Center	http://www.gsc.riken.go.jp/
J. Craig Venter Institute	http://www.jcvi.org/
Genoscope	http://www.genoscope.cns.fr/spip/

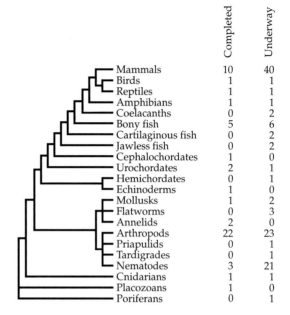

Figure 13.1 This reconstruction of the major branches of animal evolution is used to plot the numbers of taxa with complete genome sequences done and under way. The taxonomic ranks shown are arbitrary, split for illustration, but not meant to be consistent among the major groups, and taxa listed do not comprehensively cover all of life. Branch length holds no meaning. While opinions may differ on particular genomes as to whether they are complete versus needing more work, and whether they are well enough along to consider them 'under way', it is clear that there will soon be a large and phylogenetically broad sampling of genome sequences.

(4) Arthropoda is monophyletic and, within this phylum, Crustacea is united with Hexapoda to the exclusion of Myriapoda and Onychophora (Boore *et al.*, 1995, 1998); (5) Pentastomida is not a phylum, but rather a type of crustacean, and joins with Cephalocarida and Maxillopoda to the exclusion of other major crustacean groups (Lavrov *et al.*, 2004).

13.3 Nuclear genomes, a treasure-trove of phylogenetic characters

By a great margin, more DNA sequence is being generated than ever before. Facilities built and techniques developed for sequencing the human genome are now focusing on many other organisms. As recently as a year ago, the nine largest genome sequencing centres (see Table 13.1) collectively produced well over 170 billion nucleotides of DNA sequence per year: approximately 57-fold the coverage of the human genome. With next-generation sequencing platforms now in regular use, that number is exploding. Imminently there will be complete genomes of at least draft quality for many dozens of animals representing a phylogenetically diverse sample and including several equivocally placed lineages (Figure 13.1, Table 13.2).

In these genomic data are many higher-order features, beyond the linear sequences, that constitute genome-level characters that are potentially useful for phylogenetic reconstruction, including: (1) gene content, including components of multiunit complexes such as the ribosome, spliceosome, DNA replication machinery, or oxidative phosphorylation enzymes and the presence versus absence of particular biochemical pathways (e.g. de Rosa *et al.*, 1999; Fitz-Gibbon and House, 1999; House and Fitz-Gibbon, 2002; Huson and Steel, 2004; Snel *et al.*, 1999, 2005); (2) the relative arrangements of genes (Boore and Brown, 1998); (3) movements of genes among intracellular compartments (i.e. plastid, mitochondrion, nucleus) (e.g. Nugent and Palmer, 1991); (4) insertions of segments of DNA, including transposons and numts (Fukuda *et al.*, 1985; Richly and Leister, 2004); (5) variation in intron positions (e.g. Qiu *et al.*, 1998); (6) secondary structures of rRNAs or tRNAs (e.g. Murrell *et al.*,

Table 13.2 Complete nuclear genome sequencing projects completely drafted (i.e. not necessarily having every gap closed) or under way as summarized in Figure 13.1. Some of the taxa listed as under way are currently funded to only low (generally 2X) coverage. There are many other taxa not listed here whose genomes are being investigated at even lower levels of coverage.

Taxonomy	Organism	Common description
COMPLETE GENOMES		
Chordata, Mammalia	*Bos taurus*	Cow
	Callithrix jacchus	Marmoset
	Canis familiaris	Dog
	Mus musculus	Mouse
	Homo sapiens	Human
	Macaca mulatta	Rhesus macaque
	Monodelphis domestica	Opossum
	Ornithorhynchus anatinus	Duck-billed platypus
	Pan troglodytes	Chimpanzee
	Pongo pygmaeus abelii	Orangutan
Chordata, Aves	*Gallus gallus*	Red jungle fowl
Chordata, Sauria	*Anolis carolinensis*	Anole lizard
Chordata, Amphibia	*Xenopus tropicalis*	Western clawed frog
Chordata, Teleostei	*Danio rerio*	Zebrafish
	Gasterosteus aculeatus	Stickleback
	Oryzias latipes	Medakafish
	Takifugu rubripes	Japanese pufferfish
	Tetraodon nigroviridis	Green spotted pufferfish
Chordata, Cephalochordata	*Branchiostoma floridae*	Lancelet
Chordata, Urochordata	*Ciona intestinalis, C. savignyi*	Sea squirt
Echinodermata, Echinozoa	*Strongylocentrotus purpuratus*	Purple sea urchin
Mollusca, Bivalvia	*Lottia gigantea*	Owl limpet
Annelida, Oligochaeta	*Helobdella robusta*	Leech
Annelida, Polychaeta	*Capitella capitata*	None
Arthropoda, Coleoptera	*Tribolium castaneum*	Red flour beetle
Arthropoda, Diptera	*Aedes aegypti*	Yellow fever mosquito
	Anopheles gambiae	Malaria mosquito
	Culex pipiens	House mosquito
	Drosophila ananassae, D. erecta, D. grimshawi, D. melanogaster, D. mojavensis, D. persimilis, D. pseudoobscura, D. sechellia, D. simulans (8), D. virilis, D. willistoni, D. yakuba	Fruit flies
Arthropoda, Hemiptera	*Pediculus humanus corporis*	Louse
Arthropoda, Hymenoptera	*Apis mellifera*	Honeybee
	Nasonia vitripennis	Parasitic wasp
Arthropoda, Lepidoptera	*Bombyx mori*	Silkworm
Arthropoda, Crustacea	*Daphnia pulex*	Water flea
Arthropoda, Chelicerata	*Ixodes scapularis*	Deer tick
Nematoda, Chromadorea	*Caenorhabditis briggsae, C. elegans, C. remanei*	Roundworms
	Pristionchus pacificus	None
	Meloidogyne incognita	Root-knot nematode

Table 13.2   (*Continued.*)

Taxonomy	Organism	Common description
Cnidaria, Anthozoa	*Nematostella vectensis*	Sea anemone
Placozoa	*Trichoplax adhaerens*	Tablet animal
GENOMES IN PROGRESS		
Chordata, Mammalia	*Cavia porcellus*	Guinea pig
	Choloepus hoffmanni	Two-toed sloth
	Cryptomys sp.	Mole
	Cynocephalus volans	Flying lemur
	Dasypus novemcinctus	Nine-banded armadillo
	Dipodomys panamintinus	Kangaroo rat
	Echinops telfairi	Lesser hedgehog tenrec
	Elephantulus sp.	Elephant shrew
	Equus caballus	Horse
	Erinaceus europaeus	Western European hedgehog
	Felis catus	Cat
	Gorilla gorilla	Gorilla
	Loxodonta africana	African elephant
	Macaca fascicularis	Crab-eating macaque
	Macropus eugenii	Wallaby
	Manis pentadactyla	Chinese pangolin
	Microcebus murinus	Mouse lemur
	Mustela putorius furo	Ferret
	Myotis lucifugus	Little brown bat
	Nomascus leucogenys	Gibbon
	Ochotona princeps	Pika
	Oryctolagus cuniculus	European rabbit
	Otolemur garnettii	Bushbaby
	Pan paniscus	Bonobo
	Papio anubis	Baboon
	Peromyscus californicus, P. leucopus, P. maniculatus, P. polionotus	Mice
	Procavia capensis	Hyrax
	Pteropus vampyrus	Flying fox
	Saimiri sp.	Squirrel monkey
	Sorex araneus	European common shrew
	Spermophilus tridecemlineatus	Ground squirrel
	Sus scrofa	Pig
	Tarsius syrichta	Tarsier
	Tenrec ecaudatus	Common tenrec
	Tupaia belangeri	Tree shrew
	Tursiops truncatus	Dolphin
	Vicugna pacos	Alpaca

Table 13.2 (*Continued.*)

Taxonomy	Organism	Common description
Chordata, Aves	*Taeniopygia guttata*	Zebra finch
Chordata, Testudines	*Chrysemys picta*	Painted turtle
Chordata, Amphibia	*Xenopus laevis*	African clawed frog
Chordata, Coelocanthiformes	*Latimeria chalumnae*	Indonesian coelacanth
	Latimeria menadoensis	South African coelacanth
Chordata, Teleostei	*Astatotilapia burtoni*	Tilapia
	Lepisosteus oculatus	Spotted gar
	Metriaclima zebra	Tilapia
	Oreochromis niloticus	Tilapia
	Paralibidichromis chilotes	Tilapia
	Salmo salar	Atlantic salmon
Chordata, Chondrichthys	*Callorhinchus milii*	Elephant shark
	Raja erinacea	Skate
Chordata, Hyperotreti	*Eptatretus burgeri*	Hagfish
Chordata, Hyperoartia	*Petromyzon marinus*	Sea lamprey
Chordata, Urochordata	*Oikopleura dioica*	Tunicate
Hemichordata, Enteropneusta	*Saccoglossus kowalevskii*	Acorn worm
Mollusca, Gastropoda	*Aplysia californica*	Sea hare
	Biomphalaria glabrata	Snail
Platyhelminthes, Cestoda	*Echinococcus multilocularis*	Tapeworm
	Taenia solium	Pork tapeworm
Platyhelminthes, Turbellaria	*Schmidtea mediterranea*	Flatworm
Platyhelminthes, Trematoda	*Schistosoma mansoni, S. japonicum*	Blood flukes (schistosomes)
Arthropoda, Diptera	*Drosophila americana, D. auraria, D. equinoxialis, D. hydei, D. littoralis, D. mercatorum, D. mimica, D. miranda, D. novamexicana, D. repleta, D. silvestris*	Fruit flies
	Glossina morsitans	Tsetse fly
	Lutzomyia longipalpis	Sand fly
	Phlebotomus papatasi	Sand fly
Arthropoda, Hemiptera	*Acyrthosiphon pisum*	Pea aphid
	Rhodnius prolixus	Kissing bug
Arthropoda, Hymenoptera	*Nasonia giraulti, N. longicornis*	Parasitic wasps
Arthropoda, Crustacea	*Jassa slatteryi*	Amphipod
	Parhyale hawaiensis	Amphipod
Arthropoda, Chelicerata	*Limulus polyphemus*	Horseshoe crab
	Tetranychus urticae	Spider mite
Arthropoda, Myriapod	*Strigamia maritima*	Centipede
Priapula	*Priapulus caudatus*	Priapulid worm
Tardigrada	*Hypsibius dujardini*	Water bear

Table 13.2   (*Continued.*)

Taxonomy	Organism	Common description
Nematoda, Chromadorea	*Ancylostoma caninum*	Canine hookworm
	Ascaris lumbricoides	Human intestinal roundworm
	Brugia malayi	Filarial roundworm
	Caenorhabditis brenneri, C. japonica	None
	Cooperia oncophora	Intestinal worm
	Dictyocaulus viviparus	Bovine lungworm
	Haemonchus contortus	Barber pole worm
	Heterorhabditis bacteriophora	None
	Necator americanus	New World hookworm
	Nematodirus battus	Thread necked worm
	Nippostrongylus brasiliensis	Rat intestinal nematode
	Oesophagostomum dentatum	Nodule worm
	Onchocerca volvulus	River blindness roundworm
	Ostertagia ostertagi	Stomach worm
	Strongyloides ratti	Threadworm
	Teladorsagia circumcincta	Brown stomach worm
	Trichostrongylus vitrinus	Black scour worm
Nematoda, Enoplia	*Trichinella spiralis*	Trichina worm
	Trichuris muris	Whipworm
Cnidaria, Hydrozoa	*Hydra magnipapillata*	Hydra
Porifera, Demosponge	*Reniera* sp.	None

2003); (7) details of genome-level processes, such as the rearrangements that generate antibody diversity (Frieder *et al.*, 2006); and (8) deviations from the 'universal' genetic code (Telford *et al.*, 2000; Santos, 2004). Many others are likely to be found.

Of course, the reliability of these features can only be assessed by study of their consistency with other characters, and several are already suspect. Convergent gene losses, for example, may be common as organisms independently evolve smaller genomes or no longer experience selection for maintaining a particular biochemical pathway; in contrast, convergent gain of genes seems much less likely. Independent evolution of smaller genomes may also lead to parallel losses of the most expendable structures in RNA or protein genes. There is a certain time-horizon that limits the usefulness of any particular type of character; for example, once retro-elements degrade in

sequence beyond the point where the insertion can be reliably inferred to be of single origin, the insertion is no longer useful as a phylogenetic character. Certain changes in the genetic code and in tRNA secondary structures of mitochondria are known to have occurred convergently (although occasional homoplasy has not disqualified the use of either morphological characters or molecular sequence comparisons). There is also the problem in the case of closely spaced sequential internodes where random partitioning of polymorphisms, including those of genome-level characters, can lead to incorrect inference of phylogeny (e.g. Salem *et al.*, 2003). See Boore (2006) for additional caveats and precautions.

Already there have been important insights gained from comparing such features, including: (1) tarsiers have been shown to be the sister group to the clade of monkeys and apes rather than the prosimians based on patterns of short interspersed nuclear element (SINE) integration (Schmitz *et al.*, 2001); (2) patterns of SINE and long interspersed nuclear element (LINE) insertions have also supported the monophyly of toothed plus baleen whales, that hippopotamuses are the sister group to cetaceans, that camels are the most basal cetartiodactyls (Nikaido *et al.*, 1999), and that river dolphins are paraphyletic (Nikaido *et al.*, 2001); (3) animal interphylum relationships have been clarified by comparisons of the gene membership within *Hox* clusters (de Rosa *et al.*, 1999); and (4) a study of the presence of spliceosomal introns supports the monophyly of Actinopterygia and clarifies several relationships within the group, including the basal position of bichirs (Venkatesh *et al.*, 1999). For further discussion see Murphy *et al.* (2004), Okada *et al.* (2004), and Boore (2006).

13.4 What are the advantages of using these genome-level characters?

In general, these types of features would be expected to change in a saltatory, non-clocklike manner. This may seem, at first, to be wrongheaded, since great effort has been expended in many studies to identify clocklike characters, to enable accurate molecular clock estimates of time of divergence. But it is this aspect that makes these genome-level characters especially useful for addressing the most difficult branch points, those with a short time of shared history followed by a long period of divergence, as mentioned above. It is for resolving these relationships that clocklike behaviour guarantees failure, since the ratio of signal to noise will closely match the ratio of the two time periods. Rather it is the least clocklike of characters that are expected to prevail, where an occasional and abrupt change may have occurred and then remain (Figure 13.2). Admittedly, the concomitant disadvantage is that many such characters must typically be examined in order to find those that happened to have changed during the period of shared ancestry and so marking the relationship (see Boore, 2006, for further analysis and discussion).

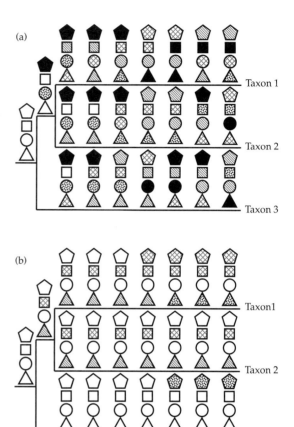

Figure 13.2 Illustration of why clocklike characters (a) may be less informative than non-clocklike characters (b) when the internode between subsequent lineage splits is short. Each of the four shapes is meant to be a character with states indicated by patterning. In (a) the circle and triangle are not informative and the square and pentagon are homoplasious. The two changes accumulated in the common ancestor of taxon 1 and 2 (for the pentagram and circle), that were at one point synapomorphies, have been erased by subsequent changes. In (b), the changes are rarer and saltatory. The pentagram and triangle are not informative and the circle is constant, but the square is informative for uniting taxa 1 and 2.

13.5 What about clades without representative genome sequences?

This enormous data set provides a new class of characters that could lead to definitive resolution of some branches of the tree of life, not only for these taxa but also for others where targeted searches for identified characters could be fruitful. As shown in Figure 13.1, whole-genome sampling will include many major lineages, but not all. Fortunately, we can use genomes in hand to identify sets of genome-level characters that can be diagnostic for the relationships of related groups without genome projects. One could then determine gene order by using Southern hybridization, for example, or probe a large DNA insert library (i.e. in bacterial artificial chromosome (BAC) or fosmid vectors) to find a clone to sequence for the region of interest of the genome. Gene rearrangements, losses, and duplications can also be identified using comparative genomic hybridization (CGH) chips with tiled large-insert clones, as has been done for a sampling of diverse human populations (Sharp *et al.*, 2005) and more broadly across the great apes (Locke *et al.*, 2003) or by using arrays of oligonucleotides (representational oligonucleotide microarray analysis (ROMA); Sebat *et al.*, 2004).

13.6 What are the main challenges before us?

First, we must increase the representation of understudied groups of animals for large-scale genomic sequencing. There is no reason to believe that taxa that have been traditionally studied intensively, i.e.

those with higher species richness, greater breadth of niche occupation, more important roles in pathogenesis, or amenability to laboratory experimentation, will be more informative toward the goals of understanding broad patterns of the evolution of animals and their genomes. Next, we need to have a codification of nomenclature for genes that is based on assessment of orthology (Dehal and Boore, 2006). The renaming of genes to indicate orthology is not feasible because it would render large bodies of literature difficult to interpret and because scientists who study model organisms (and who have largely done the naming) are invested in their parochial nomenclature. Thus, the solution must be a lexicon superimposed on these names already in place. Third, a system must be devised for codifying the genome-level characters themselves for entry into data bases and matrices for broad comparisons. Lastly, we need for the community to devise standards of interpretation and analysis, such as the use of cladistic reasoning rather than associating taxa by similarity alone (Boore, 2006). Then it seems likely that genome-level characters will provide the best data set for convincingly reconstructing relationships for some of the most hotly contended nodes in the tree of life and for establishing a framework for all organismal relationships.

CHAPTER 14

The animal in the genome: comparative genomics and evolution

Richard R. Copley

Comparisons between completely sequenced metazoan genomes have generally emphasized how similar their encoded protein content is, even when the comparison is between phyla. Given the manifest differences between phyla and, in particular, intuitive notions that some animals are more complex than others, this creates something of a paradox. Simplistic explanations have included arguments such as increased numbers of genes, greater numbers of protein products produced through alternative splicing, increased numbers of regulatory non-coding RNAs, and increased complexity of the *cis*-regulatory code. An obvious value of complete genome sequences lies in their ability to provide us with inventories of such components. Here I examine progress being made in linking genome content to the pattern of animal evolution, and argue that the gap between genome and phenotypic complexity can only be understood through the totality of interacting components.

Deus ex machina: 'A power, event, person, or thing that comes in the nick of time to solve a difficulty; providential interposition…'

Oxford English Dictionary

14.1 Introduction

Complete genome sequences provide limits to our imaginations. Even just a few years before the human genome was available in rough draft form, it was widely believed to encode at least 50,000 genes (Fields *et al.*, 1994; Editorial, 2000). In contrast, the initial publications estimated 25,000–40,000 protein-coding genes (Lander *et al.*, 2001; Venter *et al.*, 2001), and since then estimates have generally carried a downward momentum, most recently approaching 20,000 (Goodstadt and Ponting, 2006; Pennisi, 2007). Although this number is higher than the 16,000 or so found in invertebrate chordates (Dehal *et al.*, 2002) it is basically the same total as in the nematode worm *Caenorhabditis elegans* (Hillier *et al.*, 2005). Whether or not these low numbers of protein-coding genes for vertebrates stand the test of time, the sense of unease surrounding the lack of correlation between organismal complexity (often measured in number of distinct cell types) and protein-coding gene count is evident from the framing of the 'g-value paradox' by Hahn and Wray (2002), and the various explanations that have been put forward to ease it, including, for example, miRNAs (Sempere *et al.*, 2006, Heimberg *et al.*, 2008), non-protein-coding DNA (Taft *et al.*, 2007), and alternative splicing (Kim *et al.*, 2007).

Similar gene counts are, of course, a crude measure of biological complexity. There is no reason why two genomes should not encode very different sets of protein-coding genes, but still have similar overall totals. Within the field of animal evolution and the evolution of development (evo-devo), however, the g-value paradox has a particular resonance. Studies in different animal phyla have repeatedly shown the reuse of a core set of developmental genes, the so-called 'toolkit' (Carroll *et al.*, 2005), with the *HOX* genes, in particular, taking on an iconic significance. Broadly, toolkit genes come from a handful of transcription factor families, defined by the presence of particular structural domains such as the helix–turn–helix (HTH the class that includes, the homeobox genes), zinc fingers (ZnF), leucine zippers, and the helix–loop–helix (HLH). As well as transcription factors there are seven well-conserved pathways responsible for intercellular signalling (Pires-daSilva and Sommer, 2003), many of which appear to be present in sponges, the earliest branching clade of animals (Nichols *et al.*, 2006). An extreme interpretation of these data is provided by Davidson (2006): 'if we focus explicitly on the genes encoding transcription factors, and [...] signalling systems required for developmental spatial regulation, there is almost no qualitative variation among the genomes of bilaterians'.

Given all this, where in the genome do the phenotypic differences between animal taxa arise? The undoubted conservation of the protein-coding developmental genes has, particularly in the evo-devo field with its morphological concerns, focused attention on *cis*-enhancer elements affecting transcription (Carroll *et al.*, 2005; Davidson, 2006; Wray, 2007; Simpson, 2007), although there are alternative views emphasizing the importance of different kinds of regulatory elements (Alonso and Wilkins, 2005) and different protein classes, such as structural genes (Hoekstra and Coyne, 2007). Below I outline some major themes being developed by large-scale genome comparisons, principally of nematodes, insects, and vertebrates. My aim is not to present an exhaustive account, but to highlight areas where functionally relevant species-specific differences may arise, within apparently conserved systems. Although I concentrate on the evolution of the systems regulating animal development, this is not to lose sight of the things being regulated: the proteins involved in making nematode cuticles, or asynchronous flight muscles in insects, or the human brain and adaptive immune system, to name but a few, are what made it necessary to evolve those systems.

14.2 Gene duplication

Usefully summarizing the differences and similarities between more than 10,000 protein-coding genes from several species at once is not necessarily straightforward. Although pairwise similarities between sequences are easy to compute, they suffer from the imposition of arbitrary cut-offs and are less easy to interpret than measures that explicitly reflect phylogeny. Genes in different species are most obviously compared by grouping into sets of orthologues (that is, genes related by speciation events) and paralogues (genes related by intragenome duplication events). Closely related species share large numbers of orthologues: 93% of dog (*Canis familiaris*) and 82% of the marsupial *Monodelphis domesticus* gene predictions have orthologues in human (Goodstadt *et al.*, 2007). The Linnean hierarchy, however, is not necessarily a good guide of genomic relatedness by this definition of similarity. Within the nematodes 65% of *C. elegans* genes share an orthologue with *Caenorhabditis briggsae*, despite their being from the same genus (Stein *et al.*, 2003). For more distantly related genomes, orthologue counts can drop rapidly. This may be as much a sign of difficulties in reliably assigning gene phylogenies on a large scale as a real indication of the extents of the conserved cores.

Paralogues often arise via tandem duplication of genes, giving rise to localized clusters of functionally related genes. As these are the regions where gene content is evolving most rapidly between closely related species, the functions of these genes are of special interest for understanding animal-specific differences. For the most part, for any two closely related vertebrate genomes the functional classes of genes duplicated in this way are similar—olfaction and chemosensation, reproduction, and effectors of the immune response—although the duplications have occurred independently in each lineage (Emes *et al.*, 2003). These large groups of paralogues often show evidence of adaptive evolution in their amino acid sequences, suggesting that new functions have been selected for (Emes *et al.*, 2004a,b).

The recurrent nature of duplications within particular functional classes, coupled with the observed diversifying selection, suggests that they are a standard adaptive genomic response to environmental challenges. Does similar rapid duplication occur in the kinds of genes, such as transcription factors, that might be implicated in development? A growing number of examples are known. Perhaps most dramatically, in mice a set of 32 tandemly duplicated homeoboxes have arisen from apparently one or two genes in the common ancestor of humans and rodents; they are believed to play a role in germ cell development and embryonic stem cell differentiation (Maclean *et al.*, 2005; Jackson *et al.*, 2006).

Zinc finger-containing transcription factors have undergone independent rounds of gene duplication in insects and tetrapods. In insects a set of zinc fingers are found to co-occur with a zinc finger associated domain (ZAD) (Chung *et al.*, 2007); this ZAD class is found in around 100 copies in *Drosophila melanogaster* and and 150 copies in the mosquito *Anopheles gambiae*; there is only a single copy in vertebrates (Chung *et al.*, 2007). In *D. melanogaster*, many are expressed in the female germline, suggesting a role in oocyte development or embryogenesis (Chung *et al.*, 2007). An analogous story is found with Krüppel associated box (KRAB) containing zinc fingers in tetrapods. Successive independent tandem duplication events have occurred in different mammalian lineages, leading to more than 400 copies in the human genome (Huntley *et al.*, 2006). The KRAB domain itself appears to have been co-opted from a progenitor sequence conserved throughout eukaryotes (Birtle and Ponting, 2006); however, it has evolved so much as to make this similarity difficult to detect; clearly identifiable KRAB domains are specific to tetrapods. Their functions are largely unknown, and have not been tied to any general aspects of tetrapod biology. As such, why the family as a whole has expanded is a puzzle.

Nematodes too exhibit lineage-specific expansions of particular transcription factor families, most notably the nuclear hormone receptors (NHRs). The *C. elegans* genome encodes 284, many more than the 48 in human and 21 in *D. melanogaster*. The bulk of these (>200) have arisen from an apparently nematode-specific expansion of a unique gene (Lander *et al.*, 2001, Robinson-Rechavi *et al.*, 2005). Once more, the reasons for such a dramatic lineage-specific expansion of a particular transcription factor family, and any links to taxon-specific biology, are obscure, although it has been speculated that *C. elegans* relies less on combinatorial reuse of different transcription factors (Antebi, 2006). A less dramatic lineage-

specific expansion occurs in the case of the T-box-containing transcription factors: there are 21 in *C. elegans*, with 17 arising from a lineage-specific expansion when compared with *D. melanogaster* and humans. Ascertaining when and in which clades these *C. elegans* duplications took place is currently frustrated by a lack of relevant genome sequences. As a set these T-box genes map to several genomic locations, suggesting that they have arisen over a more protracted timescale than the examples discussed above; some, at least, have known roles in the development of *C. elegans* (Poole and Hobert, 2006).

14.3 The 'invention' of new genes

A number of gene families appear to be metazoan novelties, with no clear sequence similarity to other genes outside the Metazoa, but present in the more basal animal phyla such as cnidarians and sponges. These include key families involved in animal development, like T-box and SMAD transcription factors, and signalling molecules such as WNTs and FGFs (Putnam *et al.*, 2007). The most closely related non-metazoan eukaryote sequenced to date, the choanoflagellate *Monosiga brevicollis*, was reported to be missing true HOX, ETS, NHR, POU, and T-box class transcription factors, strongly suggesting their origin was co-incident with that of the metazoans (King *et al.*, 2008). Analysis of preliminary data from the sponge genome indicates that, although present, many of these gene families were much smaller in number prior to the divergence of sponges and cnidarians (Larroux *et al.*, 2008).

Was the invention of such families a prerequisite for the evolution of the Metazoa, and were analogous protein inventions required for the evolution of particular taxa, such as insects and vertebrates? At the level of three-dimensional structures (i.e. the protein fold itself), there is some reason to be sceptical that this is the case. In many cases, examination of similarities in three-dimensional protein structures shows that these genes have distant homologues in non-metazoan genomes. The MH1 (DNA-binding) domain of SMADs, for instance, is probably homologous to a family of homing endonucleases found in all kingdoms of life (Grishin, 2001); the T-box shares structural similarities indicative of homology with a variety of other transcription factors, such as STAT DNA-binding domains, which are found in other eukaryotes (Murzin *et al.*, 1995; Soler-Lopez *et al.*, 2004); and the signalling domain of metazoan hedgehog proteins shares detailed similarities with members of a family of bacterial peptidases, suggesting that they too are likely to be homologous (Murzin *et al.*, 1995). In these cases the novel families are likely to be cases of rapid sequence evolution, accompanying functional shifts, within stem lineages leading to the Metazoa. Sparse sequence sampling of non-fungal and metazoan eukaryotic genomes may contribute to the apparent co-origin of these protein domains with the animals.

As this type of domain evolution is occurring from pre-existing domain types, the process fits within a standard framework of accelerated point mutation and selection for new functions. The invention of the domain type is not a key innovation in itself; rather, it can be seen as the extension of functional diversification of subfamilies of the kind that is apparent when comparing more closely related species. The fact that so many new domain types are found to be co-incident with the origin of metazoans suggests that the selective

pressures giving rise to this kind of accelerated sequence evolution were greater in the metazoan stem lineage.

An example of a more recent domain innovation is found in the *Drosophila* gene *brinker*, which plays a key role in the establishment of dorsoventral patterning. Although the protein-coding sequence of its DNA-binding domain is well conserved in insects, using current sequence data bases it shows no significant sequence similarity to proteins from any other taxa (Figure 14.1 and Plate 9), although there is weak (non-significant) similarity to pogo-like transposases, and the structure, which is only folded when complexed with DNA, suggests similarity to various transcription factors (Cordier *et al.*, 2006).

14.4   Evolution of transcription factors: the animal in the orthologue

Lineage-specific duplication followed by sequence divergence provides one route to species-specific biology, but what scope is there for lineage-specific functional shifts within orthologous genes? In the absence of gene duplication, it is hard to imagine how the DNA specificity of a particular factor

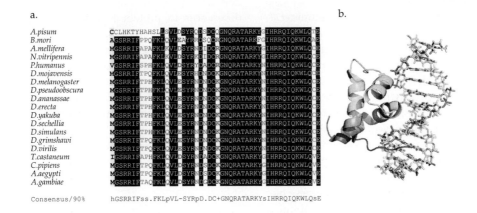

Figure 14.1   The DNA-binding domain of brinker is conserved within insects, but has no significantly similar sequences in other taxa. (a) The alignment shows the conserved core from a selection of insect species. Sequences of *Drosophila* species were taken from the UCSC web browser (http://genome.ucsc.edu/), *Anopheles* and *Aedes* from ENSEMBL (http://www.ensembl.org/), other predictions were made from sequences at the NCBI. GI accession numbers: *N. vitripennis* 146253130; *T. castaneum* 73486274; *C. pipiens* 145464888; *P. humanus* 145365328; *A. mellifera* 63051942; *B. mori* 91842977; *A. pisum* 47522326. (b) The three-dimensional structure of the aligned region when binding DNA. The structure was taken from the PDB file 2glo. (See also Plate 9.)

might be significantly changed, in such a way that it targets new genes, without deleterious consequences. The modular structure of proteins, however, suggests that other routes of functional evolution are available. A protein may have pleiotropic effects, but that is not the same as saying that every amino acid in the protein will be directly involved in all those effects. A recent illustrative example from the hox gene *Ultrabithorax*, is of an insect-specific 'QA' protein motif, found outside the homeodomain. The region is involved in limb repression; the effects of deleting the motif are strong in some tissues but close to undetectable in others (Hittinger *et al.*, 2005). Clearly, changes in the protein-coding sequences of transcription factors, apart from their more obvious DNA-binding residues, must be integrated into our understanding of the evolution of developmental regulation (Wagner and Lynch, 2008).

The majority of residues in metazoan transcription factors do not fall within regions of well-defined globular structure, with many belonging to socalled 'intrinsically disordered' regions—regions that may form a structure when complexed with other macromolecules (J. Liu *et al.*, 2006, Minezaki *et al.*, 2006). The specific sequences of these regions are typically not obviously conserved between paralogues; because they are unique to particular families they are not covered in domain data bases such as SMART and Pfam (Finn *et al.*, 2006; Letunic *et al.*, 2006). The lack of extreme conservation between distant species has sometimes masked the fact that, within closely related species, these regions are conserved. Comparisons of orthologous sequences from closely related genomes (e.g. vertebrates or drosophilids) often show that substantial proportions of these non-domain sequences are undergoing strong purifying selection—they accumulate many more synonymous nucleotide changes than non-synonymous changes—and are thus functional. For the large part, precisely what these biological functions are is unknown; two possibilities, however, suggest themselves. Firstly, they may have relatively uninteresting non-specific effects, such as facilitating folding of the major domain (for instance by reducing aggregation) or acting as spacers between globular domains. Secondly, and more interestingly from the point of view of

animal evolution, they may include short linear peptide motifs that mediate protein–protein interactions (Dyson and Wright, 2005; Neduva *et al.*, 2005; Neduva and Russell, 2005).

There are numerous examples of regulatory motifs found outside of transcription factor domains. Many hox proteins include a YPWM-like hexapeptide motif that interacts with other homeodomain-containing proteins (In der Rieden *et al.*, 2004); *Drosophila fushi tarazu* (*ftz*) orthologues have lost this motif but acquired an LXXLL motif coupled to a new role in segmentation (Lohr and Pick, 2005); and an N-terminal SSYF-like motif believed to be involved in transcriptional activation is conserved across hox orthologues and paralogues from different phyla (Tour *et al.*, 2005). Interaction motifs can be coupled with signalling pathways to create cell-type specificity. They can, for instance, be regulated by phosphorylation, such that the phosphorylation status governs what interactions can be made (e.g. Sapkota *et al.*, 2007), or alternative splicing can result in protein–protein interaction motifs being included or excluded from particular cell types, providing additional layers of regulatory complexity that are likely to be species specific (Neduva and Russell, 2005).

The challenge of identifying small regulatory motifs means that their species distributions, and how their presence might produce taxon-specific differences in protein functions, have not been well studied. Examples that tie cleanly to one taxonomic group are less common, but an interesting case has been proposed in bilaterian orthologues of the *Brachyury* gene. These possess an N-terminal motif that is not found in non-bilaterian Metazoa (Marcellini, 2006), which instead have a well-defined EH1-like motif (Copley, 2005). The bilaterian motif is believed to be responsible for an interaction with Smad1, and hence to link gastrulation to bilateral pattern formation (Marcellini, 2006).

14.5 Enhancers: transcription factor binding sites and ultraconserved regions

Theoretical considerations have led to an intense focus on transcription factor binding sites (TFBSs) as a major molecular source of morphological novelty

(Wray *et al.*, 2003; Carroll *et al.* 2005, Davidson, 2006; Wray, 2007; although see Hoekstra and Coyne, 2007, for a critique). Individual TFBSs show rapid turnover in comparisons of closely related genomes, with many being lineage specific (Dermitzakis and Clark, 2002; Moses *et al.*, 2006). This dynamic nature may not be revealed in the phenotype—patterns of gene expression may be conserved even though regulatory sequences change at the molecular level (Ludwig *et al.*, 2000; Romano and Wray, 2003; Fisher *et al.*, 2006). On the other hand, the gain and loss of individual TFBSs has been implicated in several recent cases of morphological evolution, in both vertebrates and invertebrates (reviewed in Wray, 2007, and Simpson, 2007). The relationship between individual TFBSs and enhancer function is clearly not straightforward, beyond the fact that clustering of individual binding sites can identify some enhancer regions (Markstein *et al.*, 2002). Cases of functional linkages between particular transcription factors have been proposed, for example, between Dorsal, twist, Su(H), and an unidentified motif in neurogenic ectoderm formation in Diptera (Markstein *et al.*, 2004), and even a coupling originating prior to the origin of Bilateria, of hairy and E(spl), promoting neural cell fate (Rebeiz *et al.*, 2005).

Comparisons of vertebrate genomes have revealed large regions (more than 100 nucleotides) of extreme conservation of non-coding sequences (conserved non-coding elements, CNEs) (Bejerano *et al.*, 2004). These regions are often found near transcription factors and other developmental genes (Sandelin *et al.*, 2004). Outside of the vertebrates, there is evidence for similar regions occurring near developmental genes in flies (Glazov *et al.*, 2005) and nematodes (Vavouri *et al.*, 2007). Although in many cases the conserved regions are even found near orthologous genes, there is no evidence that they are homologous; they appear to have evolved independently in each of the phyla (Vavouri *et al.*, 2007). Experimental evidence from vertebrates shows that many instances have roles as tissue-specific enhancer elements (Woolfe *et al.*, 2005, Pennacchio *et al.*, 2006).

The length, and lack of interphyla conservation of CNEs is in contrast to individual TFBSs. The DNA specificity of orthologous transcription factors is usually well conserved over large phylogenetic distances, but typical TFBSs are short, of the order of six to ten nucleotides. An obvious possibility is that longer CNEs are composed of overlapping or adjacent TFBSs. This would suggest a tight packing of transcription factor proteins on the genomic DNA of these CNEs. There is direct evidence for this: some fragments of highly conserved non-coding sequences are present in crystal structures of transcription factor complexes. An atomic model based on known crystal structures of the interferon-β enhancer, for example, shows 50 consecutive nucleotides in contact with eight different proteins; these nucleotides are well conserved in mammalian species (Panne *et al.*, 2007; see Figure 14.2 (also Plate 10) for another example). Given that such structures exist, it is not such a leap to imagine 16 proteins binding to 100 nucleotides, or even bigger complexes. This suggests a model where CNE enhancer regions controlling orthologous genes in different phyla are controlled by multiple transcription factor binding sites, although not necessarily the same transcription factors or in the same orientation. Moreover, the tight packing of transcription factors on the genomic DNA suggests that the proteins themselves may be co-adapted to interact with each other and aid the cooperative formation of enhancer complexes. Previously, Ruvinsky and Ruvkun (2003) have presented experimental evidence that enhancers and transcription factors co-evolve in this way, with neuronal and muscle-specific enhancer elements from *D. melanogaster* failing to drive expression in homologous tissue types in *C. elegans*, and Dover and co-workers (McGregor *et al.*, 2001; Shaw *et al.*, 2002) have argued for co-evolution of bicoid protein and *hunchback* regulatory regions. Wagner (2007) has proposed that the protein–protein interactions of co-adapted transcription factors may form the underpinnings of 'character identity networks'; that is, the gene regulatory networks that control the development of homologous morphological characters.

If protein–protein interactions between transcription factors are often required for the formation of enhancer complexes, close analysis of transcription factor sequence and structure may reveal evidence for co-adaptation of proteins, such as the HOX

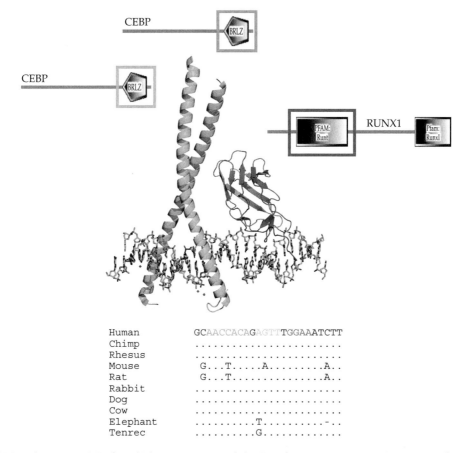

Figure 14.2 Adjacent transcription factor binding sites cause extended regions of DNA sequence conservation. Structure of CEBPβ homodimer and RUNX1 (Tahirov *et al.*, 2001). Three transcription factors (2× CEBPβ and RUNX1) bind in a region of 25 nucleotides conserved throughout placental mammals. The DNA-binding domains represented as three-dimensional structures are boxed and colour-coded in the schematic representation of the proteins. In each case, the majority of the protein is not represented in the structure; these regions could interact with other transcription factors, activators, and repressors. The human sequence coordinates are chromosome 5, bases 149,446,373–149,446,396 of the NCBI build 36. The alignment is taken from the UCSC web browser (http://genome.ucsc.edu/). (See also Plate 10.)

hexapeptide motif, through which homeotic proteins form complexes with TALE class homeodomains (LaRonde-LeBlanc and Wolberger, 2003). We might expect instances of co-adapted transcription factor combinations to be taxon specific, to match the taxon specificity of enhancer sequences.

14.6 Alternative splicing

Not all CNEs are associated with enhancer regions. There is good evidence that many are involved

in regulating alternative splicing events, including the alternative splicing of mRNAs of proteins which themselves regulate alternative splicing (Lareau *et al.*, 2007; Ni *et al.*, 2007). The presence of highly conserved control elements to regulate alternative splicing indicates that the functional consequences are of importance. Although large very conserved elements may be the exception rather than the rule, detailed comparative analyses have identified smaller conserved motifs regulating alternative splicing, for instance in nematodes

(Kabat *et al.*, 2006) and vertebrates (Sorek and Ast, 2003; Yeo *et al.*, 2005).

Alternative splicing is often touted as a mechanism by which proteomic complexity is increased. Although early reports suggested that levels of alternative splicing were comparable in vertebrates and invertebrates (Brett *et al.*, 2002), more recent studies suggest that there is indeed more alternative splicing of transcripts in vertebrates (Kim *et al.*, 2007), suggesting a link with increased phenotypic complexity. How relevant is alternative splicing for species-specific biology and morphological differences? Quantitatively, the gene products that appear to be most affected by alternative splicing are typically involved in the functioning of the nervous and immune systems (Modrek *et al.*, 2001). There are, however, ample examples of alternatively spliced transcription factors—as many as 63% of mouse transcription factors have variant exons (Taneri *et al.*, 2004). Although the differences in molecular roles of the alternatively spliced products are often unknown, the genes themselves include developmental classics such as members of Hox, SMAD, and T-box families (Noro *et al.*, 2006; Fan *et al.*, 2004; Dunn *et al.*, 2005;), although they do not necessarily present obvious morphological correlates (Yoder and Carroll, 2006). Alternative splicing of modular proteins is an obvious route through which functions can be changed by including or excluding particular combinations of domains. In this regard, it is interesting that alternative splicing often affects intrinsically disordered regions outside known protein domains (Romero *et al.*, 2006)—this again points to a critical role for finely tuned protein–protein interactions among transcriptional regulators.

There are few known cases of distant conservation of alternative splice variants of transcription factors; typically, examples are conserved within phyla at best. Widening the search to other classes of gene again suggests that splice variants are not conserved over long periods, although it should be remembered that transcript coverage of most species, from which evidence of alternative splicing is obtained, is very restricted. Perhaps the best example is currently that of fibroblast growth factor receptor 2 (FGFR2), where an exon configuration diagnostic of mutually exclusive alternative splicing is found in both vertebrates and the sea urchin *Strongylocentrotus purpuratus* (Mistry *et al.*, 2003). Examples of orthologous ion channel encoding genes showing similar alternative splicing patterns in *D. melanogaster*, *C. elegans*, and humans are likely to be cases of parallel evolution (Copley, 2004). The shared ability of vertebrates and at least insects and *C. elegans* to produce alternative transcripts in a regulated manner, alongside the absence of large numbers of conserved alternative splicing between protostomes and deuterostomes, suggests that gene products have become alternatively spliced in parallel between different lineages, while at the same time hinting that the functions performed by alternative splice variants may, over time, be replaced by different genomic solutions.

14.7 Summary

Key genetic innovations, such as alternative splicing, the invention of hox genes or the advent of micro-RNAs, have held a strong appeal for those seeking to explain animal evolution in terms of genomes. Without denying the importance of such phenomena, a more nuanced outlook is preferable. Much of the molecular complexity found in animals could have its origins in non-adaptive processes attributable to small population sizes (Lynch, 2007a,b), but this complexity may then be exploited in the service of phenotypic adaptation (Lynch, 2007a) within a framework of point mutation and selection.

Although most major classes of protein involved in animal development may be conserved throughout the Metazoa, detailed comparative analysis of these gene types reveals a more dynamic picture, with frequent gene duplication, gene loss, couplings with new motifs, and other processes such as alternative splicing and regulation by micro-RNAs, all of which are likely to be important for a full understanding of function. *Cis*-regulatory variation may well be revealed to be quantitatively the most common form of variation between species, but it seems likely that the cumulative effects of multiple *cis*-regulatory changes will have required that protein networks evolve to accommodate and correctly regulate changed enhancer structures.

Our knowledge of animal evolution and the picture presented here is currently based on a very small sampling of almost exclusively nematode, insect, and vertebrate genomes. Although this situation is beginning to change, the fact that many important functional regions, especially those that do not encode proteins, are only revealed by having sets of closely related genome sequences, and that there are 35 or so animal phyla, gives some idea of the huge scale of the challenges ahead. The rapidly falling costs of genome sequencing do, however, give grounds for optimism.

CHAPTER 15

MicroRNAs and metazoan phylogeny: big trees from little genes

Erik A. Sperling and Kevin J. Peterson

Understanding the evolution of a clade, from either a morphological or genomic perspective, first and foremost requires a correct phylogenetic tree topology. This allows for the polarization of traits so that synapomorphies (innovations) can be distinguished from plesiomorphies and homoplasies. Metazoan phylogeny was originally formulated on the basis of morphological similarity, and in some areas of the tree was robustly supported by molecular analyses, whereas in others it was strongly repudiated. Nonetheless, some areas of the tree still remain largely unknown, despite decades, if not centuries, of research. This lack of consensus may be largely due to apomorphic body plans combined with apomorphic sequences. Here, we propose that microRNAs (miRNAs) may represent a new data set that can unequivocally resolve many relationships in metazoan phylogeny, ranging from the interrelationships among genera to the interrelationships among phyla. miRNAs, small non-coding regulatory genes, show three properties that make them excellent candidates for phylogenetic markers: (1) new miRNA families are continually being incorporated into metazoan genomes through time; (2) they show very low homoplasy, with only rare instances of secondary loss, and only rare instances of substitutions occurring in the mature gene sequence; and (3) they are almost impossible to evolve convergently. Because of these three properties, we propose that miRNAs are a novel type of data that can be applied to virtually any area of the metazoan tree, to test among competing hypotheses or to forge new ones, and to help finally resolve the correct topology of the metazoan tree.

15.1 Introduction

Since the dawn of molecular phylogenetics, the relationships between animal groups, from species to the deepest nodes in Metazoa, have been the domain of ribosomal, mitochondrial, and nuclear protein-coding genes. Orthologous genes are amplified and sequenced, the sequences are aligned, and the alignment is analysed with increasingly sophisticated phylogenetic algorithms to gain an estimate of relationships. It is unarguable that our understanding of metazoan phylogeny has progressed through the use of these genes and the application of standard phylogenetic methods. Many relationships originally proposed on morphological grounds have been confirmed, while others, such as the grouping of annelids and arthropods as the Articulata, have been strongly refuted, leading to a new understanding of morphological evolution (Eernisse and Peterson, 2004; Halanych, 2004). However, many areas of the metazoan tree have remained recalcitrant, yielding trees with low statistical support or with little resemblance to any credible scenario of morphological evolution. It has often been assumed that these problems would disappear as more data (i.e. more genes and/or more taxa) were applied to the questions at hand. A sampling of the literature on multigene phylogenetics demonstrates that this has not been the case. Indeed, despite the fact that the amount of sequence data in public data bases such as NCBI's GenBank doubles every 10 months, many phylogenetic questions remain as intractable today as they were before the advent of molecular systematics. As just one example, from one of our parochial areas of interest, the interrelationships among three lophotrochozoan phyla, the nemerteans, annelids, and molluscs, still remain effectively unknown. This is despite a number of multigene studies that have recently been published including complete 18S + 28S ribosomal RNA genes (Passamaneck and Halanych, 2006), complete mitochondrial genomes (Yokobori *et al.*, 2008), multiple PCR-amplified nuclear housekeeping genes (Helmkampf *et al.*, 2008a; Peterson *et al.*, 2008), and expressed sequence tag (EST) studies (Dunn *et al.* 2008; Struck and Fisse, 2008), with all three possible arrangements of these three phyla being advocated by at least one data set (Figure 15.1).

Three problems have always plagued (and will forever plague) the field of molecular phylogenetics: differential rates of molecular evolution, long internodes caused by a recent origin of the crown group, and fast, deep radiations. Indeed, in our reading of the metazoan phylogenetic literature, a large number of questions were robustly answered in the first or second pass using 18S rDNA, and were then largely confirmed using other types of data and/or algorithms. But the remaining nodes, which are usually hampered by at least one of these three problems, have remained largely intractable despite the ever-increasing number of taxa and genes being applied. Because of this, we believe that it is not more of the same data that will

ultimately answer these questions, but new types of data.

It was originally hoped that large-scale genomic changes, such as gene rearrangements in mitochondrial genomes (Boore *et al.*, 1998) or insertion–deletion events, retroposon integrations, or gene duplications in nuclear genomes (Rokas and Holland, 2000) would provide this new data set, and provide a complementary approach to sequence-based phylogenetic estimation. These sources have provided robust support to topologies previously identified in sequence-based phylogenetic studies, notably the placement of phoronids and brachiopods within the Protostomia (Helfenbein and Boore, 2004) in the case of mitochondrial gene order, or resolution of the whale–hippo clade by retroposon analysis (Shimamura *et al.*, 1997; Nikaido *et al.*, 1999) in the case of nuclear genome changes. Ultimately, however, these structural changes have not been the panacea it was hoped they would be. The most comprehensive coding of mitochondrial gene order demonstrated that in some clades, such as the vertebrates, rearrangement was too slow or non-existent, leading to a polytomy, whereas in other clades, such as the molluscs, rearrangement was too fast, leading to a nonsensical tree (Fritzsch *et al.*, 2006). Mutational decay of the flanking regions surrounding retroposons makes their detection, at best, difficult in taxa that diverged more than about c. 50 million years ago (Ma), restricting their utility

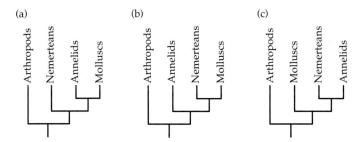

Figure 15.1   The three proposed hypotheses for the interrelationships among nemerteans, annelids, and molluscs with respect to arthropods. (a) The Neotrochozoa hypothesis (Peterson and Eernisse, 2001) posits that annelids and molluscs are each other's closest relatives with respect to nemerteans, found using morphological characters (Peterson and Eernisse, 2001) as well as an analysis using concatenated amino acid sequences of nuclear protein-coding genes (Peterson *et al.* 2008). (b) Mollusca + Nemertea was found with concatenated amino acid sequences of nuclear protein-coding genes (Hausdorf *et al.*, 2007; Helmkampf *et al.*, 2008a; Struck and Fisse, 2008), as well as concatenated amino acid sequences of mitochondrial protein-coding genes (Yokobori *et al.*, 2008). (c) Annelida + Nemertea was found using combined 18S rDNA + 28S rDNA by Passamaneck and Halanych (2006) and concatenated amino acid sequences of nuclear protein-coding genes (Dunn *et al.*, 2008).

to the post-Mesozoic portions of the phylogeny (Luo, 2000; Rokas and Holland, 2000). Moreover, the utility of these rare events has been hampered by their very rarity; although in some fortuitous examples a strong synapomorphy is captured, they are simply not present in sufficient numbers such that investigators can reliably base a research program around using them to test hypotheses concerning metazoan interrelationships.

The ultimate problem in resolving evolutionary relationships is homoplasy: similarity caused not by shared ancestry but by convergent evolution, or loss and reversion to the primitive condition. Homoplasy occurs in every data set, from examples like the torpedo shape of fish, whales, and ichthyosaurs, to the rapid gene rearrangements in molluscan mitochondrial genomes, to multiple substitutions or convergent changes in gene sequences that are limited to only four nucleotide or 20 amino acid character states. Both elevated sequence evolution in some taxa and long internodes cause homoplasy in molecular data sets, causing informative synapomophies to be eroded to misleading homoplasies. The key to resolving intractable nodes will lie in data sets that minimize homoplasy as much as possible, but whose characters arise or change at a high enough rate that they record the divergences in question. In this paper, we propose that short, highly conserved genes within the non-coding portion of the genome, specifically miRNA, may be one such data set. Not only is it almost impossible to evolve the same miRNA twice independently, but miRNAs are continually being added to metazoan genomes through time; only rarely are they secondarily lost, and nucleotide substitutions to the mature gene sequence are infrequent. Importantly, ascertaining the miRNA complement of a taxon does not require any prior knowledge of the miRNA sequences themselves, greatly facilitating their utility for attacking phylogenetic questions at all scales of the animal tree, from phyla to species.

15.2   Background

Briefly, miRNAs are small, *c.* 22 nucleotides, non-coding genes that negatively regulate protein-coding genes by binding, with imperfect complementarity, to sites in their 3′ untranslated regions (UTRs), thereby subjecting the transcript to cleavage or to blockage of its translation (Zhao and Srivastava, 2007; Filipowicz *et al.*, 2008; Hobert 2008; Stefani and Slack, 2008). The first retrospectively recognized animal miRNA, *lin-4*, was discovered in the nematode worm *Caenorhabditis elegans*, where it is involved in regulating the timing of cell cycle division in the larval worm by binding to target sites in the protein-coding gene *lin-14* and preventing its translation (Lee *et al.*, 1993; Wightman *et al.*, 1993). Because *lin-4* could not be found outside of nematodes, it was considered to be a quirk of the nematode developmental process. The wider significance of this discovery came when it was shown that this type of gene regulation exists in other systems, particularly vertebrates (Ruvkun *et al.*, 2004; Wickens and Takayama, 1994). This occurred with the finding of a second miRNA, *let-7*, which was originally discovered again in *C. elegans* (Reinhart *et al.*, 2000), but was soon found in numerous other taxa including fruitflies and vertebrates (Pasquinelli *et al.*, 2000), and quickly led to the discovery of many small regulatory RNAs subsequently named miRNAs (Lagos-Quintana *et al.*, 2001; Lau *et al.*, 2001; Lee and Ambros, 2001). *let-7* had three intriguing characteristics that held promise for a future role in phylogenetic reconstruction for these small RNA genes (Pasquinelli *et al.*, 2000). First, the mature gene product of *let-7* is unchanged in sequence between nematodes, humans, and *Drosophila*, despite a total of almost 2000 million years of independent evolution in these three taxa. Second, *let-7* was found in every protostome and deuterostome analysed, with no suggestions of secondary loss. Third, the gene was not present in any non-metazoan genome and was not detectable by Northern analysis in sponges or cnidarians, suggesting that the gene arose within Eumetazoa at the base of the nephrozoan triploblasts (i.e. protostomes and deuterostomes). Subsequent studies confirmed this pattern for *let-7* as the gene was found in, for example, chaetognaths, nemerteans, and polyclad and triclad flatworms, but not in acoel flatworms or ctenophores (Pasquinelli *et al.*, 2003).

miRNAs are defined by their mode of biogenesis, which is intimately related to their unique hairpin secondary structure (Figure 15.2) and not

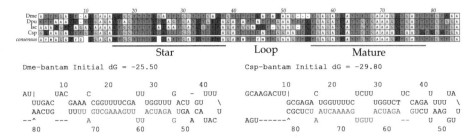

Figure 15.2 Alignment and secondary structure of representative sequences of the miRNA *bantam*. The mature sequence of the *Drosophila melanogaster* (Dme) *bantam* gene (miRBase) was used as a query against the trace archive sequences of *Daphnia pulux* (Dpu), *Ixodes scapularis* (Isc), and *Capitella* sp. (Csp) using the default settings (see Wheeler *et al.*, 2009). About 85 nucleotides of the best hits were then folded using mfold (Zuker *et al.*, 1999). Shown at the top is an alignment of these best hits using the default settings of ClustalW (MacVector, version 9.5.2), and shown below are the structures of two of these sequences, *D. melanogaster* (left) and *Capitella* (right) as determined by mfold. The mature and star sequences are shown.

by their specific nucleotide sequence. There are two components to a miRNA, the mature gene product, which is what binds to the 3′ UTRs of target genes, and the star sequence, the complement of the mature sequence, which is often degraded but is sometimes used as a gene product as well. miRNAs, which can be located either in intergenic regions or in introns, are transcribed as long primary transcripts that are capped and polyadenylated in typical Pol II fashion. However, because of the complementarity and spacing of this complementarity, the primary miRNA transcript folds into a hairpin structure, which is recognized by an enzyme complex involving at least two proteins, Drosha and Pasha, which cleave the pro-RNA into a *c.* 70 nucleotide precursor miRNA (Kim, 2005). This pre-miRNA is then exported into the cytoplasm where it is further processed by another RNAse enzyme called Dicer, and the mature gene product is then incorporated into an RNA–protein moiety that serves as the repressive entity with respect to messenger RNA translation and/or stability. Hence, miRNA biogenesis relies solely on miRNA structure and not on miRNA sequence *per se*, greatly facilitating their utility for phylogenetics because it obviates the need for a researcher to know any particular miRNA sequence (see below).

miRNAs are named in sequential order of discovery, with identical or near identical mature sequences in the same or different organism given the same number (Ambros *et al.*, 2003). miRNAs

given different numbers have different primary sequences and are assumed to have arisen independently of other named miRNAs. This can be shown using a standard maximum parsimony analysis. If the first 20 miRNAs listed for both the fly *Drosophila melanogaster* and the human are aligned (Figure 15.3, left), and analysed using bootstrap analysis (Figure 15.3, right; see legend for details) one can easily see the orthology between similarly named miRNAs in the fly and human (e.g. *let7*, miR-1), and the paralogy of similarly named miRNAs in each taxon (e.g. miR-2). Further, the unique nature of each numbered miRNA or groups of miRNAs is readily apparent as they share virtually no similarity with any other miRNA in the alignment (Figure 15.3, left) and do not cluster together in the bootstrap analysis (Figure 15.3, right).

However, phylogenetic analyses are rarely, if ever, used to help name miRNAs, and thus nomenclature problems can and do arise. For example, the two copies of miR-13 group with miR-2 (Figure 15.3, right), not unexpected given a cursory look at the alignment (Figure 15.3, left), and hence there are five, not three, copies of miR-2 in the fly genome. Even worse is when the same miRNA is given different names in different organisms. For example, Sempere *et al.* (2006) reconstructed the protostome-specific set of miRNAs to include miR-8, and the deuterostome-specific set of miRNAs to include miR-141 and miR-200, and part of the reason for this was that the seed sequences of these genes, which

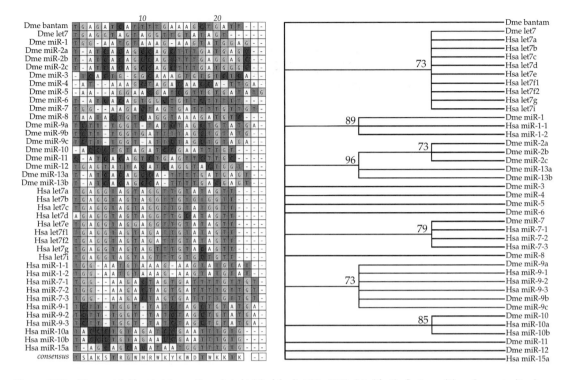

Figure 15.3 Alignment (left) and phylogenetic analysis (right) of the first 20 miRNAs listed for the fly *Drosophila melanogaster* (Dme) and the human *Homo sapiens* (Has) in miRBase. The alignment used the same parameters as Figure 15.2. Right: The 70% bootstrap tree. Sequences were analysed by PAUP, version 4.0b10 (Swofford, 2002) using maximum parsimony. Nodes found less than 70% of the time (1000 replications) were collapsed into polytomies. Note that similarly named miRNAs in the two systems cluster together, as do obvious paralogues in each system (e.g. *let-7*, miR-1). Note also that some differently numbered miRNAs (e.g. miR-2 and miR-13) group together as well, and as such constitute miRNA families, similar to, for example, the *let-7* family or miR-1 family.

are positions 2–8 of the mature gene products, were slightly different (Figure 15.4, left). And because the seed sequence is the most important area of the mature gene sequence for target recognition it is primarily used for family-level classification (Filipowicz *et al.*, 2008). But the use of only the seed sequence to name (and hence classify) miRNAs is a functional rather than a phylogenetic distinction, and in this case it is clear from a bootstrap analysis (Figure 15.4, right) that this is the same gene family, with the protostome versions called miR-8, and deuterostome versions called miR-141/200.

15.3 miRNAs as phylogenetic characters

miRNAs show three characteristics that make them outstanding candidates to arbitrate among competing phylogenetic hypotheses and to forge new ones: (1) new miRNA families are continuously being added to metazoan genomes through time; (2) once incorporated into a gene regulatory network, there are only rare instances of secondary gene loss and they show only rare nucleotide substitutions to the mature gene product; and (3) there is an infinitesimally small chance that miRNAs with the same mature sequence will evolve more than once.

15.3.1 Continuous addition of miRNA families to metazoan genomes

Sempere *et al.* (2006) showed that the miRNA repertoires of both fly and human were added sequentially through time such that each node leading to the fly, or to the human, could be characterized by

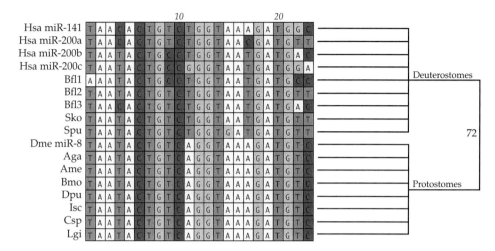

Figure 15.4   Homologous, but differently numbered, miRNAs in protostomes and deuterostomes. miR-8 in protostomes is clearly the same miRNA as miR-141 and miR-200 in deuterostomes, and indeed is supported as such in a 70% bootstrap analysis (right), despite some chordate paralogues possessing changes in the seed sequence (nucleotides 2–8) with respect to miR-8. Specifically, the human miR-141 and miR-200a, and the third paralogue found in the genomic traces of the cephalochordate *Branchiostoma floridae* (Bfl3), have T to C changes in position 4 (left). Other abbreviations: Sko, *Saccoglossus kowalevskii*; Spu, *Strongylocentrotus purpuratus*; Aga, *Anopheles gambiae*; Ame, *Apis mellifera*; Bmo, *Bombyx mori*; Lgi, *Lottia gigantea*.

a distinctive miRNA or set of miRNAs. To explore this further, we again traced the phylogenetic history of 132 uniquely numbered *D. melanogaster* miRNAs (see Heimberg *et al.*, 2008, for vertebrates), but this time used many more genomes combined with 454 sequencing of small RNA libraries (Wheeler *et al.*, 2009). We chose the arthropod example for three reasons. First, the miRNA repertoire of *D. melanogaster* is the most extensively studied of any model organism (Ruby *et al.*, 2007; Stark *et al.*, 2007a,b), and we can be confident we are examining almost every miRNA in the organism. Second, there are a large number of arthropod genomes available, including 12 from the genus *Drosophila* alone, allowing us to trace the phylogenetic acquisition over a range of taxonomic scales. And third, and most importantly for testing this data set for phylogenetic utility, there is an accepted phylogeny, allowing us to map miRNA gain (and losses, see below) against a known topology (Stark *et al.*, 2007a).

Figure 15.5 shows the phylogenetic history of all 132 *D. melanogaster* miRNAs considered, as well as the ancient miRNAs that should be present in *Drosophila*, as determined by Wheeler *et al.*, (2009),

but have been secondarily lost. Where we identify the gain of a new miRNA, it is shown in black under the node with paralogues of previously existing miRNA families underlined (see figure legend for details). Importantly, every node since the divergence between *D. melanogaster* and demosponges, where a sequenced genome is available and/or where a miRNA library has been constructed and sequenced (e.g. Priapulida; Wheeler *et al.*, 2009), is characterized by the addition of at least one novel miRNA. This often involves the innovation of new families (e.g. miR-2 at the base of Protostomia), but sometimes additionally involves the generation of a paralogue from an existing gene (e.g. miR-13 at the base of Ecdysozoa). Hence, miRNAs could be used to resolve the interrelationships of taxa at virtually every level in the taxonomic hierarchy, from species to phyla.

We emphasize that we are showing the phylogenetic history of the *D. melanogaster* miRNAs because they are well known and because the large number of genomes available enables such a study through bioinformatics alone. This does not imply that the other terminal tips will not have a similar number of miRNAs; groups such

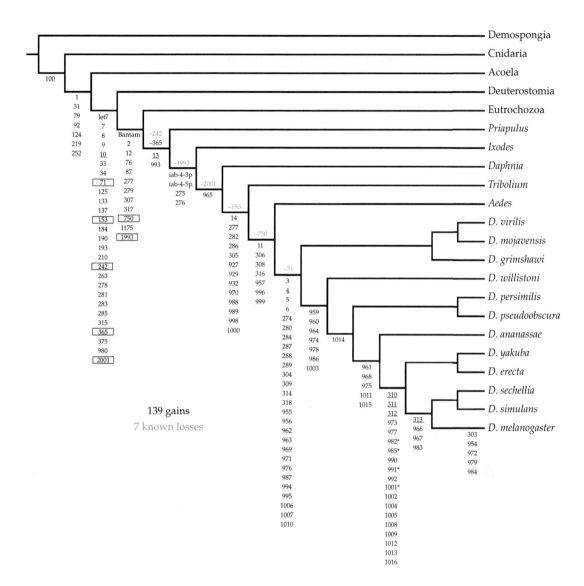

Figure 15.5   Gains and secondary losses of 132 differently numbered miRNAs in *Drosophila melanogaster*. Gains are shown in black below the node, and the seven secondary losses are shown above the node in grey (and where they were originally acquired are shown boxed below the node). Underlined miRNAs are paralogues of previously acquired miRNAs; those that are starred have slightly different seed sequences in other species of *Drosophila*, but fold properly and hence are considered gains at that point on the tree. This figure only considers gains and losses on the lineage leading to the single terminal *D. melanogaster* and does not consider those leading to other terminals, although groups like beetles or chelicerates will clearly have their own sets of clade-specific miRNAs. Results from Demospongia to *Daphnia* are taken from Wheeler *et al.* (2009); the trace archives of the remaining insects were searched using all *D. melanogaster* miRNAs as query sequences. Potential hits were folded using the program mfold and assessed using standard structural criteria (see Wheeler *et al.*, 2009, for materials and methods).

as mosquitoes or chelicerates will have their own clade-specific set of miRNAs that (we suspect) can (and hopefully will) be used to ascertain their internal phylogenetics. Indeed, Wheeler *et al.*, (2009) showed that each major lineage of metazoans, except for Deuterostomia, could be characterized by the acquisition of at least one novel miRNA family. For example, ambulacrarians were characterized by the addition of five novel miRNA families, and eleutherozoan echinoderms were characterized by the addition of 10 novel miRNA families. Even cnidarians have a novel miRNA family found only in *Hydra* and *Nematostella* and not anywhere else in the animal kingdom. And within these groups, the hemichordate worm *Saccoglossus kowalevskii* has at least an additional 34 miRNAs not found in the two echinoderms analysed, the sea urchin *Strongylocentrotus purpuratus* and the starfish *Henricia sanguinolenta*, and the hydrozoan cnidarian *Hydra* has at least an additional 17 miRNAs not found in *Nematostella* (Peterson *et al.*, unpublished data). These novel miRNAs could then be used as phylogenetic markers to explore hemichordate and hydrozoan interrelationships, respectively, assuming they show low homoplasy. Fortunately, if they are similar to virtually all other known miRNAs, this will indeed be the case.

15.3.2 Minimal secondary gene loss and rare substitutions to the mature sequence

miRNA homoplasy results from the possible combination of two factors, the first being the convergent evolution of the same miRNA in two taxa. The second is either complete loss from the genome, or nucleotide substitutions in the mature sequence that destroy the ability to recognize its true orthology. As argued below, independent evolution of miRNAs is extremely limited, but secondary gene loss and substitutions to the mature sequence can and do occur, and could obscure not only the interrelationships among the miRNAs but among the animal taxa as well. Nonetheless, in the *Drosophila* example discussed above (Figure 15.5), there are only seven secondary losses in *D. melanogaster* as compared with 139 gains—these losses are shown in grey in Figure 15.5 with their point of origin shown below the node and their inferred location of loss above the node. Note that loss can occur at any point in the evolutionary history—two of the genes not present in the fly were lost at the base of Ecdysozoa (miR-242 and miR-365), whereas one gene (miR-71) was lost in *Drosophila* after this lineage split from *Aedes* but before the

diversification of the 12 species under consideration (Figure 15.5).

If it could be shown that for most metazoan taxa miRNA gains outnumber miRNA losses by over an order of magnitude, as they do in this example, then their utility as phylogenetic markers would be unsurpassed, assuming that the mature sequence does not degrade over time. Sempere *et al.* (2006) argued that this was indeed the case, otherwise it would not be possible to map the origin of these 139 miRNAs with such minimal numbers of secondary losses, as in this example (Figure 15.5). Further, Sempere *et al.* (2006) showed that miRNAs were some of the most, if not *the* most, conserved genetic elements in the genome, with most fly and eutherian mammal miRNAs showing no substitutions to the mature sequence. But because their focus was necessarily on flies and vertebrates, it could be argued these evolutionary patterns were particular to flies and vertebrates. Subsequently, Wheeler *et al.* (2009) quantified the number and position of substitutions of all 93 shared miRNAs across 14 nephrozoan taxa, and because this study relied primarily on isolating mature sequences in small RNA libraries it was not biased towards finding only conserved miRNAs. These authors analysed 16,729 nucleotides and showed that the substitution rate of all known and novel miRNAs across these 14 taxa, whose independent evolutionary history spans over 7800 million years, is only 3.5% (567 total substitutions). When compared with 18S rDNA, one of the most conserved genes in the metazoan genome, this rate is impressively slow: aligning 18S rDNA from the same 14 taxa and removing the unalignable regions using Gblocks resulted in a substitution rate of 7.3% (Wheeler *et al.*, 2009). Hence, miRNAs evolve more than twice as slowly as the most conserved positions in a gene that is often used for reconstructing the deepest nodes in the tree of life.

15.3.3 Exceedingly small probability of the independent evolution of the same miRNA

In terms of convergent evolution, each unique 22-nucleotide sequence occurs by chance once for every 1.76×10^{13} nucleotides (4^{22}), or once for every 5864 human-genome-sized chunks of DNA queried. However, this is not an accurate estimate of the chances of two miRNAs evolving twice independently. For example, we took the (arbitrarily chosen) protostome-specific *bantam* miRNA gene (see Figure 15.2) from *D. melanogaster* and searched both protostomes and deuterostome genomes for this sequence in taxa that diverged from one another at least 500 Ma (Figure 15.7). In no case was the very same 23-nucleotide sequence found in any of these genomes (and aside from hits to *D. melanogaster* it is not found in the nucleotide data base deposited at GenBank, which consisted of 24,006,283,182 letters as of June 2008). Nonetheless, 23-nucleotide sequences were found in the two arthropods, the water flea *Daphnia* and the tick *Ixodes*, that are identical to each other but that differ from that of *D. melanogaster* by a single nucleotide at position 11. A single sequence was also found in the genomic traces of the sea urchin *S. purpuratus* that differs from the fly *bantam* sequence by a single nucleotide, at position 13; the best hits in all of the remaining deuterostomes have numerous differences, many of which are distributed in positions 2–6. The putative orthologue of *bantam* in the polychaete annelid *Capitella* shares the same nucleotide at position 11 as the water flea and the tick, but differs from all of the arthropods at positions 17, 20, and 23. Because clearly orthologous miRNAs often differ by two or three nucleotides, rather than computing the probability for 23 nucleotides, a more appropriate calculation is for the occurrence of a stretch of 19 nucleotides, which is expected every 2.75×10^{11} bases, or once in every 91 human-genome equivalents, with the possibility of a few nucleotide substitutions (see below).

On the other hand, these numbers are deceptively low because there are more constraints on a miRNA than the mature sequence of 22 nucleotides; it must also fold with a free energy value lower than about –20 kcal/mol and often lower than –25 kcal/mol. In addition, the spacing has to be such that the mature sequence, which has to be located in one of the two hairpin arms, occurs within about two nucleotides from the loop, with the entire pre-miRNA generally being from 60–80 nucleotides long. Further, the structure cannot

contain large, and in particular asymmetrical, internal loops or bulges (Ambros *et al.*, 2003). Thus, if one compares the two *bantam* miRNA sequences from *Drosophila* and the annelid *Capitella* it is obvious that these are real miRNA genes; they have the requisite free energy values and structure to be processed and thus function as bona fide miRNA genes (Figures 15.2 and 15.6). But when the deuterostome sequences are folded *in silico*, it is readily apparent that none of these are miRNAs, let alone orthologues of *bantam*. In the hemichordate (Sko), amphioxus (Bfl), and lamprey (Pma, see Figure 15.6) the free energy of these sequences are extremely high, *c.* –8 kcal/mol. In both the zebrafish (Dre, Figure 15.6) and ascidian (Cin) they have relatively low free energy values (*c.* –22 kcal/mol), but in both cases large and asymmetrical bulges and loops are present. Finally, in the sea urchin (Spu), which has the highest nucleotide similarity with the protostome sequences, not only is the free energy too high (–15 kcal/mol), but it too has large and asymmetrical bulges (Figure 15.6).

These non-folds are consistent with the observed substitution profile of the mature miRNA sequence as revealed by Wheeler *et al.* (2009). These authors found that most substitutions occurred at the 3' end of the mature sequence, but other regions of the gene, especially nucleotide 1 and nucleotide 10, showed a relatively high percentage of substitutions. Importantly, the two most infrequent places for substitutions to occur are the seed region (positions 2–8) and the 3' complementarity region spanning nucleotides 13–16, especially position 15, in concordance with the hypothesized importance of these two regions for base pairing with the 3' UTR of targets (Filipowicz *et al.*, 2008). Thus, unlike the protostome substitutions, which occur in statistically likely places (positions 11, 17, 20, and 23, see Figure 15.2), in deuterostomes, differences occur in the most conserved areas of miRNAs, positions 2–8 and 13–15. Conservation of sequence of orthologous miRNAs is explained by the constraints governing not only folding but base-pairing with targets, and these structural considerations also explain why the same miRNA gene sequence evolving twice independently is highly unlikely.

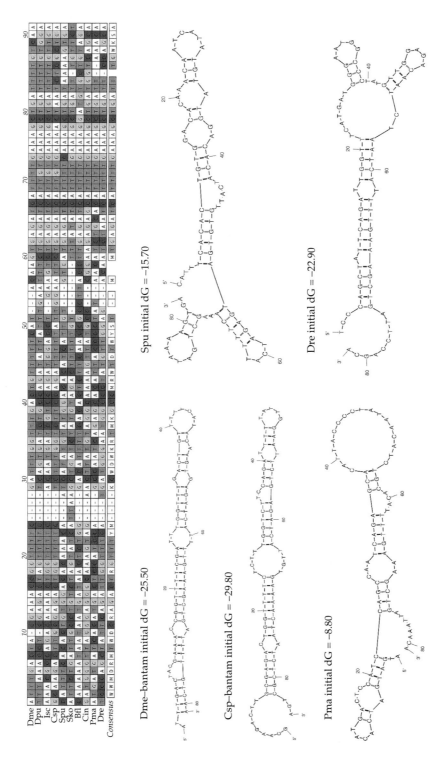

Figure 15.6 Alignment of the *bantam* gene taken from Figure 15.2 with the best hits from six different deuterostome genomes including the lamprey *Petromyzon marinus* (Pma) and the zebrafish *Danio rerio* (Dre) (other abbreviations are listed in Figure 15.4). Note that although some similarity is found in the mature sequence, especially with the sea urchin *Strongylocentrotus purpuratus* (Spu), there is no similarity in the star region, and, *contra* the protostome sequences (Dme and Csp), the deuterostome sequences do not show canonical folds (bottom), highlighting the improbability of evolving two miRNAs with the same mature sequence twice independently.

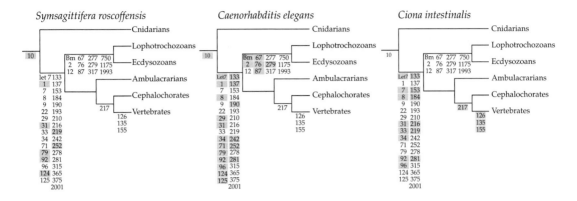

Figure 15.7   Primitive repertoire versus secondary loss of miRNAs. Shown are three taxa with reduced complements of miRNAs, the acoel flatworm *Symsagittifera roscoffensis* (Sempere *et al.*, 2007; Wheeler *et al.*, 2009); the nematode *Caenorhabditis elegans* (Ruby *et al.*, 2006), and the ascidian urochordate *Ciona intestinalis* (Norden-Krichmaer *et al.*, 2007). The miRNAs found in each taxon are shown in a grey box; the ones in black are known to characterize that particular node based on extensive comparative analyses (Wheeler *et al.*, 2009). The 37 miRNA families known to characterize vertebrates (Heimberg *et al.*, 2008) are not shown; only the three shared between vertebrates and ascidians. Note that, unlike the acoel flatworm, the nematode and the ascidian, while missing many primitive miRNAs, do possess protostome or chordate-specific miRNAs, respectively. In fact, the ascidian is grouped as the sister taxon to the vertebrates given that it shares three miRNA families with them (Heimberg *et al.*, 2008). The nematode is clearly a protostome, but cannot (at the moment) be allied with the ecdysozoans based on miRNAs, as hypothesized by numerous other data sets.

15.4   miRNAs in organisms with fast molecular evolution and frequent gene loss

We wish to take a moment to emphasize that miRNAs are not immutable, but are components of genomes that will experience some of the same processes that affect other components, especially when there is a high rate of gene loss and/or a high substitution rate. Nonetheless, the pattern that emerges from such instances will still allow an investigator to draw accurate if imprecise conclusions concerning the taxon's phylogenetic position. Ascidian urochordates, nematode worms, and acoel flatworms are all characterized by high rates of molecular evolution, and both nematodes (Copley *et al.*, 2004) and ascidians (Hughes and Friedman, 2005) are further characterized by large amounts of secondary gene loss. Both nematodes and ascidians have taken a phylogenetic 'bump up' recently with nematodes going from basal bilaterians to near relatives of arthropods (Aguinaldo *et al.*, 1997), and ascidians moving from basal chordates to the sister taxon of vertebrates (Delsuc *et al.*, 2006). Acoels, on the other hand, have followed the opposite

phylogenetic trajectory—they were originally included within the Platyhelminthes, but multiple studies with different genes have suggested that acoels and nemertodermatids form a grade or clade at the base of Bilateria (recently reviewed in Baguñà *et al.*, 2008).

Because all three of these hypotheses are controversial, they serve as useful test cases for the utility of miRNAs. Figure 15.7 shows the phylogenetic distribution of miRNAs in the acoel flatworm *Symsagittifera roscoffensis* (Wheeler *et al.*, 2009), the nematode *C. elegans* (Ruby *et al.*, 2006), and the ascidian *Ciona intestinalis* (Norden-Krichmaer *et al.*, 2007). In stark contrast to both the nematode and the ascidian, the acoel possesses only a subset of the bilaterian set of miRNAs, and no miRNAs that characterize higher clades, namely Protostomia or Platyhelminthes (Figure 15.7, left; see also Sempere *et al.*, 2007). Clearly, if acoels were in fact Platyhelminthes, or even within the protostomes or deuterostomes as suggested by recent EST studies (Philippe *et al.*, 2007; Dunn *et al.*, 2008) and have secondarily lost miRNAs so as to artificially appear to be basal bilaterians, then they have lost miRNAs in an extremely unlikely pattern. The position recovered by Dunn *et al.* (2008), which also corresponds to the traditional morphological hypothesis that allies them with the Platyhelminthes, would require the loss of 26 nephrozoan-specific and 12 protostome-specific miRNAs, in addition to some unknown set of platyzoan-specific miRNAs.

The miRNA complement of *C. elegans* is well known from deep 454 sequencing (Ruby *et al.*, 2006), and it is clear that it has lost a number of miRNAs (Figure 15.7, centre), as it possesses just over half of the reconstructed repertoire of the ancestral bilaterian miRNA-family complement (19 of 34). Nonetheless, they are clearly not basal bilaterians as they also have over half of the protostome-specific miRNAs as well (6 of 12). The ascidian *C. intestinalis* has also lost many miRNA families (Figure 15.8, right), as it only possesses 14 of the 34 miRNAs families present in the last common ancestor of protostomes and deuterostomes, but it also has the chordate-specific miRNA miR-217, and three miRNAs otherwise found only in vertebrates (Heimberg *et al.*, 2008), which is entirely consistent with the hypothesis that they, and not cephalochordates, are the sister taxon to the vertebrates (Delsuc *et al.*, 2006). Thus, it is these mosaic patterns of miRNA gene loss that characterize a secondary reduction in terms of miRNA content from primary absence and distinguishes nematodes and ascidians from acoels (Sempere *et al.*, 2007).

There is a good reason for suspecting high gene loss and/or high rates of sequence evolution in miRNA genes in these two taxa. The presence and sequence constraint of miRNA is probably dictated, to a large degree, by targeting numerous messenger RNA gene products. Several studies have shown that miRNAs can regulate up to hundreds of protein-coding genes (Lim *et al.*, 2005; Baek *et al.*, 2008; Selbach *et al.*, 2008), and because miRNAs regulate so many different transcripts, and must functionally interact with the 3′ UTR of all targets, it is difficult to lose the gene or change the primary sequence. Nematodes and ascidians have both lost a considerable fraction of their protein-coding genome, and consequently, in nematodes at least, each miRNA probably regulates, at best, only one or a few protein-coding genes (Ambros and Chen, 2007). This allows for individual miRNAs

to be more easily lost when their target messenger RNA is lost, or else to track the target site without being constrained by other targets. Thus, if a taxon is known to have a high proportion of secondary gene losses, it is likely that there will also be a relatively high number of missing and/or unrecognizable miRNAs. The pattern should nevertheless be both mosaic and random still allowing for an accurate (but possibly imprecise) placement on the metazoan tree of life.

15.5 Returning to the lophotrochozoan problem...

Ultimately, taxa such as nematodes are exceptions—most organisms have not experienced drastic gene losses. We hypothesize that lophotrochozoans in particular, which show little secondary gene loss and very little secondary modifications to their genomes (Tessmar-Raible and Arendt, 2003; Raible *et al.*, 2005), will make a near-perfect test case for miRNA phylogenetics. Returning to the problem introduced earlier in the paper, the interrelationships among nemerteans, annelids, and molluscs with respect to arthropods (Figure 15.1), the miRNAs are unequivocal. Both Eutrochozoa and Neotrochozoa are monophyletic, as nemerteans share with annelids and molluscs three unique miRNA families, one of which is the star sequence of miR-958, and annelids and molluscs share two miRNA families not found in nemerteans or any other taxon, one of which is the star sequence of an ancient miRNA family miR-133 (Wheeler *et al.*, 2009). Further, nemerteans do not share any miRNAs with either the annelids or the molluscs to the exclusion of the other, nor do they share with annelids and molluscs second copies of miR-10 and miR-22. Thus, among the three possible arrangements of these three taxa (Figure 15.1) miRNAs support the topology derived from morphological and embryological considerations (Figure 15.1a) (Peterson and Eernisse, 2001).

15.6 Methodology for miRNA phylogenetics

Because the primary sequence of the mature sequences of miRNAs is so fundamentally conserved, miRNA phylogenetics is essentially a binary system, involving simply the presence or absence of given miRNAs in different organisms. miRNAs can be identified as present in an organism by bioinformatic searches in genomes, Northern analysis, or sequencing of libraries targeting the products of Dicer cleavage, usually with new high-throughput sequencing technologies (e.g. Wheeler *et al.*, 2009). The literature on discovery and validation of miRNAs is too large to cover in this chapter, and we point the reader towards the general reviews cited above as a starting point to this literature, as well as to Ambros *et al.* (2003), which explains the requirements for annotation in miRBase, the online miRNA repository (Griffiths-Jones *et al.*, 2006).

With the exception of the phylogenetic position of acoels (Sempere *et al.*, 2007; Wheeler *et al.*, 2009), the utility of miRNAs as phylogenetic characters has not been fully tested. The main problem for miRNA-based phylogenetics is in positively demonstrating absence, which (needless to say) is far more difficult than demonstrating presence, especially in organisms without sequenced genomes. miRNAs with low expression levels will be hard to detect with both Northern analysis and libraries. As an extreme example, the miRNA *lys-6*, which was discovered in *C. elegans* by genetic screens (Johnston and Hobert, 2003) is expressed in fewer than 10 cells, and consequently has yet to be found in small RNA libraries even by extremely deep sequencing (Ruby *et al.*, 2006). Furthermore, reaction kinetics for Northern analyses indicate that only a few base changes will result in non-detection (Sempere *et al.*, 2006; Pierce *et al.*, 2008), so a negative result could be the result of a few nucleotide changes as opposed to an absence of the gene product.

Although the absence of a gene can never be proved, there are several relatively straightforward ways strongly to suggest absence. First, studies should strive to sequence libraries deeply enough to provide some confidence that the absence is real. The cost of next-generation sequencing technology is dropping quickly and as this happens the ability to sequence many organisms deeply and obtain a near complete understanding of their miRNA complement will be possible. Second, until the methodology of miRNA-based phylogenetics is more fully developed, studies should focus on understanding the relationships of non-genomic organisms to genomic organisms, rather than tackling questions where there are no genomes available at all. Genomes are never completed in the true sense of the word, but absence in both a finished genome and in a small RNA library is extremely unlikely to be a false negative. Working in a context where at least some of the organisms have sequenced genomes allows for the demonstration that the putative miRNA folds correctly in at least some of the organisms, precluding the possibility that the shared library reads are a degraded and highly conserved fragment of another gene. For non-genomic organisms, it is experimentally feasible to amplify miRNA loci using genome walking from the taxon of interest to demonstrate the necessary structural features (Wheeler *et al.*, 2009). The third, and probably most important, approach, much like any other form of phylogenetic inference, is taxon sampling. Studying more than one organism per clade of interest (especially for library construction) has the benefit of sampling two different transcriptomes which are likely to differ in their miRNA expression levels but which will help establish the polarity of individual characters and distinguish synapomorphies from plesiomorphies. As an example, miR-750 was reconstructed as a lophotrochozoan-specific miRNA by Sempere *et al.* (2007), but the presence of miR-750 in the small RNA library of the priapulid *Priapulus caudautus* (Wheeler *et al.*, 2009) demonstrated that this was in fact a protostome-specific miRNA that had been lost at the base of Insecta (Wheeler *et al.*, 2009; see Figure 15.5).

15.7 Conclusions

We see two main advantages to miRNA-based phylogeny. First, as demonstrated in Sempere *et al.*, (2006) and also in our Figure 15.5, they are applicable over a wide range of phylogenetic scales, from species divergences within a genus to phylum-level relationships at the base of Bilateria. Second, the constraints on miRNA structure means that any taxon can be queried for its complement of miRNAs, without having prior knowledge of a single miRNA sequence, simply by building a small RNA library. Given that miRNAs are continually added over time, rarely change in primary sequence, and are only rarely secondarily lost, they are potentially the near homoplasy-free data set that systematists have long wished for, and one that can be used to resolve the interrelationships among eumetazoan taxa at virtually any hierarchical level.

15.8 Acknowledgements

KJP would like to thank the National Science Foundation for funding and T. Littlewood and M. Telford for the invitation to contribute to the symposium. EAS would like to thank the Lerner-Gray fellowship from the American Museum of Natural History, the Systematics Research Fund of the Systematics Association and the Yale Enders Fund for funding. We thank D. Pisani and S. Smith for helpful suggestions on the manuscript.

CHAPTER 16

The evolution of developmental gene networks: lessons from comparative studies on holometabolous insects

Andrew D. Peel

Recent comparative studies have revealed significant differences in the developmental gene networks operating in three holometabolous insects: the beetle *Tribolium castaneum*, the parasitic wasp *Nasonia vitripennis*, and the fruit fly *Drosophila melanogaster*. In this chapter I discuss these differences in relation to divergent and convergent changes in cellular embryology. I speculate on how segmentation gene networks could have evolved to operate in divergent embryological contexts, and highlight the role that co-option might have played in this process. I argue that insects represent an important example of how diversification in life-history strategies between lineages can lead to divergence in the genetic and cellular mechanisms controlling the development of homologous adult structures.

16.1 Introduction

Arthropods are defined by a segmented body plan consisting of a series of anteroposteriorly arrayed segmental units with associated jointed appendages. The insects are traditionally viewed as one of the four major monophyletic arthropod groups, the other three being crustaceans, myriapods, and chelicerates. However, recent molecular phylogenies suggest that crustaceans are paraphyletic with respect to the insects; i.e. insects could reasonably be regarded as a monophyletic clade of terrestrial crustaceans (Carapelli *et al.*, 2007). While many insect species have retained the ancestral condition of undergoing metamorphosis from larva to adult through a series of intermediate nymphal stages (the hemimetabolous insects; Figure 16.1), the holometabolous insects undergo complete metamorphosis from larva to adult via a pupal stage (Brusca and Brusca, 2003). This is considered a derived life-history trait that arose only once during insect evolution (Brusca and Brusca, 2003) (see Figure 16.1). The vast majority of holometabolous insects belong to four speciose orders: the Diptera (two-winged flies), the Lepidoptera (butterflies and moths), the Coleoptera (beetles), and the Hymenoptera (wasps, bees, ants, etc.). We currently have a better understanding of the developmental genetic network underlying segmentation in a member of the Diptera—particularly the fruitfly *D. melanogaster*—than for any other insect, or indeed arthropod (Lawrence, 1992) (see Figure 16.2). However, a representative of the Coleoptera, the beetle *T. castaneum*, and a representative of the Hymenoptera, the parasitic wasp *N. vitripennis*, are rapidly being established as powerful insect model systems (Choe *et al.*, 2006; Brent *et al.*, 2007). Recent studies have revealed significant differences in the segmentation gene networks operating in these insects when compared with each other and with *D. melanogaster* (Schröder, 2003; Bucher and Klingler, 2004; Cerny *et al.*, 2005; Choe *et al.*, 2006; Lynch *et al.*, 2006a,b; Olesnicky *et al.*, 2006; Brent *et al.*, 2007; Choe and Brown, 2007). In this chapter, I review and discuss these differences in relation to the modes of cellular embryogenesis exhibited by these insects. Both *D. melanogaster* and *N. vitripennis* have evolved a rapid mode of development that required major changes in embryogenesis at the cellular level

(Bull, 1982; Lawrence, 1992; Davis and Patel, 2002). In contrast, *T. castaneum* has retained a more ancestral mode of cellular embryogenesis (Handel *et al.*, 2000; Davis and Patel, 2002). I speculate on how insect segmentation gene networks have evolved to operate in these divergent embryological contexts. A recent molecular phylogeny suggests that the rapid mode of cellular embryogenesis exhibited by *D. melanogaster* and *N. vitripennis* evolved convergently (Savard *et al.*, 2006). I go on to ask whether convergent changes in gene networks might have underpinned these apparent parallel transitions in cellular embryology. First I review the modes of cellular embryogenesis found within the insects, and discuss the role that life history has played in their evolution.

16.2 The influence of life-history strategy on insect embryogenesis

16.2.1 An evolutionary biologist's view on development

The principal aim of a developmental biologist is to work towards establishing a more complete picture of how the genetic information contained within an organism's genome is deployed over developmental time to transform a single cell into a functional multicellular organism. In contrast, the principal aim of many evolutionary developmental biologists is to identify—through comparative analysis of developmental data within a phylogenetic framework—the changes in developmental mechanisms

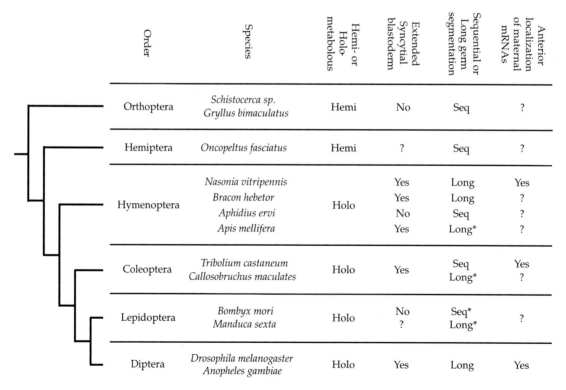

Order	Species	Hemi- or Holo-metabolous	Extended Syncytial blastoderm	Sequential or Long germ segmentation	Anterior localization of maternal mRNAs
Orthoptera	*Schistocerca sp.* *Gryllus bimaculatus*	Hemi	No	Seq	?
Hemiptera	*Oncopeltus fasciatus*	Hemi	?	Seq	?
Hymenoptera	*Nasonia vitripennis* *Bracon hebetor* *Aphidius ervi* *Apis mellifera*	Holo	Yes Yes No Yes	Long Long Seq Long*	Yes ? ? ?
Coleoptera	*Tribolium castaneum* *Callosobruchus maculates*	Holo	Yes	Seq Long*	Yes ?
Lepidoptera	*Bombyx mori* *Manduca sexta*	Holo	No ?	Seq* Long*	?
Diptera	*Drosophila melanogaster* *Anopheles gambiae*	Holo	Yes	Long	Yes

Figure 16.1 A phylogeny of the insect species discussed in this chapter with embryological features mapped on. The relative relationships of the four holometabolous insect orders follows the study by Savard *et al.* (2006). Some insects do not fit comfortably into the categories 'sequential' or 'long-germ' segmentation; for caveats in relation to the categorization of specific species (*) see Davis and Patel (2002). Character states have been left clear where there are uncertainties, i.e. when there is a lack of gene expression data and/or dye injection experiments to ascertain the existence of an extended syncytial blastoderm stage.

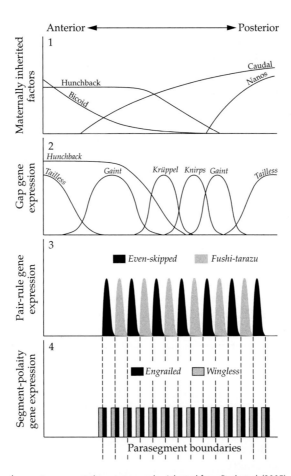

Figure 16.2   The *Drosophila melanogaster* segmentation gene cascade. Adapted from Peel *et al.* (2005).

(1) Maternal genes. Maternal transcripts of the segmentation genes *caudal* and *hunchback* are uniformly distributed, whereas maternal *bicoid* mRNA is tethered to the anterior pole of the egg. Localized at the posterior pole is a complex of maternal proteins and RNAs that includes transcripts of the gene *nanos*. On fertilization, maternal mRNAs are de-repressed, translated and Bicoid and Nanos proteins form gradients at either end of the egg. Bicoid activates zygotic *hunchback* expression and represses *caudal* translation in the anterior, whereas Nanos represses the translation of maternal *hunchback* in the posterior. As a result Hunchback protein is restricted to the anterior of the egg and protein gradients of Bicoid (decreasing posteriorly) and Caudal (decreasing anteriorly) form. In parallel, a maternally encoded terminal patterning system operates during embryogenesis; the product of *torso-like*—which is expressed within specialized follicle cells situated at both egg poles during oogenesis—catalyses the localized cleavage of a protein encoded by *trunk* within the perivitelline fluid. The trunk cleavage product acts as a ligand on the receptor tyrosine kinase encoded by *torso*, triggering a signalling cascade that regulates the zygotic expression of downstream segmentation genes, such as *tailless*, at either pole of the egg.

(2) Gap genes. The net result of maternal signalling is the activation along the anteroposterior axis of the egg of a series of zygotic gap genes (i.e. *giant*, *Krüppel*, *tailless*), thus named because their mutation leads to gaps in the region of the embryo in which they are normally expressed. The protein products of the gap genes themselves diffuse within the syncytial blastoderm, regulate each other, and thus further refine their expression. Gap genes also play an important role at this stage in regulating the expression of the Hox genes, whose proteins products confer identity to segments.

(3) Pair-rule genes. In the next tier of the *Drosophila* segmentation cascade are three genes—*even-skipped*, *runt*, and *hairy*—whose expression is driven by the maternal and gap gene transcription factor products. All three genes possess complex regulatory sequences that interpret the aperiodic expression of maternal and gap gene products and drive expression in a periodic pattern of seven stripes. These genes are collectively referred to as pair-rule genes since their mutation often leads to abnormalities in alternate segments. The three 'primary' pair-rule gene products in turn regulate expression of 'secondary' pair-rule genes, such as *fushi-tarazu*, *paired*, *sloppy-paired*, and *odd-skipped*. Black curve, *even-skipped*; grey curve, *fushi-tarazu*.

that underpin divergence in body architecture between lineages. Rather than thinking in terms of developmental time—with the rather arbitrary starting point of zygote or germ cell—evolutionary developmental biologists consider evolutionary timescales, and as such development is not viewed as a linear process but rather as continuous developmental cycles undergoing constant modification in response to selection and drift.

Natural selection can act independently on distinct stages of an organism's developmental cycle. This is obvious when considering insects. There has clearly been divergence in segment form between insect species, particularly with respect to appendage morphology; compare, for example, the sucking mouthparts of the phytophagous milkweed bug *Oncopeltus fasciatus* (Hughes and Kaufman, 2000) with the mandibles of some carnivorous beetles (Konuma and Chiba, 2007). However, it is clear that, on the whole, the basic insect segmental unit has been conserved. In contrast, oocytes and early eggs exhibit significant morphological differences, a consequence of the numerous and diverse life-history strategies that have evolved within the insects.

16.2.2   All eggs are different, but some eggs are more different than others

Evolutionary shifts in insect life-history strategies often correlate with changes in cellular modes of embryogenesis. This was dramatically illustrated in a study by Grbic and Strand (1998) on two parasitic wasps belonging to the hymenopteran family Braconidae. *Bracon hebetor* is an ectoparasite that lays yolky eggs on the integument of moth larvae. In the lineage leading to *Aphidius ervi*, however, there has been a transition to an endoparasitic life history; *A. ervi* lays a single yolkless egg into the haemocoel of an aphid host. Grbic and

Strand (1998) studied the cellular embryology of these insects and found significant differences. In the eggs of *B. hebetor*, the cellularization of early cleavage nuclei is delayed until after they form a blastoderm, and all segments develop more or less simultaneously. In contrast, in *A. ervi* eggs, complete cytokinesis (the formation of cell membranes) occurs from the fourth round of nuclear divisions onwards, the early embryo ruptures from the chorion within the host haemocoel, and segments form one by one in an anterior to posterior progression. One can only speculate on why the transition to an endoparasitic life history required such dramatic changes in cellular embryology, but is seems likely they are associated with the transition from receiving nutrients in the form of maternal yolk to the use of nutrients available from the haemolymph of the unfortunate host.

Similarly dramatic cellular transitions in embryogenesis have occurred within non-parasitic insect lineages (for an in-depth review see Davis and Patel, 2002). Although the precise ecological reasons remain speculative, it seems likely that in many cases these transitions occurred in response to selection for increases in the speed of embryogenesis. Here I discuss two specific cellular adaptations and how they might have facilitated the faster development of an insect segmented body plan.

The timing of cellularization
In most insect species early nuclear divisions are superficial; the formation of cell membranes around early cleavage nuclei is delayed until they have migrated to the egg surface and formed the blastoderm. For example, dye injection experiments have demonstrated this to be the case in the locust *Schistocerca gregaria* (Ho *et al.*, 1997) (Figure 16.1). However, this delay is particularly pronounced in some holometabolous insect lin-

---

(4) Segment-polarity genes. The pair-rule gene products activate the final tier in the *Drosophila* segmentation gene cascade, the segment polarity genes. These are the genes encoding proteins that actually initiate the formation of segment boundaries, and, as the name suggests, confer polarity to segments. Segment polarity genes are expressed in a series of 14 stripes, with odd and even stripes regulated by a different combination of the pair-rule proteins. The boundary between the expression of two of these genes, *engrailed* and *wingless*, becomes the parasegmental boundary, whereas segment boundaries form later, posterior to *engrailed* expression. Black bar, *engrailed*; grey bar, *wingless*.

eages, creating an extended syncytial blastoderm stage: examples of such insects include *N. vitripennis* (Bull, 1982), *T. castaneum* (Handel *et al.*, 2000), and *D. melanogaster* (Lawrence, 1992) (Figure 16.1). Within a syncytium, gradients of patterning molecules can form quickly across a field of nuclei, without the need for complex intercellular signalling pathways.

The allocation of cells to segments
The temporal dynamics by which cells are allocated to segments varies across insect species. In insects exhibiting primitive modes of development, anterior segments are patterned through the subdivision of blastoderm nuclei/cells, while posterior segments are patterned sequentially after the blastoderm stage, within a posteriorly located cellular zone of extension. Examples of such insects include the hemimetabolous insects *Gryllus bimaculatus*, *Schistocerca* sp., and *O. fasciatus* and the holometabolous insect *T. castaneum* (see Davis and Patel, 2002) (Figure 16.1). I shall refer to these as 'sequentially segmenting' insects. In many insect lineages there has been an increase in the number of anterior segments patterned through subdivision in the blastoderm (Davis and Patel, 2002); this has occurred, for example, in some coleopteran lineages (Patel *et al.*, 1994). In many holometabolous insects this trend has reached its logical extreme, and all segments form through early subdivision of embryonic blastoderm nuclei. These insects are said to exhibit 'long-germ' embryogenesis, since the embryonic germ rudiment typically occupies almost the entire length of the egg. Examples of such insects include *D. melanogaster* (Lawrence, 1992) and *N. vitripennis* (Bull, 1982) (see below and Figures 16.1 and 16.2).

16.3   Molecular transitions underlying the evolution of long-germ embryogenesis

During long-germ embryogenesis in *D. melanogaster*, a cascade of transcription factors acts within a syncytium to divide the embryo into progressively smaller domains such that segments develop more or less simultaneously. The *D. melanogaster* segmentation gene cascade is briefly outlined in

Figure 16.2, but for a more thorough understanding the reader is referred to Lawrence (1992).

In order to identify the changes in gene networks that underpinned the evolution of long-germ embryogenesis, a good understanding of the segmentation mechanisms operating in insects that have retained sequential segmentation is required. One such insect is the beetle *T. castaneum* (Handel *et al.*, 2005). Recent studies on this holometabolous insect have revealed significant differences in the genetic circuitry underlying segmentation when compared with *D. melanogaster* (Schröder, 2003; Bucher and Klingler, 2004; Bucher *et al.*, 2005; Cerny *et al.*, 2005; Choe *et al.*, 2006; Choe and Brown, 2007; Schinko *et al.*, 2008). Together, comparative studies on *T. castaneum*, *N. vitripennis*, and *D. melanogaster* suggest that the evolution of long-germ embryogenesis required distinct molecular transitions to occur in concert at either pole of the egg (Choe *et al.*, 2006; Brent *et al.*, 2007).

16.3.1   Molecular transitions at the anterior egg pole: the evolution of maternally encoded anterior patterning gradients

In *T. castaneum*, head and thoracic segments are patterned through the subdivision of blastoderm nuclei located near the posterior egg pole; the anterior blastoderm forms extra-embryonic tissue (Handel *et al.*, 2000) (Figure 16.3a,d). In insects exhibiting long-germ embryogenesis, however, head and thoracic segments are patterned further towards the anterior egg pole (Figure 16.3b,c,e,f). It has been proposed that this spatial shift in anterior patterning required the evolution of an instructive anterior patterning gradient to complement the action of existing posterior determinants (Lynch *et al.*, 2006a). The localization of maternal mRNAs to the anterior pole of the oocyte is observed in *T. castaneum*, *N. vitripennis*, and *D. melanogaster* (Lawrence, 1992; Bucher *et al.*, 2005; Olesnicky and Desplan, 2007) (Figure 16.1). In *N. vitripennis* and *D. melanogaster*, mRNA and/or translated protein from anterior and posterior sources of maternal mRNAs form largely non-overlapping, and opposing, instructive patterning gradients (Lawrence, 1992; Lynch *et al.*, 2006a; Olesnicky *et al.*, 2006; Brent

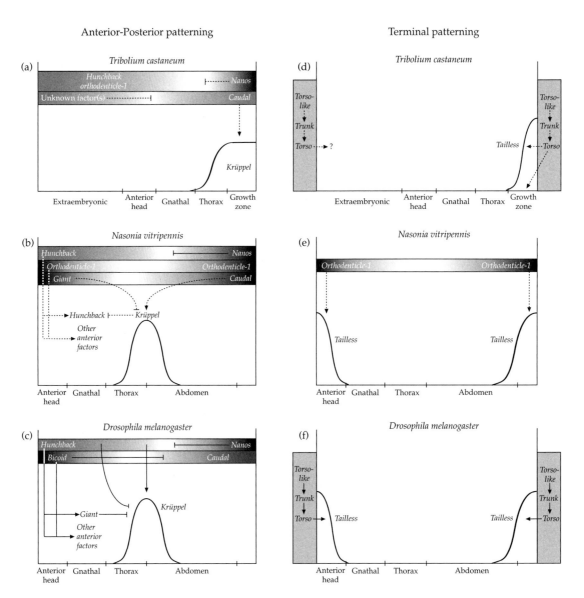

Figure 16.3 A schematic representation of the variation in maternal patterning observed between the holometabolous insects *Tribolium castaneum*, *Nasonia vitripennis*, and *Drosophila melanogaster* (as described in the text and summarized in Table 16.1). Particular focus is placed on variation in the maternal regulation of the zygotically expressed central gap gene *Krüppel* (a)–(c) and terminal gap gene *tailless* (d)–(f). Note that in *T. castaneum Krüppel* is expressed like, but does not function as, a canonical gap gene (see text and Cerny *et al.*, 2005), and *tailless* is not expressed in the anterior (see text and Schröder *et al.*, 2000, and Schoppmeier and Schröder, 2005). Maternally expressed genes are shown within shaded rectangles, above or beside representations of zygotically expressed genes: gradients of shading within the rectangles depict expression gradients (not to scale). The mRNA of maternal genes written in bold is known to be anteriorly and/or posteriorly tethered during oogenesis. Genetic interactions depicted for *T. castaneum* and *N. vitripennis* are determined from RNA interference experiments and so cannot be assumed to reflect direct regulation (hence the dotted lines). Recent models suggest that the establishment of the *Krüppel* gap domain in *D. melanogaster* can largely be explained by both positive (at low concentrations) and negative (at high concentrations) regulation by maternal *hunchback* (Papatsenko and Levine, 2008).

et al., 2007) (see Table 16.1 and Figure 16.3b,c,e,f). In *T. castaneum*, on the contrary, neither of the anteriorly localized mRNAs identified to date play a significant role in anterior–posterior patterning (Bucher *et al.*, 2005), and the maternal mRNAs of two genes known to be important anterior determinants in *T. castaneum*—*hunchback* and *orthodenticle-1*—are initially distributed uniformly in the egg (Wolff *et al.*, 1995; Schröder, 2003; Schinko *et al.*, 2008) (see Table 16.1 and Figure 16.3a,d). It is possible that an anteriorly localized maternal mRNA, whose protein product forms an instructive patterning gradient exists in *T. castaneum* but has been overlooked. However, it is tempting to speculate that there is an association between the retention of sequential segmentation in *T. castaneum*, and the lack of an instructive anterior patterning gradient. It will be interesting to determine whether an instructive anterior patterning gradient has evolved in those beetle lineages in which there has been an increase in the number of segments patterned in the blastoderm prior to gastrulation (Patel *et al.*, 1994).

16.3.2 Molecular transitions at the posterior egg pole: changes in the regulation of pair-rule gene homologues

The primary pair-rule genes (*even-skipped*, *hairy*, and *runt*) are the first genes within the *D. melanogaster* segmentation cascade to be expressed in a periodic pattern of stripes (Jaynes and Fujioka, 2004) (see Figure 16.2). The primary pair-rule genes activate a suite of secondary pair-rule genes, that includes *paired*, *sloppy-paired-1* and *-2*, and *odd-skipped* (Jaynes and Fujioka, 2004). They are collectively referred to as pair-rule genes since their mutation often leads to abnormalities in alternate segments. Recent work on *T. castaneum* has revealed divergent regulatory interactions between the homologues of *D. melanogaster* pair-rule genes. Choe *et al.* (2006) showed that it is the homologues of *D. melanogaster even-skipped*, *D. melanogaster runt* and *D. melanogaster odd-skipped* that comprise the primary tier of pair-rule genes in *T. castaneum*. The authors disrupted the expression of each primary pair-rule gene in turn using parental RNA interference (RNAi) and then examined the expression of the remaining genes in knockdown embryos. Surprisingly, rather than canonical pair-rule phenotypes, the individual knockdown of each of these genes result in asegmental phenotypes in which all but a few anterior segments are deleted. Although direct regulatory interactions were not proven, the results suggested to the authors that *T. castaneum even-skipped* activates *T. castaneum runt*, which in turn activates *T. castaneum odd-skipped*, which completes a regulatory cycle by repressing *T. castaneum even-skipped*. On the basis of these data, it was proposed that the genes comprise a regulatory gene circuit, each cycle of which sequentially patterns pairs of segments via the downstream regulation of a secondary tier of pair-rule gene homologues composed of *T. castaneum paired* and *T. castaneum sloppy-paired* (note that *T. castaneum hairy* does not appear to play a role in trunk segmentation; Choe and Brown, 2007).

This model and the data it is based on raise an interesting question. How could a regulatory circuit of transcription factors—that by definition must operate intracellularly—pattern segments within a cellularized zone of extension? One possibility I suggest is that the proposed transcription factor circuit—or perhaps just some components of it—constitute an intracellular molecular oscillator with an analogous role to the molecular oscillators that control sequential segmentation of the pre-somitic mesoderm during vertebrate development (see reviews by Pourquié, 2004, and Gridley, 2006). The vertebrate segmentation clock relies on the Notch intercellular signalling pathway, both as a component of the molecular oscillator in some cases (Pourquié, 2004; Gridley, 2006) and to coordinate oscillations amongst neighbouring cells (Masamizu *et al.*, 2006). Work on myriapods and chelicerates has shown that some pair-rule gene homologues and members of the Notch intercellular signalling pathway are expressed in a dynamic fashion during sequential segmentation in a manner reminiscent of that seen during vertebrate sequential segmentation (Stollewerk *et al.*, 2003; Chipman *et al.*, 2004; Schoppmeier and Damen, 2005), leading to the exciting hypothesis that a segmentation clock, analogous if not homologous

Table 16.1 A summary of some *Tribolium castaneum* and *Nasonia vitripennis* segmentation gene homologues shown by recent studies to exhibit differences in expression and function when compared to *Drosophila melanogaster* (see also Figures 16.2 and 16.3).

Gene	T. castaneum	N. vitripennis
bicoid[1]	bicoid not present in the genomes of insects outside the cychlorrhaphan flies Evolved from a zerknüllt (zen)–like precursor Orthodenticle-1 appears to play an analogous role to *D. melanogaster* Bicoid in these species	
orthodenticle-1[2–4]	mRNA is maternally inherited—unlike in *D. melanogaster* where expression is purely zygotic—but not localized to the egg poles as in *N. vitripennis* Maternal plus zygotic expression becomes anteriorly restricted Functions with Hunchback to pattern the anterior of the embryo	Unlike in *D. melanogaster*, maternal mRNA is localized to the anterior and posterior poles (via distinct mechanisms at either pole) On fertilization, mRNA is released and forms an opposing anterior and posterior mRNA and, through translation, protein gradients Anterior gradient functions with maternal Hunchback to activate anterior gap genes: *empty spiracles*, *giant*, and (zygotic) *hunchback* Posterior gradient functions with Caudal to pattern posterior segments Largely conserved zygotic head gap gene role
giant[4–6]	No maternal expression as is the case in *D. melanogaster* Expressed in two zygotic gap-like domains as in *D. melanogaster*, except that the posterior domain is positioned much more to the anterior Anterior domain controls segment identity via the regulation of Hox genes, but unlike in *D. melanogaster*, it is not required for segment formation Required for the formation of all thoracic and abdominal segments, not just the segments in which it is expressed	Unlike in *D. melanogaster*, maternal mRNA is localized to the anterior during oogenesis On fertilization, mRNA is released and forms an anterior mRNA and, through translation, a protein gradient Represses central gap gene *Krüppel* in anterior, preventing repression of anterior gap gene *hunchback* by *Krüppel*, and thus plays a permissive role in anterior development A similar zygotic gap gene role to *D. melanogaster*
cauda1[3, 4,7–10]	Maternal mRNA initially uniformly distributed as in *D. melanogaster* Posterior protein gradient forms through translational repression by an unknown factor/factors (i.e. not Bicoid) Expressed in the cellularized growth zone Unlike in *D. melanogaster*, required for the formation of all but the most anterior few segments	Unlike in *D. melanogaster*, maternal mRNA localized to posterior during oogenesis On fertilization, mRNA is released and forms a posterior mRNA and, through translation, a protein gradient Functions with Orthodenticle-1 to pattern posterior via activation of posterior gap genes Influence extends further to the anterior than in *D. melanogaster* and includes activation of the central gap gene *Krüppel*
tailless and the terminal patterning system[3,11–13]	tailless expressed by small group of cells at the posterior pole of the blastoderm. In contrast to *D. melanogaster*, there is no expression at the anterior pole of the blastoderm Terminal patterning system (*torso* and *torso-like*) required for sequential segmentation and formation of anterior extra-embryonic tissue	tailless expression is activated in anterior and posterior by Orthodenticle-1 (i.e. there is no evidence for a terminal patterning system in *N. vitripennis*) Unlike *D. melanogaster*, the anterior domain is not required for segmentation Posterior domain has more extensive influence on posterior patterning than in *D. melanogaster*

References: 1, Stauber *et al.* (2002); 2, Schröder (2003); 3, Lynch *et al.* (2006a); 4, Olesnicky and Desplan (2007); 5, Bucher and Klingler (2004); 6, Brent *et al.* (2007); 7, Schulz *et al.* (1998); 8, Wolff *et al.* (1998); 9, Copf *et al.* (2004); 10, Olesnicky *et al.* (2006); 11, Schröder *et al.* (2000); 12, Schoppmeier and Schröder (2005); 13, Lynch *et al.* (2006b).

to that operating in vertebrates, controls sequential segmentation in these arthropods (Peel and Akam, 2003; Stollewerk *et al.*, 2003). However, there is no evidence, as yet, for the involvement of the Notch signalling pathway during insect sequential segmentation. Wingless signalling also plays a central role in the vertebrate segmentation clock (Pourquié, 2004; Gridley, 2006). Perhaps wingless signalling forms the basis of a possible segmentation clock in insects (Miyawaki *et al.*, 2004) or alternatively other signalling pathways might be involved.

In *D. melanogaster* the periodic expression of primary pair-rule genes is, somewhat curiously, activated by an aperiodic series of anteroposteriorly restricted domains of gap gene expression (Figure 16.2). Gap genes play an additional role in *D. melanogaster* development; they regulate the anteroposteriorly restricted domains of Hox gene expression that confer identity to segments (Irish *et al.*, 1989). In *T. castaneum*, most *D. melanogaster* gap gene homologues are expressed in restricted anteroposterior domains, consistent with a gap gene function, and in a roughly conserved anteroposterior order, albeit shifted towards the anterior (Schröder, 2003; Bucher and Klingler, 2004; Cerny *et al.*, 2005). However, the knockdown by RNAi of at least two of the *D. melanogaster* gap gene homologues—*Krüppel* and *giant*—does not result in canonical gap gene phenotypes: i.e. the loss of the segments in and around their domains of expression (Bucher and Klingler, 2004; Cerny *et al.*, 2005). Instead, these segments take on abnormal identities as a result of the misexpression of Hox genes (Bucher and Klingler, 2004; Cerny *et al.*, 2005). This might imply that one of the ancestral roles of gap gene homologues in insects was to position Hox gene domains correctly, and that they were only later recruited to pattern pair-rule genes (Peel and Akam, 2003). Under this model, gap gene recruitment is correlated with the transition to activating pair-rule stripes simultaneously in a syncytium, where control by intercellular signalling becomes redundant and where a spatial rather than temporal regulatory input is required (Peel and Akam, 2003). Presumably transcription factors expressed at the right time and place in the posterior blastoderm were co-opted to regulate progressively more posterior pair-rule stripes, thus explaining the complex nature of the regulatory sequence of *D. melanogaster* primary pair-rule genes. And perhaps the co-option of *D. melanogaster* gap gene homologues was favoured, since they had already evolved a spatially and temporally corresponding role in Hox gene regulation.

16.4 Molecular transitions underlying the convergent evolution of long-germ embryogenesis

The four major holometabolous insect orders all contain species exhibiting long-germ embryogenesis, for example the dipteran *D. melanogaster*, the lepidopteran *Manduca sexta*, the coleopteran *Callosobruchus maculates*, and the hymenopterans *Apis mellifera*, *N. vitripennis*, and *Bracon hebetor* (Grbic and Strand, 1998; see also Davis and Patel, 2002). However, the Lepidoptera, Coleoptera, and Hymenoptera also contain species that have retained, or re-evolved in the case of parasitic hymenopterans (Grbic and Strand, 1998), differing degrees of sequential segmentation; for example, the lepidopteran *Bombyx mori*, the coleopteran *T. castaneum*, and the hymenopteran *A. ervi* (Grbic and Strand, 1998). This has led to the idea that long-germ development evolved multiple times independently during the holometabolous insect radiation (see Davis and Patel, 2002). This scenario now seems more likely due to a recent re-evaluation of holometabolous insect phylogeny. The general consensus surrounding the relationship of the major holometabolous insect orders used to be that the Diptera and Lepidoptera are sister groups, and that the Coleoptera form the basal branch in the tree (Whiting, 2002). However, a recent molecular phylogenetic study supports a reversal in the position of the Hymenoptera and Coleoptera, such that the Hymenoptera now form the basal branch (Savard *et al.*, 2006). Thus, the Diptera and Hymenoptera are now separated by an order that contains many species with clear sequential segmentation; the Coleoptera.

Work by the Desplan and Pultz laboratories has begun to reveal the molecular basis to segmentation in *N. vitripennis*, which has an embryonic fate map almost identical to that of *D. melanogaster* (Bull, 1982; Brent *et al.*, 2007). If *N. vitripennis* and

D. melanogaster did evolve long-germ embryogenesis independently, as the latest molecular phylogenies suggest, comparisons between these two species are a first step to determining the extent to which underlying gene network changes were also convergent.

As in *D. melanogaster* (Figure 16.2), the segmentation cascade operating in *N. vitripennis* can be divided into four distinct tiers of maternal, gap, pair-rule, and segment-polarity genes. The major genetic differences identified so far between *D. melanogaster* and *N. vitripennis* are found right at the top of the segmentation cascade and relate to changes in the maternal contribution to patterning. These differences are summarized in Table 16.1 and Figure 16.3 and are discussed below.

In *D. melanogaster,* anterior development is largely under the control of the Bicoid morphogen gradient (Lawrence, 1992) (Figures 16.2 and 16.3c). However, *bicoid* is known to be an invention of the higher Diptera (Stauber *et al.*, 2002). In *N. vitripennis* anterior patterning is accomplished by two distinct anterior gradients of patterning molecules (Lynch *et al.*, 2006a; Brent *et al.*, 2007) (Figure 16.3b). In contrast to *D. melanogaster,* where their expression is purely zygotic, maternal mRNAs of both *orthodenticle-1* and *giant* are tethered to the anterior pole during oogenesis in *N. vitripennis* (for details see Lynch *et al.*, 2006a, Brent *et al.*, 2007, and Olesnicky and Desplan, 2007). Following fertilization these mRNAs are translated to form anterior gradients of Orthodenticle-1 and Giant protein. The maternal Orthodenticle-1 gradient functions to activate anterior segmentation genes, such as the gap genes *empty spiracles*, (zygotic) *giant*, and *hunchback* (Lynch *et al.*, 2006a), while the maternal Giant gradient functions to set the anterior expression boundary of the central gap gene *Krüppel* (Brent *et al.*, 2007) (Figure 16.3b). The repressive role of maternal Giant is permissive for anterior development, since, in its absence, *Krüppel* expression spreads anteriorly to repress the anterior gap gene *hunchback* (Brent *et al.*, 2007) (Figure 16.3b).

The influence of Orthodenticle-1 and Giant on embryonic patterning does not extend as far to the posterior as does Bicoid in *D. melanogaster* (Lynch *et al.*, 2006a; Brent *et al.*, 2007). Instead, the influence of the posterior determinant *caudal* extends

further towards the anterior; in *D. melanogaster hunchback* regulates (probably both positively and negatively; Papatsenko and Levine, 2008) (Figure 16.3c) the central gap domain of *Krüppel*, but in *N. vitripennis* it is *caudal* that activates this gap domain (Olesnicky *et al.*, 2006) (Figure 16.3b). Indeed, posterior patterning in *N. vitripennis* exhibits significant differences when compared with *D. melanogaster*. A posterior gradient of Caudal is established in *N. vitripennis*—in the absence of translational repression by Bicoid (Figure 16.2 and 16.3c)—via the tethering of *caudal* mRNA to the posterior pole during oogenesis (Olesnicky *et al.*, 2006; Olesnicky and Desplan, 2007) (Figure 16.3b). The mRNA of *orthodenticle-1* is also tethered to the posterior pole during oogenesis, such that, together, gradients of Caudal and Orthodenticle-1 protein control posterior patterning (Lynch *et al.*, 2006a; Olesnicky *et al.*, 2006) (Figure 16.3b).

The large degree of variation between *D. melanogaster* and *N. vitripennis* in maternal patterning suggests that the changes in the gene network underlying the putative independent evolution of long-germ embryogenesis within the hymenopteran and dipteran lineages might have been very different. However, many of the observed differences could be attributed to the evolution of a maternal Bicoid gradient within the higher Diptera (Stauber *et al.*, 2002). Indeed, *bicoid* and *orthodenticle* are both homeobox-containing genes, and Bicoid is predicted to have usurped the role of Orthodenticle as an anterior determinant (as exemplified by *T. castaneum*; Schröder, 2003; Schinko *et al.*, 2008) and *N. vitripennis* (Lynch *et al.*, 2006a) *orthodenticle-1* (Table 16.1, Figures 16.3a,b) through gaining affinity for Orthodenticle DNA-binding sites via the convergent acquisition of a lysine residue at position 50 in its homeodomain (Treisman *et al.*, 1989). Further data from additional long-germ dipterans (e.g. *Anopheles gambiae*) and hymenopterans (e.g. *A. mellifera*) will be required to map divergent (and convergent) gene network changes to the dipteran and/or hymenopteran lineages. For example, it is interesting that regulation by zygotic *giant* is involved in the maintenance and refinement of the *Krüppel* gap domain in *D. melanogaster* (Papatsenko and Levine, 2008) (Figure 16.3c). Was *giant* recruited to play an earlier role in the establishment of the

Krüppel gap domain within the lineage leading to *N. vitripennis*, or was a role for *giant* at the maternal level a character that has been lost during the dipteran radiation within the lineage leading to *D. melanogaster*?

There does, however, already appear to be a strong case for differences in the evolution of the terminal patterning system during the putative independent evolution of long-germ development within the hymenopteran and dipteran lineages. In *D. melanogaster* a terminal patterning system acts to determine the most anterior and posterior regions of the embryo (Lawrence, 1992) (Figure 16.2). The product of the gene *torso-like* is maternally restricted to the egg poles where it cleaves the protein encoded by the gene *trunk* (Casali and Casanova, 2001). The Trunk C-terminal then acts as a ligand, binding to the receptor tyrosine kinase encoded by *torso* and triggering a signalling cascade that regulates the zygotic expression of downstream segmentation genes, such as *tailless*, at either pole of the egg (Casali and Casanova, 2001) (see Table 16.1 and Figure 16.3f). In *T. castaneum*, both *torso* and *torso-like* are required at the posterior pole for the activation and/or maintenance of sequential segmentation (Schoppmeier and Schröder, 2005) (Figure 16.3d). However, in *N. vitripennis* the expression of *tailless* has been shown to be dependent on the anterior and posterior gradients of maternal Orthodenticle-1 and thus perhaps not a terminal patterning system (Lynch *et al.*, 2006b) (Figure 16.3e). The absence of a *D. melanogaster*-like terminal patterning system in hymenopterans is further supported by the failure to find homologues of *torso* or *trunk* in the *A. mellifera* genome (Dearden *et al.*, 2006).

The apparent lack of a *Drosophila*-like terminal patterning system in long-germ hymenopteran insects is intriguing, particularly since one appears to be required for sequential segmentation in *T. castaneum* (Schoppmeier and Schröder, 2005; Lynch *et al.*, 2006b) (Figure 16.3e). Studies on more holometabolous insects, as well as non-holometabolous insects, will be required to determine if a terminal patterning system is a character that has been lost within the hymenopteran lineage or gained in the lineage leading to the other major holometabolous insect orders.

One possibility is that in some of the insect lineages that underwent the transition to long-germ embryogenesis (i.e. in dipteran lineages) a terminal patterning system that ancestrally played an important role in initiating or maintaining sequential segmentation was co-opted to activate posterior segmentation genes (Schoppmeier and Schröder, 2005), whereas in other lineages undergoing parallel transitions (i.e. in hymenopteran lineages) this did not occur.

16.5   General conclusions

16.5.1   The role of co-option in the evolution of segmentation gene networks

A theme emerging from comparative studies on insects is the role that co-option plays in evolution, at different levels of complexity. At the regulatory sequence level, it seems possible, if not likely, that some gap gene homologues—expressed at the right time and in the right place due to an ancestral role in Hox gene regulation—have been co-opted into regulating pair-rule gene homologues, perhaps via the simple acquisition of binding sites (Bucher and Klingler, 2004; Cerny *et al.*, 2005; Choe *et al.*, 2006). At the protein level, Bicoid—or perhaps more accurately Zerknüllt—was co-opted to an anterior patterning role within the higher Diptera via a simple coding mutation that allowed it to recognize the regulatory targets of an existing anterior determinant; Orthodenticle (Stauber *et al.*, 2002; Lynch *et al.*, 2006a). At the intracellular level, existing cytoskeletal machinery may have been co-opted during the evolution of instructive anterior patterning gradients (Bucher *et al.*, 2005). All these cases are consistent with a long series of simple modifications to developmental gene circuits having, over evolutionary time, underpinned diversification at the cellular level and above. The results of a recent whole-genome study are consistent with the widespread occurrence of gene co-option during insect evolution. Dearden *et al.* (2006) looked at the presence/absence of homologues of *D. melanogaster* developmental genes in the *A. mellifera* genome. They found that, of the developmental genes involved in processes that are known to be divergent between these two species (i.e. sex determination, dosage compensation, meiosis, and germ-cell development), a significant number of those conserved in *A. mellifera* ($c = 19.03$; $P < 0.001$, $n = 78$) have multiple (i.e. pleiotropic) functions in *D. melanogaster*. This suggests either that the homologues of genes with multiple functions in *D. melanogaster* have been lost less frequently in the honeybee lineage and/or that genes with ancestral conserved functions have been frequently co-opted into new roles in the fruitfly lineage (Dearden *et al.*, 2006).

16.5.2 The evolution of developmental gene networks in relation to adult morphology

The comparative studies reviewed in this chapter clearly demonstrate that genetic networks controlling the development of conserved adult structures (i.e. homologous insect segmental units) can diverge significantly over time due to lineage-specific transitions in cellular embryology associated with changes in life-history strategy. They also suggest that distinct changes in gene networks might underlie convergent transitions in modes of cellular embryogenesis. This implies that over the course of hundreds of millennia, developmental gene networks and the adult morphology they pattern can become 'decoupled'. Assigning homology, or otherwise, to adult morphological features based on comparative developmental genetic data alone is therefore risky. Accurately reconstructing the evolution of animal body plans will require a holistic approach, which includes adequate and intelligent sampling of species (i.e. perhaps less focus on species exhibiting highly derived modes of embryogenesis), a more thorough understanding of the embryological contexts in which gene networks operate, and a better appreciation of how evolutionary changes in life-history strategy (i.e. changes in species ecology) can influence the evolution of development.

CHAPTER 17

Conserved developmental processes and the evolution of novel traits: wounds, embryos, veins, and butterfly eyespots

Patrícia Beldade and Suzanne V. Saenko

17.1 Introduction

The origin and diversification of novel traits is one of the most exciting topics in evolutionary developmental biology (evo-devo). The genetic and developmental mechanisms that underlie morphological innovations are, however, poorly understood, and the fact that no unambiguous definition of novelty exists does not make things easier (see Moczek, 2008; Pigliucci, 2008). While some authors suggest that only a structure for which no homologue can be found in the ancestral species or in the same organism can be considered a morphological novelty (e.g. Müller and Wagner, 1991), others emphasize its ecological importance and define novelty as a new trait that enables new functions and opens up new adaptive zones (Mayr, 1960). The animal kingdom has numerous examples of such adaptive morphological innovations. Feathers of birds, spinnerets of spiders, and carapaces of turtles are unique traits that have played a crucial role in the diversification of these lineages. Analysing the genetic and developmental underpinnings of novel traits, however, can be a challenge when they are not represented in model organisms and the comparative method, so successful in evo-devo, is harder to apply. The co-option of existing genes, pathways, or organs in the evolution of novel traits offers the opportunity to overcome this limitation.

17.1.1 Co-option in the evolution of novelties

Among the different mechanisms that have been proposed to explain the origin of morphological novelties, the co-option, or recruitment, of pre-existing features into performing novel functions has received a great deal of attention (e.g. True and Carroll, 2002; Sanetra et al., 2005). This phenomenon seems to be prevalent and includes the co-option of tissues and organs as well as of single genes and whole developmental pathways, often with modification of components therein. Avian feathers, for example, have evolved from primitive feather-like epithelial outgrowths used for thermoregulation and/or camouflage in non-avian dinosaurs (Prum, 1999), and insect wings and spider spinnerets are both derived from the respiratory organs of the common arthropod ancestor (Damen et al., 2002). The development of horns in a number of beetle species seems to rely on redeployment of the arthropod limb patterning genes Distal-less and aristaless (Moczek and Nagy, 2005). The Wnt signalling pathway, involved in various developmental processes in vertebrates (Logan and Nusse, 2004), has been implicated in the evolution of the turtle shell (Kuraku et al., 2005).

Studies in butterflies and moths provide some spectacular examples of pathways that are shared across insects and have been co-opted in the evolution of colour patterns. The remarkably diverse lepidopteran wing patterns are built up from a mosaic of thousands of flattened pigmented scales produced by wing epidermal cells (Figure 17.1a and Plate 11). It has been proposed that these scales are homologous to insect sensory bristles, and have evolved through the recruitment of bristle-patterning genes of the *Achaete-Scute* complex, followed by acquisition of target genes responsible for typical scale morphology (Galant *et al.*, 1998). The variable colour patterns generated by those wing scales have themselves evolved through the co-option of a number of genetic pathways. The pigments that colour butterfly scales, for instance, derive from the redeployment of several genes from the ommochrome synthesis pathway (e.g. *vermilion, cinnabar*), known to function in insect eye pigmentation (Reed and Nagy, 2005; Reed *et al.*, 2008). The patterns made by the spatial arrangement of coloured scales, on the other hand, rely on genetic pathways (e.g. the Hedgehog and Wingless pathways), involved in embryonic and wing development in butterflies and other insects (Carroll *et al.*, 1994; Monteiro *et al.*, 2006). Such co-option of genetic pathways offers the potential to use extensive knowledge gathered from work on classical model organisms to dissect the formation of butterfly-specific colour patterns.

17.1.2 Butterfly eyespots as an example of an evolutionary novelty

The study of butterfly eyespots, characteristic pattern elements composed of concentric rings of different colours (Figure 17.1b), has started to shed light on how novel patterns have arisen and diversified in the Lepidoptera. Eyespots probably evolved from primitive, uniformly coloured spots through the recruitment and modification of conserved developmental genes and pathways, acquisition of signalling activity, and further diversification of colour schemes under the influence of natural selection (Brunetti *et al.*, 2001; Monteiro, 2008). Their ecological significance in predator avoidance and sexual selection is well documented (Stevens, 2005; Costanzo and Monteiro, 2007), as is the spectacular diversity in eyespot morphology. Variation

in eyespot number, position, shape, size, or colour composition is found not only across species and among individuals of the same species, but often also between different wing surfaces of the same individual butterfly (Nijhout, 1991). Eyespot development is amenable to detailed characterization, ranging from the genetic pathways involved in establishing the pattern, to the molecular and cellular interactions underlying pattern specification, to the biochemical networks involved in pigment production (Beldade and Brakefield, 2002; McMillan *et al.*, 2002).

Models of eyespot formation involve the production and diffusion of one or more signalling molecules from a central organizer, the focus, and the response of the surrounding epithelial cells to the signal(s) in a threshold-like fashion, eventually leading to the production of rings of different pigments (Nijhout, 1980; Dilão and Sainhas, 2004). The organizer properties of the focus are supported by experiments in which transplantation of the focal cells into a different position on the early pupal wing induces the formation of an ectopic eyespot around the transplanted tissue (Nijhout, 1980; French and Brakefield, 1995). The molecular identity of the signal is not known, but both Wingless and Decapentaplegic have been proposed as candidate morphogens (Monteiro *et al.*, 2006). A number of genes have also been implicated in the determination of eyespot centres and colour rings, including members of the Hedgehog pathway (Keys *et al.*,1999), the receptor gene *Notch* (Reed and Serfas, 2004), and the transcription factor-encoding genes *Distal-less, engrailed*, and *spalt* (Carroll *et al.*, 1994; Brakefield *et al.*, 1996; Brunetti *et al.*, 2001). The latter three, for example, are expressed in scale-building cells in association with the different colour rings of the eyespots of several butterfly species (Figure 17.1h). Despite the fact that these and other genes have been implicated in eyespot development, we still know little about the interactions between them, how they regulate pigment synthesis, or the extent to which they contribute to phenotypic variation in eyespot morphology (see Beldade and Saenko, 2009).

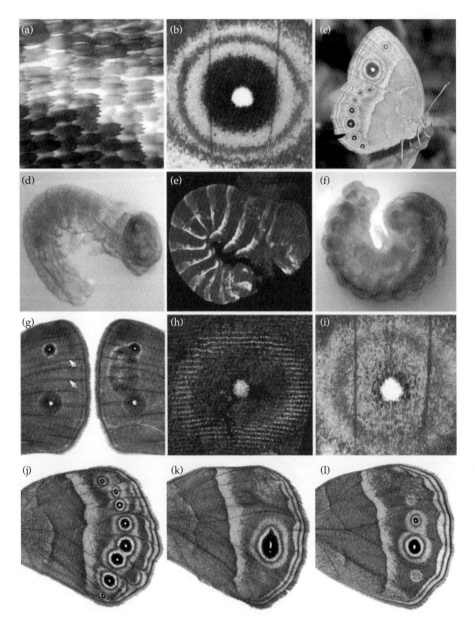

Figure 17.1   Conserved developmental processes implicated in butterfly eyespot formation. Coloured scales covering butterfly wings (a), and eyespot patterns formed by these scales (b), are key innovations in the lepidopteran lineage and are represented in the laboratory-tractable system, *Bicyclus anynana* (c, ventral view of female at rest). The formation of eyespots in *B. anynana* shares genetic commonalities with different conserved developmental processes such as embryonic development (d–f), wound healing (g), and wing vein patterning (j–l). Embryonic development in *B. anynana* has been characterized in wild-type and pleiotropic eyespot mutants (Saenko *et al.*, 2008): wild-type embryo after completion of blastokinesis (d), the characteristic expression of the segment polarity gene *engrailed* at that stage (e), and a homozygous *Goldeneye* embryo of the same age that has failed to undergo blastokinesis (f). Expression pattern of *engrailed* in pupal wings in association with the gold ring of the presumptive adult eyespot (h). This expression is altered in *Goldeneye* eyespots (Brunetti *et al.*, 2001; Saenko *et al.*, 2008) in a manner that matches the change in adult eyespot colour-composition (i). Damage with a fine needle applied to the distal part of the developing pupal wing (arrows in left panel) results in formation of ectopic eyespots around the wound site (right panel) (g). Wing venation mutants often affect eyespot patterns (all photos show the ventral surface of the hindwing): the additional vein in *extra veins* mutants can lead to the formation of an extra eyespot (j), while partial vein loss in *Cyclops* (k) and vestigial venation in *veinless* (l) mutants typically result in changes in eyespot size, number, and/or shape. (See also Plate 11.)

17.1.3 *Bicyclus anynana* as an emerging 'eyespot evo-devo' model

The tropical Nymphalid *B. anynana* (Figure 17.1c) has been established as a laboratory system for studying the reciprocal interactions between evolutionary and developmental processes underlying ecologically relevant phenotypic variation, with emphasis on wing patterns (Brakefield *et al.*, 2009). This system combines knowledge of ecology, often minimal for classical genetic model species, with experimental tractability, including recently developed genomic resources (Beldade *et al.*, 2008) and transgenic techniques (Ramos and Monteiro, 2007), and allows an integrated analysis, from the developmental and genetic basis of eyespot formation to the molecular underpinnings of variation in natural populations.

Here we focus on butterfly eyespots as an example of a morphological novelty and review recent findings relating to the co-option of conserved developmental processes in the evolution of eyespot patterns. First, we discuss experimental evidence for the similarities between eyespot patterning and wound repair. We then give examples of pleiotropic mutations isolated in *B. anynana* laboratory populations which affect not only eyespots but also either embryonic or wing vein development, and discuss how studies of such mutations may help elucidate the genetic pathways involved in eyespot formation and variation. A detailed analysis of such conserved genetic networks, extensively studied in the model organism *Drosophila melanogaster*, in the context of eyespot formation will be invaluable for our understanding of the evolutionary origin and diversification of butterfly eyespots.

17.2 Wounds and eyespots

The ability to repair wounded tissue is a fundamental property of all multicellular organisms, and a key topic of current research (see Gurtner *et al.*, 2008). Here we review evidence suggesting that some components of this process are involved in eyespot formation.

17.2.1 Damage-induced eyespot formation in *B. anynana*

Local damage to pupal wing tissue has long been known to disturb colour patterns in many lepidopterans, and has been used to study the mechanisms of pattern formation in butterflies and moths (e.g. Kühn and Von Englehardt, 1933). In eyespot-bearing butterfly species, for example, damage with a very fine needle applied to the presumptive eyespot in early pupae can completely eliminate eyespots in the adults (Nijhout, 1980; French and Brakefield, 1992). Also, damage to other locations on the pupal wing epidermis can result in the formation of an ectopic eyespot around the wound site (French and Brakefield, 1995). In *B. anynana*, for example, rings of black and/or gold, typically poorly defined and less symmetrical than those of the native eyespots, are found around the healed wound (Figure 17.1g). Interestingly, this type of damage produces eyespots only on the distal area of the wing, and only if applied during a very narrow time window (Brakefield and French, 1995), which more or less corresponds to the period when 'colour-ring genes', *Distal-less*, *engrailed*, and *spalt*, are upregulated in the presumptive eyespot area (Monteiro *et al.*, 2006). The mechanisms by which the genes and pathways of the damage response machinery might contribute to the formation of ectopic eyespots are as yet unclear, but insights from studies in model organisms might provide some clues.

17.2.2 Genetic mechanisms of wound repair

Comparative studies on the genetic and cellular mechanisms of wound repair and regeneration in representatives of various animal phyla have suggested their evolutionary conservation (see Woolley and Martin, 2000). For example, some steps in the wound healing process are regulated by the same transcription factor, Grainyhead, in flies and mice (Mace *et al.*, 2005; Ting *et al.*, 2005). Also, wound healing seems to recapitulate some aspects of embryonic morphogenesis, such as dorsal closure in flies and eyelid fusion in mice, raising the possibility of co-option of genetic pathways active during embryogenesis into wound repair processes during adult life (see Martin and Parkhurst, 2004).

The mechanism by which damage results in the formation of ectopic eyespots is an intriguing question. It is known that candidate genes for eyespot signalling also perform functions related to wound healing. Studies in fly larvae, for example, have shown that wounds act as sources of short- and long-range signalling molecules and activate downstream pathways in a gradient-like manner, as has been proposed for eyespot development. In *Drosophila*, the Jun N-terminal kinase (JNK) pathway is activated in a gradient centred around the wound (Galko and Krasnow, 2004) and up-regulates the transcription factor AP-1 which, in turn, leads to induction of *decapentaplegic*, a transforming growth factor (TGF)-β family member and one of the candidate eyespot-inducing signals (Monteiro *et al.*, 2006). Some evidence also exists for the involvement of the Wnt proteins in wound repair in mammals (e.g. Okuse *et al.*, 2005), whereas their insect homologue, Wingless, is a candidate morphogen in eyespot formation (Monteiro *et al.*, 2006).

To investigate whether the same genetic mechanisms that underlie wound healing are implicated in damage-induced eyespot formation in butterflies, Monteiro *et al.* (2006) used immunostaining to detect the expression of known 'eyespot genes' around damage sites in developing pupal wings. They showed that these genes were upregulated in the epidermal cells surrounding the wound (Monteiro *et al.*, 2006) and proposed that the formation of both native and ectopic eyespots relies on at least some identical molecules. Their data, however, do not distinguish between the possibility that the rings of expression of the tested eyespot genes around wound sites are induced by known candidate eyespot morphogens (Monteiro *et al.*, 2006), or whether they result from some other, as yet unidentified, signals produced by wounded cells. Future work will focus on the identification of genes and pathways that are upregulated upon wing damage and lead to the formation of ectopic eyespots.

17.3 Embryos and eyespots

Among the spontaneous mutants maintained in laboratory populations of *B. anynana* (Brakefield *et al.*, 2009), some have been characterized that are embryonic lethal in the homozygous state and have a dramatic effect on eyespot morphology in heterozygotes (Saenko *et al.*, 2008). Comparative analysis of disturbed embryonic development in these mutants with similar phenotypes described in model insects might help identify genes involved in eyespot formation. The mechanisms of early embryonic development are well studied in the dipteran *D. melanogaster* (Peel *et al.*, 2005), and are becoming increasingly well understood in representatives of other insect orders, such as the coleopteran *Tribolium castaneum* and the hemipteran *Oncopeltus fasciatus* (reviewed in Liu and Kaufman, 2005), the hymenopteran *Nasonia vitripennis* (e.g. Pultz *et al.*, 2005; Lynch *et al.*, 2006a), and the lepidopterans *Bombyx mori* (e.g. Nagy, 1995) and *Manduca sexta* (e.g. Kraft and Jackle, 1994). To the extent that the genetic mechanisms of embryogenesis are conserved across insects (see Peel *et al.*, 2005; Damen, 2007; see also Chapter 16), a comparison of disturbed embryonic development in *B. anynana* eyespot mutants with mutants in insect model species can help identify signalling pathways and/or specific genes involved in eyespot formation and variation.

Embryonic development in *B. anynana* (described in Saenko *et al.*, 2008) is characterized by a long-germ mode of segmentation, as in most other lepidopterans and dipterans (Kraft and Jackle, 1994), but differs from that of *D. melanogaster* in some important aspects. For example, at about half way through embryonic development, lepidopterans go through a characteristic movement within the egg that results in the reversal from a ventral to a dorsal flexion, called blastokinesis (Broadie *et al.*, 1991; Figure 17.1d). Other aspects of embryonic development, such as segment patterning by segment polarity and Hox genes (Figure 17.1e) and limb patterning by *Distal-less*, are similar in *B. anynana* and other insects (Saenko *et al.*, 2008). This suggests that the direct comparison of disrupted embryonic development in pleiotropic *B. anynana* eyespot mutants with described mutants in model insects could be a useful approach for identifying new pathways and candidate genes for eyespot development (Saenko *et al.*, 2008).

17.3.1 The homozygous lethal mutation Goldeneye

Goldeneye is one such pleiotropic mutation that is embryonic lethal in homozygotes (Saenko *et al.*, 2008) and affects eyespot colour composition in heterozygous adults (Brunetti *et al.*, 2001). In *Goldeneye* butterflies, the scales that typically form the black inner ring of *B. anynana* eyespots are replaced with gold-coloured scales characteristic of the outer ring (Figure 17.1i), and the expression of genes associated with the different colour rings changes accordingly (Brunetti *et al.*, 2001). Analysis of embryonic development in *Goldeneye* homozygotes has shown that these embryos do not undergo blastokinesis, subsequently become shorter and thicker than wild-type embryos of the same age, and end up dying shortly after the time at which blastokinesis would normally have occurred (Saenko *et al.*, 2008; Figure 17.1f).

The study of disturbed embryogenesis in *Goldeneye* has suggested that genes involved in blastokinesis may play a role in eyespot formation (Saenko *et al.*, 2008). Unfortunately, the specific genetic regulation of embryonic movements in insects, including blastokinesis in the Lepidoptera, is poorly understood, despite the fact that several mutations have been identified that disturb these processes (e.g. Ueno *et al.*, 1995; Schock and Perrimon, 2002; Van der Zee *et al.*, 2005; Panfilio *et al.*, 2006). Even though the *Goldeneye* embryonic phenotype did not show resemblance to described mutants in other species, and despite the fact that we do not know whether embryonic movements are regulated by similar mechanisms across insect lineages, analysis of candidate genes affecting this process seems like a valuable starting point for further genetic dissection of variation in eyespot morphology.

17.3.2 Conservation versus divergence in insect embryonic development

The strategy of comparing disturbed embryonic phenotypes between *B. anynana* eyespot mutants and mutants in genetic model systems will be useful only to the extent that the genetic mechanisms of embryonic development are conserved across insects. Recent studies extending the analysis of insect embryonic development outside *D. melanogaster* (see Peel *et al.*, 2005, Damen, 2007; Peel, 2008) have shown that while some aspects of embryonic development are indeed highly conserved (e.g. the functions of segment polarity and Hox genes), others appear to be unexpectedly diverged (e.g. the functions of gap and pair-rule genes; see Chapter 16).

The analysis of embryonic lethality in three other pleiotropic *B. anynana* mutants, in which development appears to be disturbed during the segmented germband stage, is more promising for identifying mutated genes (P. Beldade and S. V. Saenko, unpublished data). Unlike blastokinesis, this stage of embryogenesis is highly conserved among arthropods (e.g. Farzana and Brown, 2008), and the genes and developmental pathways that regulate it have been studied in greater detail in model organisms (see Galis *et al.*, 2002). Still, whatever the embryonic stage affected in any pleiotropic eyespot mutant, because direct comparison of disturbed eyespot phenotypes with 'eyespot mutants' in model species is impossible (model insects do not have eyespots!), comparative analysis of embryonic phenotypes in such mutants remains a valuable first approach.

17.4 Veins and eyespots

Models for butterfly wing pattern formation propose an important role for wing veins and the wing margin (see Nijhout, 1991). An association between wing venation and either patterns of colourful stripes and bands or eyespot formation has been supported by the phenotypic characterization of spontaneous venation mutants in *Papilio* and *Heliconius* species (Koch and Nijhout, 2002; Reed and Gilbert, 2004) and in *B. anynana* (Saenko *et al.*, 2008), respectively. Models of eyespot formation have suggested that wing veins act as sources of diffusible molecules involved in the determination of eyespot-organizing centres, the foci (see Nijhout *et al.*, 2003; Evans and Marcus, 2006), but this role, as well as the nature or even the existence of the proposed diffusible signals, has not yet been shown experimentally.

17.4.1 Parallels between fruitfly and butterfly vein development

The mechanisms of vein patterning in *Drosophila* are fairly well understood (reviewed in Blair, 2007), as is the role of veins in the distribution of melanin precursors in newly eclosed fruitflies (True *et al.*, 1999). As is often the case for work in non-model systems, this knowledge is an invaluable starting point for our understanding of the mechanisms behind vein establishment and its role in pattern formation in butterfly wings.

Unsurprisingly, positional specification in butterfly wing discs seems to be achieved in a manner very similar to that described for *Drosophila*. Division of the developing wing discs into antero-posterior and dorsoventral compartments is marked by the expression of the genes *engrailed* and *apterous*, respectively, and proximal–distal patterning is regulated by *Distal-less* and *wingless* (Carroll *et al.*, 1994). The signalling pathways that are involved in the positioning and differentiation of longitudinal and cross veins in *Drosophila* (reviewed in Marcus, 2001, and Crozatier *et al.*, 2004) might also be conserved between the lineages of Diptera and Lepidoptera (De Celis and Diaz-Benjumea, 2003). Functional analysis of homologues of known *Drosophila* vein patterning genes during butterfly wing development will be instrumental in our understanding of vein establishment and role in butterfly wings.

17.4.2 Wing venation and eyespot patterns in *B. anynana* mutants

Study of *B. anynana* mutants with disturbed venation has started to explore the functional relationship between wing veins and eyespot formation, and has suggested that eyespot patterning depends on normal formation of veins and tracheae (Saenko *et al.*, 2008). Mutations that lead to the addition of veins, such as *extra veins* (Figure 17.1j) can also lead to the appearance of extra eyespots, presumably when the ectopic vein bisects an existing

eyespot signalling centre or because the additional vein itself acts as an inducer of eyespot formation. Conversely, partial or complete loss of wing veins often leads to formation of smaller, fewer and/or misshaped eyespots, as in the *Cyclops* (Figure 17.1k) and *veinless* (Figure 17.1l) mutants of *B. anynana* (Saenko *et al.*, 2008).

Surgical manipulations on developing wings of *B. anynana* venation mutants have provided the first insights into the mechanisms by which loss of veins can interfere with eyespot formation. The absence of eyespots on the dorsal surface of *veinless* individuals, investigated by transplanting the signalling focus of a wild-type pupa into a *veinless* wing, has shown that vestigial venation leads to impaired determination of the eyespot focus or lack of focal signal (Saenko *et al.*, 2008). The molecular mechanisms of such relationship are yet to be explored.

17.5 Concluding remarks

Here we have reviewed knowledge of the genetic and developmental mechanisms of eyespot formation, and discussed new approaches to the study of these lineage-specific structures, based on the commonalities with conserved developmental processes such as wound healing, embryonic development, or vein patterning.

How much the study of laboratory populations can tell us about variation in natural populations remains a crucial issue in evo-devo. In particular, the extent to which the mutants of large effect identified in the laboratory are relevant for natural variation within and across species is still debated (see Haag and True, 2001). Whilst it seems unlikely that pleiotropic mutations with negative effects on other traits (e.g. embryonic lethality in *Goldeneye*, or fragile wings in mutants with vestigial venation) will contribute to variation in natural populations, it is possible that the same loci harbour other, less deleterious, alleles relevant to variation in eyespot patterns. Future work will explore the extent to which loci identified in laboratory eyespot mutants contribute to variation segregating in natural populations and to variation across species.

17.6 Acknowledgements

We thank Max Telford and Tim Littlewood for organizing the Novartis Foundation Symposium on Animal Evolution and inviting us to contribute to this book. We also thank Paul Brakefield, Vernon French, and Antónia Monteiro for many inspiring discussions on butterfly wing patterns.

CHAPTER 18

Reassembling animal evolution: a four-dimensional puzzle

Maximilian J. Telford and D. Timothy J. Littlewood

18.1 Introduction

Drawing from the latest literature and the contributions in this volume, we consider some of the recent progress made in the study of animal evolution and the hurdles that remain. Each of the disciplines considered—palaeontology, evo-devo, phylogenetics, and the incorporation of genomic data—have made major contributions to our understanding of how animals have diversified. Together, these pursuits are resulting in a return to whole-organism biology where the link between genotype and phenotype is considered in the context of changing physical and biological environments. The modern approach integrates across all these sometimes disparate disciplines, with the aim of reconciling available evidence to describe the patterns and processes that have led to the existing diversity of animal life.

Arguably, there is one underlying common quest that unites the goals of individual researchers: the search for homology—recognizing it, defining it, and using it. Whether it is establishing shared common ancestry of form or function, similar challenges face those contemplating strings of nucleotides, protein structure, gene expression, biochemical pathways, organs systems, or fossilized microstructures. As we move towards a greater understanding of evolution and the biological entities undergoing selection, it is the study of homology that allows us to detect patterns and interpret processes.

Gaps in our knowledge can be daunting. At best they define the limits of our ignorance, and at worst they prevent any meaningful or confident interpretation of available information. We consider how some of the major gaps are being addressed with the renaissance of whole-organism biology, the development of improved models, and the advent of new technologies.

18.2 Phylogenies and phylogenetics

Since the first credible molecular estimate of animal relationships was published by Field *et al.* (1988) there have been a number of significant changes in our understanding of the evolution of the animal kingdom. The largest shift has been from the widely held assumption of gradualism, whereby morphologically simpler animals such as flatworms were placed towards the base of the tree, and complex features such as coeloms and segments were thought to be homologous and to define major groups of animals higher up the tree. The tree widely accepted today has its roots firmly in Field *et al.*'s study, and subsequent studies adding to the sampling of small subunit (SSU) ribosomal RNA gene (rDNA) sequences; the major revolutions have, until recently, almost all come from efforts using SSU rDNA. Terms such as Ecdysozoa and Lophotrochozoa draw upon shared morphological features, but their roots stem from SSU rDNA. The new animal phylogeny, hand in hand with comparative developmental studies of homologous gene expression, has forced a reassessment of the evolution and homology of many characteristics of animals; a recognition of the pervasive effects of the loss of characters and secondary simplification of body plans (Copley *et al.*, 2004; Jenner, 2004c) as is apparent in the flatworms.

While there has been enormous progress in our understanding of metazoan phylogeny leading to broad agreement over the outline of the animal tree (Halanych, 2004; Telford, 2006), there remain a number of hotly contested questions in metazoan phylogeny; with inevitability, the outstanding questions are the hardest to answer and the difficulties encountered are likely to stem from multiple sources. The first major source of difficulty occurs when the living phyla emerged in an explosive radiation leaving little chance for the fixation of informative substitutions; such a situation is exemplified by the difficulty of resolving relationships between the lophotrochozoan clades (Dunn *et al.*, 2008). The second important source of difficulty arises when living exemplars are the result of unusual patterns of genomic evolution that violate assumptions of models used to reconstruct trees, resulting in inaccuracies in their placement on the tree (Philippe and Telford, 2006). This is undoubtedly seen in the case of the acoel flatworms, chaetognaths, myzostomids, gnathostomulids, and various other 'Problematica'.

The tendency for phylogeneticists to contradict each other over the placement of problematic groups may be rather frustrating to outsiders but is inevitable. First, all animals that have ever been described have also been positioned somewhere on a phylogenetic tree. Any progress to be made inevitably involves changing this position and hence introduces contradiction. Secondly, and alluded to above, all the easily solved aspects of the tree were answered 10 or 20 years ago, meaning anything currently worth studying is by definition problematic. A reliable phylogeny is fundamental to comparative biology and to our understanding of evolution, and progress continues.

The progress currently being made stems from the combination of four approaches; much larger data sets (phylogenomics) which avoid stochastic error from limited samples; data from additional representatives of problematic taxa to avoid or reduce systematic error; alternative sources of data (e.g. microRNAs) and, potentially, other rare genomic changes which it is hoped are resistant to homoplastic evolution (Rokas and Holland, 2000;

Boore, 2006); and finally, improved methods of tree reconstruction that more accurately model the underlying process of molecular evolution so reducing further the possibility of stochastic error (Philippe and Telford, 2006). The biggest contributors to progress in terms of data are the new, cheap technologies for DNA sequencing. We are not far from the day when any given species (with a 'normal' sized genome) will have its genome completely sequenced for less than the sum that a single gene may have cost 25 years ago. This will provide the greatest possible source of data for phylogenetic analysis and the resolution of any remaining errors will be the province of the model makers.

18.3 Palaeontology

The frustrations inherent in reconstructing the phylogeny of living animals are echoed by the problems of palaeontology. Many fossils are hard to decipher, especially for outsiders, and confusion is exacerbated by the vehement disagreements over their interpretation by the experts. As an example, the Lower Cambrian *Emmonaspis cambrensis* has been linked with graptolites (hemichordates), chordates and arthropods, and even with Ediacaran frond-like organisms since its description in 1886 (Conway Morris, 1993b). Beyond the well-known problems of preservation and interpretation (Budd and Jensen, 2000), the most interesting fossils—those in the stem lineages of living taxa with the potential to show the order of acquisition of clade synapomorphies—are the hardest to interpret and to relate to modern groups by their very lack of synapomorphies.

Despite the undoubted problems of palaeontology, fossils are unique in their ability to inform us about certain aspects of evolution (Smith, 1994). While comparisons of living taxa within an accurate phylogenetic framework give tremendous insight into the pattern of evolution, this approach remains limited by the fact that most of the steps of evolution leading to living clades are absent. As an example, it seems clear that the closest relatives of the arthropods are to be found amongst the cycloneuralian worms. It is not clear, however, how much a comparison of priapulids and arthropods will tell us about the stages by which segments and jointed appendages were acquired in the arthropod stem; in such a case, fossils can be of enormous importance.

The importance of studying fossil lineages for our understanding of the evolution of crown groups has been discussed. Stem-lineage fossils make an important contribution in several ways; they break long branches leading to crown groups and show intermediate character states; they may reveal unsuspected character homologies or indeed convergent evolution between extant groups; they can highlight character loss in certain groups; and, finally, they provide the sole means to calibrate evolutionary trees by giving minimum divergence times of living clades. Fossils are also able to provide the ecological background to specific evolutionary events, perhaps most spectacularly the great extinctions and the invasion of new habitats such as the land. All of this information is provided uniquely by fossils; it is vital that evolutionary biologists do not damn fossil evidence too readily based on the difficulties inherent in the field. Palaeontologists themselves recognize the problems they face, and efforts are being made to strengthen the objectivity of fossil interpretation and to understand the limits of inference; e.g. in calibrating trees (Drummond *et al.*, 2006; Marshall, 2008), and the interpretation of biological evidence for historical events (Budd and Jensen, 2000; Domazet-Los *et al.*, 2007; Peterson *et al.*, 2007; Donoghue and Purnell, 2009). Newly discovered deposits, new tools to visualize internal and microscopic features, new methods of detecting and characterizing biomolecules, and simply returning repeatedly to problematic taxa in the light of new evidence will keep the study of fossils alive.

18.4 Developmental evolution

A phylogenetic tree can describe the relationships of species of living and fossil taxa; mapping the characteristics of those taxa onto the framework of the tree permits us to track the evolution of those characters, showing in which groups—and even at what time—key morphological novelties have evolved. While this combination of a dated phylogenetic framework and the distribution of

characters provides a historical description or pattern of character evolution, to understand morphological novelty and how such morphological change has occurred at the level of the genome and the embryo (the process of morphological evolution) we need to study the genetics behind changes in ontogeny (see, for example, Moczek, 2008).

The birth of modern developmental evolutionary biology came 25 years ago with the molecular cloning of the homeobox motif from *Drosophila* homeotic genes (Carrasco *et al.*, 1984; McGinnis *et al.*, 1984) alongside the amazing discovery that the same motif (and indeed the same genes) existed in vertebrates with conserved functions. Comparative molecular genetic analyses of development have since changed our view of the evolution of developmental mechanisms and the origins of novel morphology, revealing surprising conservation and providing an alternative to phylogenetic proximity for determining homology. The promise of current evo-devo research is to expand the focus of research to new groups of organisms. While a great deal of progress continues to be made using comparisons of expression patterns (using *in situ* hybridization) for detecting similarity of function of homologous genes and identifying homology of characters, the export of genomics and true functional studies (e.g. RNA interference and transgenesis) to animals not previously considered model organisms is extremely exciting (see, for example, Abzhanov *et al.*, 2008, and Vera *et al.*, 2008).

By expanding beyond the traditional model organisms, practitioners of developmental evolutionary biology are able to build on the discoveries of the phylogeneticists and palaeontologists to address some of the more intriguing questions in morphological evolution. Current questions revealed by the new animal phylogeny and palaeontological discoveries include the origins of arthropods from the cycloneuralian worms such as priapulids and kinorhynchs, the unexpected relationship of the deuterostome-like brachiopods to lophotrochozoans such as annelids and molluscs, and the possible origins of bilaterians from animals resembling the acoel flatworms.

In addition to investigating specifics such as those questions mentioned above, another focus of developmental evolutionary studies is the generalities of the genetics behind morphological evolution. A current debate concerns the relative importance of changes in regulatory DNA versus coding DNA of genes (Carroll, 2008; Stern and Orgogozo, 2008; Wagner and Lynch, 2008). One thing on which both sides seem to agree, however, and perhaps this realization is more fundamental than scoring points, is that changes of small effect predominate. *Cis*-regulatory changes are common due to the possibility of making subtle changes in independent enhancers, and coding changes occur where their pleiotropic effects are minimized. There is nothing new under the sun, however (*Ecclesiastes* 1:9–14), and this debate harks back, of course, to R. A. Fisher's analogy of the focusing of a microscope using small adjustments (Fisher, 1930).

18.5   Mind the gaps

Addressing what is missing in the study of animal evolution is unavoidable and necessary, not least because it demonstrates openness, attempts to define the limits of our knowledge, and indicates possible directions for future research. The influence of missing empirical information can be substantial, and assessing the impact of missing fossils, missing taxa, and missing data is almost a discipline itself in systematics. What is not known can influence estimates of tree topology and stability and the biological inferences we are prepared to make (see Wiens, 2006; Geuten *et al.*, 2007; Fitzhugh, 2008). In phylogeny, should missing features be scored as losses or simply missing data, and when are multiple related missing features indicative of single losses (e.g. the deletion of strings of nucleotides or the loss of entire organs systems)? In palaeontology and evo-devo, when can absence of evidence be used as evidence of absence?

Incomplete information necessarily pushes us either towards caution, in the fear that any inferences from gappy data may be deemed premature, or towards bravery (perhaps even foolhardiness) as the constant need to take stock of available evidence forces phylogenetic estimates, character mapping, taxonomic revisions, recalibrated histories, and the desire to provide a narrative that explains biodiversity through space and time. Diligent

researchers are keen to indicate the strength of their arguments by circumscribing the limits and possible influence of what is not known, at the risk of undermining any conclusions drawn from what is known. In contrast, selective sampling can provide more robust arguments and may obviate the need to consider uncertainty or less compelling scenarios. Though we do not set out to sample selectively, the nature of certain data sets puts us firmly at the mercy of exemplars. Just as the early days of SSU rDNA estimates of animal phylogeny relied on single taxa as representatives of entire phyla, we have seen phylogenomic analyses suffering from over-representation of taxonomically biased model organisms or unbalanced data sets as more or fewer expressed sequence tags (ESTs) are recruited for analysis from unrelated research. Using all available evidence from GenBank to estimate animal interrelationships would be cumbersome and unwise, but that is not to say we should not consider all the available data for statements on homology, and sample them for balanced representative data sets.

Balancing taxon and character sampling is difficult, and has been the focus of empirical and theoretical studies (e.g. Graybeal, 1998; Pollock *et al.*, 2002), but there is little doubt that with each new data set we are liable to repeat the mistakes of insufficient or biased sampling. In many cases we simply do not know that our sampling is insufficient or biased, or may not be able address any shortfalls until new data sets become available. Many gaps in phylogenetic data sets await attention on key taxa for known characters that need to be scored. Meanwhile, expert morphologists and taxonomists are declining in number, character coding is frequently controversial, archival specimens may not be available or suitable for sampling the missing data, and the animals may be difficult to sample, being rare, cryptic, geographically isolated, elusive, or extinct. We need to live with gaps but also to recognize the need to address them when the opportunity arises.

The age of genomics arrived with the expectation that knowledge of complete genetic blueprints would provide a surfeit of phylogenetic information for robust tree reconstruction. This has yet to occur, since our efforts to uncover form, function, and homology have been achieved for very few components of genomes (Kuzniar *et al.*, 2008). For animal evolutionary biologists the era of post-genomics is a long way off, not just because of the lack of understanding of available genomes, but also because of the lack of characterized genomes themselves. Sampling systematically across the animal tree of life is an important strategy in developing comparative genomic data sets, but until now evolutionary biologists have rarely dictated sampling priorities. Furthermore, even a cursory look at the revolutions in molecular systematics show how sampling just a few key taxa can upset the entire understanding of animal evolution. For example, it was preliminary molecular systematic surveys of flatworms that highlighted the phylogenetic uniqueness of acoelomorph flatworms (Carranza *et al.*, 1997; Littlewood *et al.*, 1999) and that led ultimately to their current status, their distinctness from the Platyhelminthes and their importance as links to our deep bilaterian past (Baguñà *et al.*, 2008; Hejnol and Martindale, 2008b). Undoubtedly, denser sampling of animal genomes will provide more surprises.

Whilst evolutionary biologists are constantly concerned with homology either implicitly or explicitly (see recent review by Szucsich and Wirkner, 2007), large-scale data sets are moving us away from an intimate understanding of all the statements of homology that we make or rely upon. To some, this may appear to be neglecting our responsibility as those whose task it is to detect, highlight, and interpret the evidence for shared ancestry. Recently there has been a shift from poring over nucleotide and amino acid alignments with reference to secondary structures, open reading frames, and function, where indels (insertion/ deletion markers) might be placed judiciously and exclusion sets chosen carefully, to a need for automation in order to harness considerable volumes of data (Wong *et al.*, 2008). A plethora of data requires the building and implementation of bioinformatic pipelines to make many of these decisions for us, swiftly, consistently (with given criteria), and routinely in the hope that we are minimizing noise and maximizing signal. Whilst these routines and algorithms might be borne of an understanding of the underlying data, such automated efforts do not negate the need to make evolutionary sense of the biological data, and we must be wary of opening new gaps in our understanding.

18.6 Learning from the past and taking advantage of the present

In an era dominated by unprecedented access to information, we have an opportunity for embracing considerable bodies of primary data, meta-data, and the thoughts and arguments of generations of researchers. Global efforts to digitize literature and specimens, internet tools that mine, parse, and link databases, and concerted global efforts by a generation of researchers willing to synthesize existing information are generating new understanding, whilst complementary efforts by others to generate primary data continue unabated. Indeed, the increase in rate at which gene sequence data can now be generated with second-generation sequencing is phenomenal, and third-generation sequencing, now on the horizon, promises orders of magnitude more data (Shendure and Ji, 2008). The information revolution is vast in scale and breadth and brings with it new powers and challenges, not least for bioinformaticians (Helaers *et al.*, 2008; Pop and Salzberg, 2008). New ways of studying genomes and inferring historical events challenge underlying philosophies and resurrect arguments against phenetics, but there is little doubt that presence/absence of genes, gene networks and biochemical pathways, relative arrangement of genes, and so on, provide an entirely new vocabulary with which to consider the past (Boore, 2006; Ding *et al.*, 2008; Dulith *et al.*, 2008).

Although we strive for pragmatic approaches to the onslaught of information, and welcome the opportunities to bring disparate fields back into the fold, caution is always at the back of our minds. For example, although we might expect to be able to access information at the click of a mouse, at what point should we select the following without a second thought: a gene sequence with no associated voucher specimen, a distribution map based on inaccurate identifications or DNA barcodes, a tree topology based on data we have not seen, a cluster of genes we have not verified as being orthologous, a supertree? Clearly, no individual can make all these decisions independently and it is as a community that we police ourselves, and the data we choose to accept as fit for purpose. Systematics continues to be about maximizing the signal and minimizing the noise, but there is a constant battle against a modern trend towards 'one-gene-fits-all' approaches, the undermining of systems that 'ain't broke and don't need fixing' (e.g. the Linnean system for classification), a lack of rigour in the understanding or implementing the tools (and underlying philosophies) of the trade, and false claims as to how we will have catalogued or barcoded every species on the planet and resolved the position of every twig of the tree of life within the next 25 years. Rhetoric aside, there has been no better time to study animal evolution.

References

Abedin, M. and King, N. (2008) The premetazoan ancestry of cadherins. *Science* **319**, 946–948.

Abouheif, E., Zardoya, R., and Meyer, A. (1998) Limitations of metazoan 18S rRNA sequence data: implications for reconstructing a phylogeny of the animal kingdom and inferring the reality of the Cambrian explosion. *Journal of Molecular Evolution* **47**, 394–405.

Abzhanov, A., Extavour, C.G., Groover, A., Hodges, S.A., Hoekstra, H.E., Kramer, E.M. *et al.* (2008) Are we there yet? Tracking the development of new model systems. *Trends in Genetics* **24**, 353–360.

Acampora, D., Annino, A., Tuorto, F., Puelles, E., Lucchesi, W., Papalia, A. *et al.* (2005) *Otx* genes in the evolution of the vertebrate brain. *Brain Research Bulletin* **66**, 410–420.

Ackermann, C., Dorresteijn, A., and Fischer, A. (2005) Clonal domains in postlarval *Platynereis dumerilii* (Annelida: Polychaeta). *Journal of Morphology* **266**, 258–280.

Adamska, M., Matus, D.Q., Adamski, M., and Green, K. (2007a) The evolutionary origin of hedgehog proteins. *Current Biology* **17**, R836–R837.

Adamska, M., Degnan, S.M., Green, K.M., Adamski, M., Craigie, A., Larroux, C. *et al.* (2007b) Wnt and TGF-β expression in the sponge *Amphimedon queenslandica* and the origin of metazoan embryonic patterning. *PLoS ONE* **10**, e1031.

Adell, T., Nefkens, I., and Müller, W.E.G. (2003) Polarity factor 'Frizzled' in the demosponge *Suberites domuncula*: identification, expression and localization of the receptor in the epithelium/pinacoderm. *FEBS Letters* **554**, 363–368.

Adell, T., Thakur, A.N., and Müller, W.E.G. (2007) Isolation and characterization of Wnt pathway-related genes from Porifera. *Cell Biology International* **31**, 939–949.

Adoutte, A., Balavoine, G., Lartillot, N., and de Rosa, R. (1999) Animal evolution. The end of the intermediate taxa? *Trends in Genetics* **15**, 104–108.

Adoutte, A., Balavoine, G., Lartillot, N., Lespinet, O., Prud'homme, B., and de Rosa, R. (2000) The new animal phylogeny: reliability and implications. *Proceedings of the National Academy of Sciences USA* **97**, 4453–4456.

Aguinaldo, A.M., Turbeville, J.M., Linford, L.S., Rivera, M.C., Garey, J.R., Raff, R.A. *et al.* (1997) Evidence for a clade of nematodes, arthropods and other moulting animals. *Nature* **387**, 489–493.

Ahlrichs, W. (1995) *Ultrastruktur und Phylogenie von Seison nebaliae (Grube 1859) und Seison annulatus (Claus 1876). Hypothesen zu phylogenetischen Verwandtschaftsverhältnissen innerhalb der Bilateria.* Cuvillier Verlag, Göttingen.

Åkesson, B. (1967) The embryology of the Polychaete *Eunice kobiens. Acta Zoologica* **48**, 141–192.

Aldridge, R.J., Xian-Guang, H., Siveter, D.J., and Gabbott, S.E. (2007) The systematics and phylogenetic relationships of vetulicolians. *Palaeontology* **50**, 131–168.

Allen, J.D. and Pernet, B. (2007) Intermediate modes of larval development: bridging the gap between planktotrophy and lecithotrophy. *Evolution & Development* **9**, 643–653.

Alonso, C.R. and Wilkins, A.S. (2005) The molecular elements that underlie developmental evolution. *Nature Reviews Genetics* **6**, 709–715.

Alwes, F. and Scholtz, G. (2004) Cleavage and gastrulation of the euphausiacean *Meganyctiphanes norvegica* (Crustacea, Malacostraca). *Zoomorphology* **123**, 125–137.

Ambros, V. and Chen, X. (2007) The regulation of genes and genomes by small RNAs. *Development* **134**, 1635–1641.

Ambros, V., Bartel, B., Bartel, D.P., Burge, C.B., Carrington, J.C., Chen, X. *et al.* (2003) A uniform system for microRNA annotation. *RNA* **9**, 277–279.

Amthor, J.E., Grotzinger, J.P., Schröder, S., Bowring, S.A., Ramezani, J., Martin, M.W. *et al.* (2003) Extinction of *Cloudina* and *Namacalathus* at the Precambrian-Cambrian boundary in Oman. *Geology* **31**, 431–434.

Anbar, A.D. and Knoll, A.H. (2002) Proterozoic ocean chemistry and evolution: a bioinorganic bridge? *Science* **297**, 1137–1142.

Anderson, F.E. and Swofford, D.L. (2004) Should we be worried about long-branch attraction in real data sets?

Investigations using metazoan 18S rDNA. *Molecular Phylogenetics and Evolution* **33**, 440–451.

Antcliffe, J.B. and Brasier, M.D. (2007) *Charnia* and sea pens are poles apart. *Journal of the Geological Society* **164**, 49–51.

Antebi, A. (2006) Nuclear hormone receptors in *C. elegans* (January 03, 2006). In: *WormBook*, The *C. elegans* Research Community (eds), doi/10.1895/wormbook.1.64., http://www.wormbook.org/.

Arduini, P., Pinna, G., and Teruzzi, G. (1981) Megaderaion sinemuriense, a new fossil enteropneust of the Sinemurian. *Atti della Societa Italiana de Scienze Naturale e del Museo Civico di Storia Naturale de Milano* **122**, 104–108.

Arenas-Mena, C., Cameron, A.R., and Davidson, E.H. (2000) Spatial expression of Hox cluster genes in the ontogeny of a sea urchin. *Development* **127**, 4631–4643.

Arendt, D. (2004) Comparative aspects of gastrulation. In: C.D. Stern (ed.) *Gastrulation*, pp. 679–693. Cold Spring Harbor Laboratory Press, New York.

Arendt, D. and Nübler-Jung, K. (1994) Inversion of dorso-ventral axis? *Nature* **371**, 26.

Arendt, D. and Nübler-Jung, K. (1996) Common ground plans in early brain development in mice and flies. *BioEssays* **18**, 255–259.

Arendt, D. and Nübler-Jung, K. (1997) Dorsal or ventral: similarities in fate maps and gastrulation patterns in annelids, arthropods and chordates. *Mechanisms of Development* **61**, 7–21.

Arendt, D. and Nübler-Jung, K. (1999) Comparison of early nerve cord development in insects and vertebrates. *Development* **126**, 2309–2325.

Arendt, D., Technau, U., and Wittbrodt, J. (2001) Evolution of the bilaterian larval foregut. *Nature* **409**, 81–85.

Arendt, D., Tessmar-Raible, K., Snyman, H., Dorresteijn, A.W., and Wittbrodt, J. (2004) Ciliary photoreceptors with a vertebrate-type opsin in an invertebrate brain. *Science* **306**, 869–871.

Aris-Brosou, S. and Yang, Z.H. (2003) Bayesian models of episodic evolution support a late Precambrian explosive diversification of the Metazoa. *Molecular Biology and Evolution* **20**, 1947–1954.

Aris-Brosou, S. and Yang, Z. (2002) Effects of models of rate evolution on estimation of divergence dates with special reference to the metazoan 18S ribosomal RNA phylogeny. *Systematic Biology* **51**, 703–714.

Aronowicz, J. and Lowe, C.J. (2006) Hox gene expression in the hemichordate *Saccoglossus kowalevskii* and the evolution of deuterostome nervous systems. *Integrative and Comparative Biology* **46**, 890–901.

Aruga, J., Odaka, Y.S., Kamiya, A., and Furuya, H. (2007) *Dicyema* Pax6 and Zic: tool-kit genes in a highly simplified bilaterian. *BMC Evolutionary Biology* **7**, 201.

Ayala, F.J., Rzhetsky, A., and Ayala, F.J. (1998) Origin of the metazoan phyla: molecular clocks confirm paleontological estimates. *Proceedings of the National Academy of Sciences USA* **95**, 606–611.

Baek, D., Villén, J., Shin, C., Camargo, F.D., Gygi, S.P., and Bartel, D.P. (2008) The impact of microRNAs on protein output. *Nature* **455**, 64–71.

Baguñà, J. and Riutort, M. (2004) The dawn of bilaterian animals: the case of acoelomorph flatworms. *BioEssays* **26**, 1046–1057.

Baguñà, J., Martinez, P., Paps, J., and Riutort, M. (2008) Back in time: a new systematic proposal for the Bilateria. *Philosophical Transactions of the Royal Society B: Biological Sciences* **363**, 1481–1491.

Bailey, J.V., Joye, S.B., Kalanetra, K.M., Flood, B.E., and Corsetti, F.A. (2007a) Evidence of giant sulphur bacteria in Neoproterozoic phosphorites. *Nature* **445**, 198–201.

Bailey, J.V., Joye, S.B., Kalanetra, K.M., Flood, B.E., and Corsetti, F.A. (2007b) Palaeontology—undressing and redressing Ediacaran embryos—reply. *Nature* **446**, E10–E11.

Balfour, F.M. (1883) Anatomy and development of *Peripatus capensis*. *Quarterly Journal of the Microscopical Society* **23**, 213–259.

Bapteste, E., Susko, E., Leigh, J., Ruiz-Trillo, I., Bucknam, J., and Doolittle, W.F. (2008) Alternative methods for concatenation of core genes indicate a lack of resolution in deep nodes of the prokaryotic phylogeny. *Molecular Biology and Evolution* **25**, 83–91.

Barfod, G.H., Albarede, F., Knoll, A.H., Xiao, S.H., Telouk, P., Frei, R. *et al.* (2002) New Lu–Hf and Pb–Pb age constraints on the earliest animal fossils. *Earth and Planetary Science Letters* **201**, 203–212.

Barolo, S. and Posakony, J.W. (2002) Three habits of highly effective signaling pathways: principles of transcriptional control by developmental cell signaling. *Genes and Development* **16**, 1167–1181.

Barrington, E.J.W. (1965) *The biology of Hemichordata and Protochordata*. W. H. Freeman, San Francisco.

Bateson, W. (1884) Early stages in the development of *Balanoglossus* (sp. incert.). *Quarterly Journal of Microscopical Science* **24**, 208–236.

Bateson, W. (1885) Later stages in the development of *Balanoglossus kowalevskii* with a suggestion as to the affinities of the Enteropneusta. *Quarterly Journal of Microscopical Science* **25**, 81–128.

Bateson, W. (1886) The ancestry of the Chordata. *Quarterly Journal of Microscopical Science* **26**, 535–571.

Baurain, D., Brinkman, H., and Philippe, H. (2007) Lack of resolution in the animal phylogeny: closely spaced cladogeneses or undetected systematic errors? *Molecular Biology and Evolution* **24**, 6–9.

Bejerano, G., Pheasant, M., Makunin, I., Stephen, S., Kent, W.J., Mattick, J.S. *et al.* (2004) Ultraconserved elements in the human genome. *Science* **304**, 1321–1325.

Beklemishev, W.N. (1969) *Principles of comparative anatomy of invertebrates.* University of Chicago Press, Edinburgh.

Beldade, P. and Brakefield, P.M. (2002) The genetics and evo-devo of butterfly wing patterns. *Nature Reviews Genetics* **3**, 442–452.

Beldade, P. and Saenko, S.V. (2009) Evolution and development of butterfly wing patterns: focus on *Bicyclus anynana* eyespots. In: M. Goldsmith and F. Marec (eds) *Molecular biology and genetics of the Lepidoptera.* CRC Press, Boca Raton, FL. (In press, ISBN 9781420060140.)

Beldade, P., McMillan, W.O., and Papanicoloau, A. (2008) Butterfly genomics eclosing. *Heredity* **100**, 150–157.

Bengtson, S. (1992) The cap-shaped Cambrian fossil *Maikhanella* and the relationship between coeloscleritophorans and mollusks. *Lethaia* **25**, 401–420.

Bengtson, S. (1994) The advent of animal skeletons. In: S. Bengtson (ed.) *Early life on Earth. Nobel Symposium No. 84*, pp. 412–425.Columbia University Press, New York.

Bengtson, S. (2000) Teasing fossils out of shales with cameras and computers. *Palaeontologia Electronica* **3**, 14 pp. (http://palaeo-electronica.org/2000_1/fossils/issue1_00.htm).

Bengtson, S. (2002) Origins and early evolution of predation. In: M. Kowalewski and P.H. Kelley (eds) *The fossil record of predation. The Paleontological Society Papers* **8**, 289–317.

Bengtson, S. (2006) Mineralized skeletons and early animal evolution. In: D.E.G. Briggs (ed.) *Evolving form and function: fossils and development. Proceedings of a symposium honoring Adolf Seilacher for his contributions to paleontology in celebration of his 80th birthday*, pp. 101–124. Peabody Museum of Natural History, Yale University, New Haven.

Bengtson, S. and Budd, G. (2004) Comment on 'Small bilaterian fossils from 40 to 55 million years before the Cambrian'. *Science* **306**, 1291.

Bengtson, S. and Conway Morris, S. (1984) A comparative study of Lower Cambrian *Halkieria* and Middle Cambrian *Wiwaxia. Lethaia* **17**, 307–329.

Bengtson, S. and Urbanek, A. (1986) *Rhabdotubus*, a Middle Cambrian rhabdopleurid hemichordate. *Lethaia* **19**, 293–308.

Bengtson, S. and Zhao, Y. (1997) Fossilized metazoan embryos from the earliest Cambrian. *Science* **277**, 1645–1648.

Benton, M.J. and Donoghue, P.C.J. (2007) Paleontological evidence to date the tree of life. *Molecular Biology and Evolution* **24**, 26–53.

van den Biggelaar, J.A.M., Edsinger-Gonzales, E., and Schram, F.R. (2002) The improbability of dorso-ventral axis inversion during animal evolution, as presumed by Geoffroy Saint Hilaire. *Contributions to Zoology* **71**, 29–36.

Birtle, Z. and Ponting, C.P. (2006) Meisetz and the birth of the KRAB motif. *Bioinformatics* **22**, 2841–2845.

Blair, J.E. and Hedges, S.B. (2005) Molecular phylogeny and divergence times of deuterostome animals. *Molecular Biology and Evolution* **22**, 2275–2284.

Blair, J.E., Ikeo, K., Gojobori, T., and Hedges, S.B. (2002) The evolutionary position of nematodes. *BMC Evolutionary Biology* **2**, 7.

Blair, S.S. (2007) Wing vein patterning in *Drosophila* and the analysis of intercellular signaling. *Annual Reviews of Cell and Developmental Biology* **23**, 293–319.

Blake, D.B. and Guensberg, T.E. (2005) Implications of a new Early Ordovician asteroid (Echinodermata) for the phylogeny of asterozoans. *Journal of Paleontology* **79**, 395–399.

Bleidorn, C., Vogt, L., and Bartolomaeus, T. (2003) New insights into polychaete phylogeny (Annelida) inferred from 18S rDNA sequences. *Molecular Phylogenetics and Evolution* **29**, 279–288.

Bleidorn, C., Eeckhaut, I., Podsiadlowski, L., Schult, N., McHugh, D., Halanych, K.M. *et al.* (2007) Mitochondrial genome and nuclear sequence data support Myzostomida as part of the annelid radiation. *Molecular Biology and Evolution* **24**, 1690–1701.

Bolker, J.A. (1995) Model systems in developmental biology. *BioEssays* **17**, 451–455.

Boore, J.L. (1999) Animal mitochondrial genomes. *Nucleic Acids Research* **27**, 1767–1780.

Boore, J.L. (2006) The use of genome-level characters for phylogenetic reconstruction. *Trends in Ecology and Evolution* **21**, 439–446.

Boore, J.L. and Brown, W.M. (1998) Big trees from little genomes: mitochondrial gene order as a phylogenetic tool. *Current Opinion in Genetics and Development* **8**, 668–674.

Boore, J.L. and Brown, W.M. (2000) Mitochondrial genomes of *Galathealinum, Helobdella*, and *Platynereis*: sequence and gene arrangement comparisons indicate that Pogonophora is not a phylum and Annelida and Arthropoda are not sister taxa. *Molecular Biology and Evolution* **17**, 87–106.

Boore, J.L. and Staton, J. (2002) The mitochondrial genome of the sipunculid *Phascolopsis gouldii* supports its association with Annelida rather than Mollusca. *Molecular Biology and Evolution* **19**, 127–137.

Boore, J.L., Collins, T.M., Stanton, D., Daehler, L.L., and Brown, W.M. (1995) Deducing the pattern of arthropod phylogeny from mitochondrial DNA rearrangements. *Nature* **376**, 163–165.

Boore, J.L., Lavrov D., and Brown, W.M. (1998) Gene translocation links insects and crustaceans. *Nature* **392**, 667–668.

Borchiellini, C., Manuel, M., Alivon, E., Boury-Esnault, N., Vacelet, J., and Le Parco, Y. (2001) Sponge paraphyly and the origin of Metazoa. *Journal of Evolutionary Biology* **14**, 171–179.

Borchiellini, C., Chombard, C., Manuel, M., Alivon, E., Vacelet, J., and Boury-Esnault, N. (2004) Molecular phylogeny of Demospongiae: implications for classification and scenarios of character evolution. *Molecular Phylogenetics and Evolution* **32**, 823–837.

Botting, J.P. and Butterfield, N.J. (2005) Reconstructing early sponge relationships by using the Burgess Shale fossil *Eiffelia globosa*, Walcott. *Proceedings of the National Academy of Sciences USA* **102**, 1554–1559.

Bottjer, D.J., Davidson, E.H., Peterson, K.J., and Cameron, R.A. (2006) Paleogenomics of echinoderms. *Science* **314**, 956–960.

Bourlat, S.J., Nielsen, C., Lockyer, A.E., Littlewood, D.T.J., and Telford, M.J. (2003) *Xenoturbella* is a deuterostome that eats molluscs. *Nature* **424**, 925–928.

Bourlat, S.J., Juliusdottir, T., Lowe, C.J., Freeman, R., Aronowicz, J., Kirschner, M. *et al.* (2006) Deuterostome phylogeny reveals monophyletic chordates and the new phylum Xenoturbellida. *Nature* **444**, 85–88.

Bourlat, S.J., Nielsen, C., and Telford, M.J. (2008) Testing the new animal phylogeny: a phylum level molecular analysis of the animal kingdom. *Molecular Phylogenetics and Evolution* **49**, 23–31.

Boury-Esnault, N., Ereskovsky, A., Bezac, C., and Tokina, D. 2003. Larval development in the Homoscleromorpha (Porifera, Demospongiae). *Invertebrate Biology* **122**, 187–202.

Boute, N., Exposito, J.Y., Boury-Esnault, N., Vacelet, J., Noro, N., Miyazaki, K. *et al.* (1996) Type IV collagen in sponges, the missing link in basement membrane ubiquity. *Biology of the Cell* **88**, 37–44.

Boveri, T.H. (1899) Die Entwicklung von *Ascaris megalocephala* mit besonderer Rücksicht auf die Kernverhältnisse. *Festschrift für C. v. Kupffer*, pp. 383–430. Gustav Fischer Verlag, Jena.

Bowers-Morrow, V.M., Ali, S.O., and Williams, K.L. (2004) Comparison of molecular mechanisms mediating cell contact phenomena in model developmental systems: an exploration of universality. *Biological Reviews* **79**, 611–642.

Bowring, S., Myrow, P., Landing, E., Ramezani, J., and Grotzinger, J. (2003) Geochronological constraints on terminal Neoproterozoic events and the rise of metazoans. *Geophysical Research Abstracts* **5**, 13219.

Boyer, B.C., Henry, J.J., and Martindale, M.Q. (1998) The cell lineage of a polyclad turbellarian embryo reveals close similarity to coelomate spiralians. *Developmental Biology* **204**, 111–123.

Brakefield, P.M. and French, V. (1995) Eyespot development on butterfly wings: the epidermal response to damage. *Developmental Biology* **168**, 98–111.

Brakefield, P.M., Gates, J., Keys, D., Kesbeke, F., Wijngaarden P.J., Montelro, A. *et al.* (1996) Development, plasticity and evolution of butterfly eyespot patterns. *Nature* **384**, 236–242.

Brakefield, P.M., Beldade, P., and Zwaan, B.J. (2009) The African butterfly *Bicyclus anynana*: a model for evolutionary genetics and evo-devo. In: R.R. Behringer, A.D. Johnson and R.E. Krumlauf (eds), *Emerging model organisms: a laboratory manual*, Vol. 1, pp. 291–330. Cold Spring Harbor Laboratory Press, New York.

Brambell, F.W.A. and Cole, C.A. (1939) The preoral ciliary organ of the Enteropneusta; its occurrence, structure and possible phylogenetic significance. *Proceedings of the Zoological Society of London B* **109**, 181–193.

Brasier, M., Green, O., and Shields, G. (1997) Ediacarian sponge spicule clusters from southwestern Mongolia and the origins of the Cambrian fauna. *Geology* **25**, 303–306.

Brent, A.E., Yucel, G., Small, S., and Desplan, C. (2007) Permissive and instructive anterior pattering rely on mRNA localization in the wasp embryo. *Science* **315**, 1841–1843.

Brett, D., Pospisil, H., Valcarcel, J., Reich, J., and Bork, P. (2002) Alternative splicing and genome complexity. *Nature Genetics* **30**, 29–30.

Briggs, D.E.G. (2003) The role of decay and mineralization in the preservation of soft-bodied fossils. *Annual Reviews in Earth and Planetary Sciences* **31**, 275–301.

Briggs, D.E.G. and Fortey, R.A. (2005) Wonderful strife: systematics, stem groups, and the phylogenetic signal of the Cambrian radiation. *Paleobiology* **31**, 94–112.

Briggs, D.E.G., Fortey, R.A., and Wills, M.A. (1992) Morphological disparity in the Cambrian. *Science* **256**, 1670–1673.

Briggs, D.E.G., Erwin, D.H., and Collier, F.J. (1995) *Fossils of the Burgess Shale*. Smithsonian Institution Press, Washington, DC.

Brinkmann, H. and Philippe, H. (1999) Archaea sister group of Bacteria? Indications from tree reconstruction artifacts in ancient phylogenies. *Molecular Biology and Evolution* **16**, 817–825.

Brinkmann, H. and Philippe, H. (2007) The diversity of eukaryotes and the root of the eukaryotic tree. *Advances in Experimental and Medical Biology* **607**, 20–37.

Brinkmann, H. and Philippe, H. (2008) Animal phylogeny and large-scale sequencing: progress and pitfalls. *Journal of Systematics and Evolution* **46**, 274–286.

Briscoe, J., Sussel, L., Serup, P., Hartigan-O'Connor, D., Jessell, T.M., Rubenstein, J.L. *et al.* (1999) Homeobox gene *Nkx2.2* and specification of neuronal identity by graded Sonic hedgehog signalling. *Nature* **398**, 622–627.

Broadie, K.S., Bate, M., and Tublitz, N.J. (1991) Quantitative staging of embryonic development of the tobacco hawk moth *Manduca sexta*. *Roux's Archives of Developmental Biology* **199**, 327–334.

Brocco, S.L., O'Clair, R.M., and Cloney, R.A. (1974) Cephalopod integument: the ultrastructure of Kölliker's organs and their relationship to setae. *Cell and Tissue Research* **151**, 293–308.

Bromham, L. (2006) Molecular dates for the Cambrian explosion: is the light at the end of the tunnel an oncoming train. *Palaeontologia Electronica* **9.1**, 2E.

Bromham, L.D. and Degnan, B.M. (1999) Hemichordates and deuterostome evolution: robust molecular phylogenetic support for a hemichordate + echinoderm clade. *Evolution & Development* **1**, 166–171.

Brown, F.D., Prendergast, A. and Swalla, B.J. 2008. Man is but a worm: chordate origins. *Genesis* **46**, 605–613.

Brunetti, C.R., Selegue, J.E., Monteiro, A., French, V., Brakefield, P.M., and Carroll, S.B. (2001) The generation and diversification of butterfly eyespot colour patterns. *Current Biology* **11**, 1578–1585.

Brusca, R.C. and Brusca, G.J. (1990) *Invertebrates*, 1st edn. Sinauer Associates, Sunderland, MA.

Brusca, R.C. and Brusca, G.J. (2003) *Invertebrates*, 2nd edn. Sinauer Associates, Sunderland, MA.

Bucher, G. and Klingler, M. (2004) Divergent segmentation mechanism in the short germ insect *Tribolium* revealed by *giant* expression and function. *Development* **131**, 1729–1740.

Bucher, G., Farzana, L., Brown, S.J., and Klingler, M. (2005) Anterior localization of maternal mRNAs in a short germ insect lacking *bicoid*. *Evolution & Development* **7**, 142–149.

Budd, G.E. (1996) The morphology of *Opabinia regalis* and the reconstruction of the arthropod stem-group. *Lethaia* **29**, 1–14.

Budd, G.E. (1998) Arthropod body-plan evolution in the Cambrian with an example from anomalocaridid muscle. *Lethaia* **31**, 197–210.

Budd, G.E. (2001a) Tardigrades as 'stem-group arthropods': the evidence from the Cambrian fauna. *Zoologischer Anzeiger* **240**, 265–279.

Budd, G.E. (2001b) Why are arthropods segmented? *Evolution & Development* **3**, 332–342.

Budd, G.E. (2002) A palaeontological solution to the arthropod head problem. *Nature* **417**, 271–275.

Budd, G.E. (2003) The Cambrian fossil record and the origin of the phyla. *Integrative and Comparative Biology* **43**, 157–165.

Budd, G.E. (2004) Lost children of the Cambrian. *Nature* **427**, 205–207.

Budd, G.E. and Jensen, S. (2000) A critical reappraisal of the fossil record of the bilaterian phyla. *Biological Reviews of the Cambridge Philosophical Society* **75**, 253–295.

Budd, G.E. and Jensen, S. (2003) The limitations of the fossil record and the dating of the origin of the Bilateria. In: P.C.J. Donoghue and M.P. Smith (eds) *Telling evolutionary time: molecular clocks and the fossil record*, pp. 166–189. Taylor and Francis, London.

Budd, G. and Telford, M. (2005) Evolution: along came a sea spider. *Nature* **437**, 1099–1102.

Bull, A.L. (1982) Stages of living embryos in the jewel wasp *Mormoniella* (*Nasonia*) *vitripennis* (Walker). *International Journal of Morphology and Embryology* **11**, 1–23.

Bullock, T.H. (1945) The anatomical organization of the nervous system of enteropneusta. *Quarterly Journal of Microscopical Science* **86**, 55–112.

Bullock, T.H. and Horridge, G.A. (1965) *Structure and function in the nervous systems of invertebrates*. W. H. Freeman, San Francisco.

Burger, G., Forget, L., Zhu, Y., Gray, M.W., and Lang, B.F. (2003) Unique mitochondrial genome architecture in unicellular relatives of animals. *Proceedings of the National Academy of Sciences USA* **100**, 892–897.

Butterfield, N.J. (1990) A reassessment of the enigmatic Burgess Shale fossil *Wiwaxia corrugata* (Matthew) and its relationship to the polychaete *Canadia spinosa* Walcott. *Paleobiology* **16**, 287–303.

Butterfield, N.J. (1997) Plankton ecology and the Proterozoic–Phanerozoic transition. *Paleobiology* **23**, 247–262.

Butterfield, N.J. (2001) Ecology and evolution of Cambrian plankton. In: A.Y. Zhuravlev and R. Riding (eds) *The ecology of the Cambrian radiation*, pp. 200–216.Columbia University Press, New York.

Butterfield, N.J. (2006) Hooking some stem-group 'worms': fossil lophotrochozoans in the Burgess Shale. *BioEssays* **28**, 1161–1166.

Butterfield, N.J. (2007) Macroevolution and macroecology through deep time. *Palaeontology* **50**, 41–55.

Butterfield, N.J. (2008) An early Cambrian radula. *Journal of Paleontology* **82**, 543–554.

Butterfield, N.J., Knoll, A.H., and Swett, K. (1990) A bangiophyte red alga from the Proterozoic of arctic Canada. *Science* **250**, 104–107.

Butterfield, N.J., Balthasar, U., and Wilson, L.A. (2007) Fossil diagenesis in the Burgess Shale. *Palaeontology* **50**, 537–543.

Calver, C.R., Black, L.P., Everard, J.L., and Seymour, D.B. (2004) U–Pb zircon age constraints on late Neoproterozoic glaciation in Tasmania. *Geology* **32**, 893–896.

Cameron, C.B. (2002) Particle retention and flow in the pharynx of the enteropneust worm *Harrimania planktophilus*: the filter-feeding pharynx may have evolved before the chordates. *Biological Bulletin* **202**, 192–200.

Cameron, C.B. and Mackie, G.O. (1996) Conduction pathways in the nervous system of *Saccoglossus* sp. (Enteropneusta). *Canadian Journal of Zoology* **74**, 15–19.

Cameron, C.B., Garey, J.R., and Swalla, B.J. (2000) Evolution of the chordate body plan: new insights from phylogenetic analyses of deuterostome phyla. *Proceedings of the National Academy of Sciences USA* **97**, 4469–4474.

Cameron, R.A., Rowen, L., Nesbitt, R., Bloom, S., Rast, J.P., Berney, K. *et al.* (2006) Unusual gene order and organization of the sea urchin hox cluster. *Journal of Experimental Zoology (Molecular and Developmental Evolution)* **306B**, 45–58.

Canestro, C., Bassham, S., and Postlethwait, J. (2005) Development of the central nervous system in the larvacean *Oikopleura dioica* and the evolution of the chordate brain. *Developmental Biology* **285**, 298–315.

Canfield, D.E. (1998) A new model for Proterozoic ocean chemistry. *Nature* **396**, 450–453.

Canfield, D.E. and Teske, A. (1996) Late Proterozoic rise in atmospheric oxygen concentration inferred from phylogenetic and sulphur-isotope studies. *Nature* **382**, 127–132.

Canfield, D.E., Poulton, S.W., and Narbonne, G.M. (2007) Late-Neoproterozoic deep-ocean oxygenation and the rise of animal life. *Science* **315**, 92–95.

Carapelli, A., Lio, P., Nardi, F., van der Wath, E., and Frati, F. (2007) Phylogenetic analysis of mitochondrial protein coding genes confirms the reciprocal paraphyly of Hexapoda and Crustacea. *BMC Evolutionary Biology* **7**(Suppl. 2), S8.

Carlson, S.J. (1995) Phylogenetic relationships among extant brachiopods. *Cladistics* **11**, 131–197.

Caron, J.B. (2006) *Banffia constricta*, a putative vetulicolid from the Middle Cambrian Burgess Shale. *Transactions of the Royal Society of Edinburgh: Earth Sciences* **96**, 95–111.

Caron, J.B., Scheltema, A., Schander, C., and Rudkin, D. (2006) A soft-bodied mollusc with radula from the Middle Cambrian Burgess Shale. *Nature* **442**, 159–163.

Caron, J.-B., Scheltema, A., Schander, C., and Rudkin, D. (2007) Reply to Butterfield on stem-group 'worms': fossil lophotrochozoans in the Burgess Shale. *BioEssays* **29**, 200–202.

Carranza, S., Baguñà, J., and Riutort, M. (1997) Are the Platyhelminthes a monophyletic primitive group? An assessment using 18S rDNA sequences. *Molecular Biology and Evolution* **14**, 485–497.

Carrasco, A.E., McGinnis, W., Gehring, W.J., and De Robertis, E.M. (1984) Cloning of an *X. laevis* gene expressed during early embryogenesis coding for a peptide region homologous to *Drosophila* homeotic genes. *Cell* **37**, 409–414.

Carrera, M.G. and Botting, J.P. (2008) Evolutionary history of Cambrian spiculate sponges: implications for the Cambrian evolutionary fauna. *Palaios* **23**, 124–138.

Carroll, S.B. (2005) *Endless forms most beautiful: the new science of evo devo and the making of the animal kingdom*, 1st edn. W. W. Norton & Co., New York.

Carroll, S.B. (2008) Evo-devo and an expanding evolutionary synthesis: a genetic theory of morphological evolution. *Cell* **134**, 25–36.

Carroll, S.B., Gates, J., Keys, D.N., Paddock, G.E., Panganiban, G.E., Selegue, J.E. *et al.* (1994) Pattern formation and eyespot determination in butterfly wings. *Science* **265**, 109–114.

Carroll, S.B., Grenier, J.K., and Weatherbee, S.D. (2001) *From DNA to diversity: molecular genetics and the evolution of animal design*, 1st edn. Blackwell Science, Malden, MA.

Carroll, S.B., Grenier, J.K., and Weatherbee, S.D. (2005) *From DNA to diversity: molecular genetics and the evolution of animal design*, 2nd edn. Blackwell Publishing, Oxford.

Cartwright, P. and Collins, A. (2007) Fossils and phylogenies: integrating multiple lines of evidence to investigate the origin of early major metazoan lineages. *Integrative and Comparative Biology* **47**, 744–751.

Casali, A. and Casanova, J. (2001) The spatial control of Torso RTK activation: a C-terminal fragment of the Trunk protein acts as a signal for Torso receptor in the *Drosophila* embryo. *Development* **128**, 1709–1715.

Castro, L.F., Rasmussen, S.L., Holland, P.W., Holland, N.D., and Holland, L.Z. (2006) A *Gbx* homeobox gene in amphioxus: insights into ancestry of the ANTP class and evolution of the midbrain/hindbrain boundary. *Developmental Biology* **295**, 40–51.

Catling, D.C., Glein, C.R., Zahnle, K.J., and McKay, C.P. (2005) Why O_2 is required by complex life on habitable planets and the concept of planetary 'oxygenation time'. *Astrobiology* **5**, 415–438.

Cavalier Smith, T. (1998) A revised six-kingdom system of life. *Biological Reviews* **73**, 203–266.

Cavalier Smith, T., Chao, E.E., Boury Esnault, N., and Vacelet, J. (1996). Sponge phylogeny, animal monophyly, and the origin of the nervous system: 18S rRNA evidence. *Canadian Journal of Zoology* **74**, 2031–2045.

de Ceccatty, M.P. (1974) Coordination in sponges. The foundations of integration. *American Zoologist* **14**, 895–903.

Celerin, M.R., Schisler, N.J., Day, A.W., Stetler-Davidson, W.G., and Laudenbach, D.E. (1996) Fungal fimbrae are composed of collagen. *EMBO Journal* **15**, 4445–4453.

Cereijido, M., Contreras, R.G., and Shoshani, L. (2004) Cell adhesion, polarity, and epithelia in the dawn of metazoans. *Physiological Reviews* **84**, 1229–1262.

Cerfontaine, P. (1906) Recherches sur le développement de l'*Amphioxus*. *Archives de Biologie* **22**, 229–418.

Cerny, A.C., Bucher, G., Schröder, R., and Klingler, M. (2005) Breakdown of abdominal patterning in the *Tribolium Kruppel* mutant *jaws*. *Development* **132**, 5353–5363.

Chan, K.M.A. and Levin, S.A. (2005) Leaky prezygotic isolation and porous genomes: rapid introgression of maternally inherited DNA. *Evolution* **59**, 720–729.

Chapman, A. (2005) *Numbers of living species in Australia and the world.* Report for the Department of the Environment and Heritage Canberra, Australia, pp. 1–64.

Chea, H.K., Wright, C.V., and Swalla, B.J. (2005) Nodal signaling and the evolution of deuterostome gastrulation. *Developmental Dynamics* **234**, 269–278.

Cheesman, S.E., Layden, M.J., Von Ohlen, T., Doe, C.Q., and Eisen, J.S. (2004) Zebrafish and fly Nkx6 proteins have similar CNS expression patterns and regulate motoneuron formation. *Development* **131**, 5221–5232.

Chen, J.Y. and Chi, H.M. (2005) Precambrian phosphatized embryos and larvae from the Doushantuo Formation and their affinities, Guizhou (SW China). *Chinese Science Bulletin* **50**, 2193–2200.

Chen, J.Y., Dzik, J., Edgecombe, G.D., Ramskold, L., and Zhou, G.Q. (1995) A possible Early Cambrian chordate. *Nature* **377**, 720–722.

Chen, J.Y., Huang, D.Y., and Li, C.W. (1999) An early Cambrian craniate-like chordate. *Nature* **402**, 518–522.

Chen, J.Y., Oliveri, P., Li, C.W., Zhou, G.Q., Gao, F., Hagadorn, J.W. *et al.* (2000) Precambrian animal

diversity: putative phosphatized embryos from the Doushantuo Formation of China. *Proceedings of the National Academy of Sciences USA* **97**, 4457–4462.

Chen, J.Y., Huang, D.Y., Peng, Q.Q., Chi, H.M., Wang, X.Q., and Feng, M. (2003) The first tunicate from the Early Cambrian of South China. *Proceedings of the National Academy of Sciences USA* **100**, 8314–8318.

Chen, J.Y., Bottjer, D.J., Oliveri, P., Dornbos, S.Q., Gao, F., Ruffins, S. *et al.* (2004) Small bilaterian fossils from 40 to 55 million years before the Cambrian. *Science* **305**, 218–222.

Chen, J.Y., Schopf, J.W., Bottjer, D.J., Zhang, C.Y., Kudryavtsev, A.B., Tripathi, A.B. *et al.* (2007). Raman spectra of a lower Cambrian ctenophore embryo from southwestern Shaanxi, China. *Proceedings of the National Academy of Sciences USA* **104**, 6289–6292.

Chipman, A.D., Arthur, W., and Akam, M. (2004) A double segment periodicity underlies segment generation in centipede development. *Current Biology* **14**, 1250–1255.

Choe, C.P. and Brown, S.J. (2007) Evolutionary flexibility of pair-rule patterning revealed by functional analysis of secondary pair-rule genes, *paired* and *sloppy-paired* in the short-germ insect, *Tribolium castaneum*. *Developmental Biology* **302**, 281–294.

Choe, C.P., Miller, S.C., and Brown, S.J. (2006) A pair-rule gene circuit defines segments sequentially in the short-germ insect *Tribolium castaneum*. *Proceedings of the National Academy of Sciences USA* **103**, 6560–6564.

Chung, H.R., Lohr, U., and Jackle, H. (2007) Lineage-specific expansion of the zinc finger associated domain ZAD. *Molecular Biology and Evolution* **24**, 1934–1943.

Ciccarelli, F.D., Doerks, T., von Mering, C., Creevey, C.J., Snel, B., and Bork, P. (2006) Toward automatic reconstruction of a highly resolved tree of life. *Science* **311**, 1283–1287.

Clarke, B., Johnson, M.S., and Murray, J. (1996) Clines in the genetic distance between two species of island land snails: how 'molecular leakage' can mislead us about speciation. *Philosophical Transactions of the Royal Society B: Biological Sciences* **351**, 773–784.

Clausen, S. and Smith, A.B. (2005) Palaeoanatomy and biological affinities of a Cambrian deuterostome (Stylophora). *Nature* **438**, 351–354.

Clayton, D.A. (1992) Transcription and replication of animal mitochondrial DNAs. *International Review of Cytology* **141**, 217–232.

Cobbett, A., Wilkinson, M., and Wills, M.A. (2007) Fossils impact as hard as living taxa in parsimony analysis of morphology. *Systematic Biology* **56**, 753–766.

Cohen, B.L. (2000) Monophyly of brachiopods and phoronids: reconciliation of molecular evidence with

Linnaean classification (the subphylum Phoroniformea nov.). *Proceedings of the Royal Society B: Biological Sciences* **267**, 225–231.

Cohen, B.L. and Weydmann, A. (2005) Molecular evidence that phoronids are a subtaxon of brachiopods (Brachiopoda: Phoronata) and that genetic divergence of metazoan phyla began long before the early Cambrian. *Organisms Diversity and Evolution* **5**, 253–273.

Cohen, B.L., Gawthrop, A.B., and Cavalier-Smith, T. (1998) Molecular phylogeny of brachiopods and phoronids based on nuclear-encoded small subunit ribosomal RNA gene sequences. *Philosophical Transactions of the Royal Society B: Biological Sciences* **353**, 2039–2061.

Cohn, M.J. and Tickle, C. (1999) Developmental basis of limblessness and axial patterning in snakes. *Nature* **399**, 474–479.

Colgan, D.J., Hutchings, P.A., and Braune, M. (2006) A multigene framework for polychaete phylogenetic studies. *Organisms Diversity and Evolution* **6**, 220–235.

Collin, R. (2004) Phylogenetic effects, the loss of complex characters, and the evolution of development in calyptraeid gastropods. *Evolution* **58**, 1488–1502.

Collins, A.G. (1998) Evaluating multiple alternative hypotheses for the origin of Bilateria: an analysis of 18S rRNA molecular evidence. *Proceedings of the National Academy of Sciences USA* **95**, 15458–15463.

Condon, D., Zhu, M.Y., Bowring, S., Wang, W., Yang, A.H., and Jin, Y.G. (2005) U–Pb ages from the Neoproterozoic Doushantuo Formation, China. *Science* **308**, 95–98.

Conklin, E.G. (1897) The embryology of *Crepidula*. *Journal of Morphology* **13**, 1–226.

Conklin, E.G. (1902) The embryology of *Terebratulina septentrionalis* Couthouy. *Proceedings of the American Philosophical Society* **16**, 41–76.

Conway Morris, S. (1979) The Burgess Shale (Middle Cambrian) fauna. *Annual Reviews in Ecology and Systematics* **10**, 327–349.

Conway Morris, S. (1985) The Middle Cambrian metazoan *Wiwaxia corrugata* (Matthew) from the Burgess Shale and *Ogygopsis* Shale, British Columbia, Canada. *Philosophical Transactions of the Royal Society of London B, Biological Sciences* **307**, 507–582.

Conway Morris, S. (1993a) The fossil record and the early evolution of the Metazoa. *Nature* **361**, 219–225.

Conway Morris, S. (1993b) Ediacaran-like fossils in the Cambrian Burgess Shale-type faunas of North America. *Palaeontology* **36**, 593–635.

Conway Morris, S. (1998a) *The crucible of creation*. Oxford University Press, Oxford.

Conway Morris, S. (1998b) The question of metazoan monophyly and the fossil record. In: W.E.G. Muller (ed.) *Molecular evolution: towards the origin of the Metazoa*, pp. 1–19. Springer, Berlin.

Conway Morris, S. (2000) The Cambrian 'explosion': slow-fuse or megatonnage? *Proceedings of the National Academy of Sciences USA* **97**, 4426–4429.

Conway Morris, S. (2003a) The Cambrian 'explosion' of metazoans and molecular biology: would Darwin be satisfied? *International Journal of Developmental Biology* **47**, 505–515.

Conway Morris, S. (2003b) *Life's solution: inevitable humans in a lonely universe*. Cambridge University Press, Cambridge.

Conway Morris, S. (2006) Darwin's dilemma: the realities of the Cambrian 'explosion'. *Philosophical Transactions of the Royal Society B: Biological Sciences* **361**, 1069–1083.

Conway Morris, S. and Caron, J.-B. (2007) Halwaxiids and the early evolution of the lophotrochozoans. *Science* **315**, 1255–1258.

Conway Morris, S. and Collins, D.H. (1996) Middle Cambrian ctenophores from the Stephen Formation, British Columbia, Canada. *Philosophical Transactions of the Royal Society B: Biological Sciences* **351**, 279–308.

Conway Morris, S. and Peel, J.S. (1995) Articulated Halkieriids from the Lower Cambrian of North Greenland and their role in early protostome evolution. *Philosophical Transactions of the Royal Society B: Biological Sciences* **347**, 305–358.

Cook, C.E., Yue, Q., and Akam, M. (2005) Mitochondrial genomes suggest that hexapods and crustaceans are mutually paraphyletic. *Proceedings of the Royal Society B: Biological Sciences* **272**, 1295–1304.

Copf, T., Schroder, R., and Averof, M. (2004) Ancestral role of *caudal* genes in axis elongation and segmentation. *Proceedings of the National Academy of Sciences USA* **101**, 17711–17715.

Copley, R.R. (2004) Evolutionary convergence of alternative splicing in ion channels. *Trends in Genetics* **20**, 171–176.

Copley, R.R. (2005) The EH1 motif in metazoan transcription factors. *BMC Genomics* **6**, 169.

Copley, R.R., Aloy, P., Russell, R.B., and Telford, M.J. (2004) Systematic searches for molecular synapomorphies in model metazoan genomes give some support for Ecdysozoa after accounting for the idiosyncrasies of *Caenorhabditis elegans*. *Evolution & Development* **6**, 164–169.

Cordier, F., Hartmann, B., Rogowski, M., Affolter, M., and Grzesiek, S. (2006) DNA recognition by the brinker repressor—an extreme case of coupling between binding and folding. *Journal of Molecular Biology* **361**, 659–672.

Cornell, R.A. and Ohlen, T.V. (2000) Vnd/nkx, ind/gsh, and msh/msx: conserved regulators of dorsoventral neural patterning? *Current Opinions in Neurobiology* **10**, 63–71.

Costanzo, K. and Monteiro, A. (2007) The use of chemical and visual cues in female choice in the butterfly *Bicyclus anynana*. *Proceedings of the Royal Society B: Biological Sciences* **274**, 845–851.

Coyne, J.A. and Orr, H.A. (2004) *Speciation.* Sinauer Associates, Sunderland, MA.

Crozatier, M., Glise, B., and Vincent, A. (2004) Patterns in evolution: veins of the *Drosophila* wing. *Trends in Genetics* **20**, 498–505.

Damen, W. (2007) Evolutionary conservation and divergence of the segmentation process in arthropods. *Developmental Dynamics* **236**, 1379–1391.

Damen, W.G., Saridaki, T., and Averof, M. (2002) Diverse adaptations of an ancestral gill: a common evolutionary origin for wings, breathing organs, and spinnerets. *Current Biology* **12**, 1711–1716.

Darras, S. and Nishida, H. (2001) The BMP/CHORDIN antagonism controls sensory pigment cell specification and differentiation in the ascidian embryo. *Developmental Biology* **236**, 271–288.

Darwin, C. (1859) *On the origin of species.* John Murray, London (republished by Harvard University Press, Cambridge, MA, 1964).

Dautert, E. (1929) Die Bildung der Keimblätter von *Paludina vivipara. Zoologische Jahrbücher, Abteilung für Anatomie und Ontogenie der Tiere* **50**, 433–496.

Davidson, E.H. (2006) *The regulatory genome: gene regulatory networks in development and evolution.* Academic Press, London.

Davidson, E.H., Peterson, K.J., and Cameron, R.A. (1995) Origin of adult bilaterian body plans: evolution of developmental regulatory mechanisms. *Science* **270**, 1319–1325.

Davis, G.K. and Patel, N.H. (2002) Short, long, and beyond: molecular and embryological approaches to insect segmentation. *Annual Review of Entomology* **47**, 669–699.

Davis, G.K., D'Alessio, J.A., and Patel, N.H. (2005) Pax3/7 genes reveal conservation and divergence in the arthropod segmentation hierarchy. *Developmental Biology* **285**, 169–184.

Dearden, P.K., Wilson, M.J., Sablan, L., Osborne, P.W., Havler, M., McNaughton, E. *et al.* (2006) Patterns of conservation and change in honey bee developmental genes. *Genome Research* **16**, 1376–1384.

De Celis, J.F. and Diaz-Benjumea, F.J. (2003) Developmental basis for vein pattern variations in insect wings. *International Journal for Developmental Biology* **47**, 653–663.

Dehal, P. and Boore, J.L. (2006) A phylogenomic gene cluster resource: the phylogenetically inferred groups (PhIGs) database. *BMC Bioinformatics* **7**, 201.

Dehal, P., Satou, Y., Campbell, R.K., Chapman, J., Degnan, B., De Tomaso, A. *et al.* (2002) The draft genome of *Ciona intestinalis*: insights into chordate and vertebrate origins. *Science* **298**, 2157–2167.

Dellaporta, S.L., Xu, A., Sagasser, S., Jakob, W., Moreno, M.A., Buss, L.W. *et al.* (2006) Mitochondrial genome of *Trichoplax adhaerens* supports Placozoa as the basal lower metazoan phylum. *Proceedings of the National Academy of Sciences USA* **103**, 8751–8756.

Delsuc, F., Philips, M.J., and Penny, D. (2003) Comment on 'Hexapod origins: monophyletic or paraphyletic?' *Science* **301**, 1482–1483.

Delsuc, F., Brinkmann, H., and Philippe, H. (2005) Phylogenomics and the reconstruction of the tree of life. *Nature Reviews Genetics* **6**, 361–375.

Delsuc, F., Brinkmann, H., Chourrout, D., and Philippe, H. (2006) Tunicates and not cephalochordates are the closest living relatives of vertebrates. *Nature* **439**, 965–968.

Denes, A.S., Jekely, G., Steinmetz, P.R., Raible, F., Snyman, H., Prud'homme, B. *et al.* (2007) Molecular architecture of annelid nerve cord supports common origin of nervous system centralization in Bilateria. *Cell* **129**, 277–288.

Derelle, R. (2007) L'apport des Cténaires à la compréhension de l'évolution des Métazoaires: positionnement par la phylogénomique et recherche de gènes de développement chez *Pleurobrachia pileus. Doctoral thesis.* Université de Pierre et Marie Curie, Paris (unpublished).

Dermitzakis, E.T. and Clark, A.G. (2002) Evolution of transcription factor binding sites in mammalian gene regulatory regions: conservation and turnover. *Molecular Biology and Evolution* **19**, 1114–1121.

De Robertis, E.M. and Sasai, Y. (1996) A common plan for dorsoventral patterning in Bilateria. *Nature* **380**, 37–40.

Deutsch, J.S. (2008) Do acoels climb up the 'Scale of Beings'? *Evolution & Development* **10**, 135–140.

Dictus, W.J. and Damen, P. (1997) Cell-lineage and clonal-contribution map of the trochophore larva of *Patella vulgata* (Mollusca). *Mechanisms of Development* **62**, 213–226.

Dilão, R. and Sainhas, J. (2004) Modeling butterfly wing eyespot patterns. *Proceedings of the Royal Society B: Biological Sciences* **271**, 1565–1569.

Dilly, P.N., Welsch, U., and Storch, V. (1970) The structure of the nerve fiber layer and neurocord in the enteropneusts *Zeitschrift für Zellforschung und Mikroskopische Anatomie* **103**, 129–148.

Ding, G., Yu, Z., Zhao, J., Li, Y., Xing, X., Wang, C. *et al.* (2008) Tree of life based on genome context networks. *PLoS ONE* **3**, e3357.

Dohle, W. (1997) Are the insects more closely related to the crustaceans than to the myriapods? *Entomologica Scandinavica* **51**(Suppl.), 7–16.

Dohle, W. (2001) Are the insects terrestrial crustaceans? A discussion of some new facts and arguments and the proposal of a proper name 'Tetraconata' for the monophyletic unit Crustacea + Hexapoda. *Annales de la Societé Entomologique de France* **37**, 85–103.

von Döhren, J. and Bartolomaeus, T. (2007) Ultrastructure and development of the rhabdomeric eyes in *Lineus viridis* (Heteronemertea, Nemertea). *Zoology* **110**, 430–438.

Dohrmann, M., Janussen, D., Reitner, J., Collins, A.G., and Wörheide, G. (2008) Phylogeny and evolution of glass sponges (Porifera, Hexactinellida). *Systematic Biology* **57**, 388–405.

Dohrn, A. (1875) *Der Ursprung der Wirbelthiere und das Princip des Functionswechsels. Genealogische Skizzen.* Wilhem Engelmann, Leipzig.

Domazet-Los, T., Brajkovic, J., and Tautz, D. (2007) A phylostratigraphy approach to uncover the genomic history of major adaptations in metazoan lineages. *Trends in Genetics* **23**, 533–539.

Dominguez, P., Jacobson, A.G., and Jefferies, R.P. (2002) Paired gill slits in a fossil with a calcite skeleton. *Nature* **417**, 841–844.

Dong, X.P., Donoghue, P.C.J., Cheng, H., and Liu, J.B. (2004) Fossil embryos from the Middle and Late Cambrian period of Hunan, South China. *Nature* **427**, 237–240.

Donoghue, P.C.J. (2007) Embryonic identity crisis. *Nature* **445**, 155–156.

Donoghue, P.C.J. and Benton, M.J. (2007) Rocks and clocks: calibrating the tree of life using fossils and molecules. *Trends in Ecology and Evolution* **22**, 424–431.

Donoghue, P.C.J. and Purnell, M.A. (2009) Distinguishing heat from light in debate over controversial fossils. *BioEssays* **31**, 178–189.

Donoghue, P.C.J., Forey, P.L., and Aldridge, R.J. (2000) Conodont affinity and chordate phylogeny. *Biological Reviews* **75**, 191–251.

Donoghue, P.C.J., Bengtson, S., Dong, X.P., Gostling, N.J., Huldtgren, T., Cunningham, J.A. *et al.* (2006a) Synchrotron X-ray tomographic microscopy of fossil embryos. *Nature* **442**, 680–683.

Donoghue, P.C.J., Kouchinsky, A., Waoszek, D., Bengtson, S., Dong, X., Val'kov, A. K. *et al.* (2006b) Fossilized embryos are widespread but the record is temporally and taxonomically biased. *Evolution & Development* **8**, 232–238.

Dopazo, H., Santoyo, J., and Dopazo, J. (2004) Phylogenomics and the number of characters required for obtaining an accurate phylogeny of eukaryote model species. *Bioinformatics* **20**(Suppl. 1), i116–i121.

Dornbos, S.Q., Bottjer, D.J., Chen, J.Y., Gao, F., Oliveri, P., and Li, C.W. (2006) Environmental controls on the taphonomy of phosphatized animals and animal embryos from the Neoproterozoic Doushantuo Formation, southwest China. *Palaios* **21**, 3–14.

Douzery, E.J.P., Snell, E.A., Bapteste, E., Delsuc, F., and Philippe, H. (2004) The timing of eukaryotic evolution: does a relaxed molecular clock reconcile proteins and fossils? *Proceedings of the National Academy of Sciences USA* **101**, 15386–15391.

Dove, H. and Stollewerk, A. (2003) Comparative analysis of neurogenesis in the myriapod *Glomeris marginata* (Diplopoda) suggests more similarities to chelicerates than to insects. *Development* **130**, 2161–2171.

Droser, M.L., Gehling, J.G., and Jensen, S.R. (2005) Ediacaran trace fossils: true and false. In: D.E.G. Briggs (ed.) *Evolving form and function: fossils and development*, pp. 125–138. Yale Peabody Museum Publications, New Haven, CT.

Drummond, A.J. and Rambaut, A. (2006) *BEAST, version 1.4.* http://beast.bio.ed.ac.uk/ (accessed 2 February 2009).

Drummond, A.J., Ho, S.Y.W., Phillips, M.J., and Rambaut, A. (2006) Relaxed phylogenetics and dating with confidence. *PLoS Biology* **4**, e88.

Duboc, V. and Lepage, T. (2006) A conserved role for the nodal signaling pathway in the establishment of dorso-ventral and left-right axes in deuterostomes. *Journal of Experimental Zoology (Molecular and Developmental Evolution)* **306B**, 1–13.

Duboc, V., Rottinger, E., Besnardeau, L., and Lepage, T. (2004) Nodal and *BMP2/4* signaling organizes the oral-aboral axis of the sea urchin embryo. *Developmental Cell* **6**, 397–410.

Duboc, V., Rottinger, E., Lapraz, F., Besnardeau, L., and Lepage, T. (2005) Left–right asymmetry in the sea urchin embryo is regulated by nodal signaling on the right side. *Developmental Cell* **9**, 147–158.

Dulith, B.E., Snel, B., Ettema, T.J.G., and Huynen, M.A. (2008) Signature genes as a phylogenomic tool. *Molecular Biology and Evolution* **25**, 1659–1667.

Dunn, C.W., Hejnol, A., Matus, D.Q., Pang, K., Browne, W.E., Smith, S.A. *et al.* (2008) Broad phylogenomic sampling improves resolution of the animal tree of life. *Nature* **452**, 745–749.

Dunn, E.F., Moy, V.N., Angerer, L. M., Angerer, R.C., Morris, R.L., and Peterson, K.J. (2007) Molecular paleoecology: using gene regulatory analysis to address the origins of complex life cycles in the Late Precambrian. *Evolution & Development* **9**, 10–24.

Dunn, N.R., Koonce, C.H., Anderson, D.C., Islam, A., Bikoff, E.K., and Robertson, E.J. (2005) Mice exclusively expressing the short isoform of Smad2 develop normally and are viable and fertile. *Genes and Development* **19**, 152–163.

Dunne, J.A., Williams, R.J., Martinez, N.D., Wood, R.A., and Erwin, D.H. (2008) Compilation and network analyses of Cambrian food webs. *PLoS Biology* **6**, e102.

Dyson, H.J. and Wright, P.E. (2005) Intrinsically unstructured proteins and their functions. *Nature Reviews Molecular Cell Biology* **6**, 197–208.

Dzik, J. (2002) Possible ctenophoran affinities of the Precambrian 'sea-pen' *Rangea*. *Journal of Morphology* **252**, 315–334.

Dzik, J. (2003) Anatomical information content in the Ediacaran fossils and their possible zoological affinities. *Integrative and Comparative Biology* **43**, 114–126.

Dzik, J. (2005) Behavioral and anatomical unity of the earliest burrowing animals and the cause of the 'Cambrian explosion'. *Paleobiology* **31**, 503–521.

Dzik, J. and Krumbiegel, G. (1989) The oldest 'onychophoran' *Xenusion*: a link connecting phyla? *Lethaia* **22**, 169–182.

Economou, A.D. (2008) *Phylogenetic and developmental studies of the evolution of an insect novelty*. PhD thesis, University College London.

Edgecombe, G., Richter, S., and Wilson, G. (2003) The mandibular gnathal edges: homologous structures throughout Mandibulata? *African Invertebrates* **44**, 115–135.

Editorial (2000) The nature of the number. *Nature Genetics* **25**, 127–128.

Eeckhaut, I. and Lanterbecq, D. (2005) Myzostomida: a review of the phylogeny and ultrastructure. *Hydrobiologia* **535/536**, 253–275.

Eeckhaut, I., McHugh, D., Mardulyn, P., Tiedemann, R., Monteyne, D., Jangoux, M. *et al.* (2000) Myzostomida: a link between trochozoans and flatworms? *Proceedings of the Royal Society B: Biological Sciences* **267**, 1383–1392.

Eernisse, D.J. and Peterson, K.J. (2004) The history of animals. In: J. Cracraft and M.J. Donoghue (eds) *Assembling the tree of life*, pp. 197–208. Oxford University Press, Oxford.

Eernisse, D.J., Albert, J.S., and Anderson, F.E. (1992) Annelida and Arthropoda are not sister taxa: a phylogenetic analysis of spiralian metazoan morphology. *Systematic Biology* **41**, 305–330.

Ehlers, U. (1985) *Das phylogenetische System der Plathelminthes*. Gustav Fischer, Stuttgart.

Eibye-Jacobsen, D. (2004) A reevaluation of *Wiwaxia* and the polychaetes of the Burgess Shale. *Lethaia* **37**, 317–335.

Eisig, H. (1898) Zur Entwicklungsgeschichte der Capitelliden. *Mitteilungen aus der Zoologischen Station zu Neapel* **13**, 1–292.

Elliott, G.R.D. and Leys, S.P. (2007) Coordinated contractions effectively expel water from the aquiferous system of a freshwater sponge. *Journal of Experimental Biology* **210**, 3736–3748.

Emes, R.D., Goodstadt, L., Winter, E.E., and Ponting, C.P. (2003) Comparison of the genomes of human and mouse lays the foundation of genome zoology. *Human Molecular Genetics* **12**, 701–709.

Emes, R.D., Beatson, S.A., Ponting, C.P., and Goodstadt, L. (2004a) Evolution and comparative genomics of odorant- and pheromone-associated genes in rodents. *Genome Research* **14**, 591–602.

Emes, R.D., Riley, M.C., Laukaitis, C.M., Goodstadt, L., Karn, R.C., and Ponting, C.P. (2004b) Comparative evolutionary genomics of androgen-binding protein genes. *Genome Research* **14**, 1516–1529.

Ericson, J., Rashbass, P., Schedl, A., Brenner-Morton, S., Kawakami, A., van Heyningen, V. *et al.* (1997) Pax6 controls progenitor cell identity and neuronal fate in response to graded Shh signaling. *Cell* **90**, 169–180.

Eriksson, B.J. and Budd, G.E. (2000) Onychophoran cephalic nerves and their bearing on our understanding of head segmentation and stem-group evolution of Arthropoda. *Arthropod Structure and Development* **29**, 197–209.

Eriksson, B.J., Tait, N.N., and Budd, G.E. (2003) Head development in the onychophoran *Euperipatoides kanangrensis* with particular reference to the central nervous system. *Journal of Morphology* **255**, 1–23.

Erpenbeck, D. and Wörheide, G. (2007) On the molecular phylogeny of sponges (Porifera). *Zootaxa* **1668**, 107–126.

Erwin, D.H. (2006) The developmental origins of animal bodyplans. In: S.H. Xiao and A.J. Kaufman (eds) *Neoproterozoic geobiology and paleobiology*, pp. 159–197. Plenum, New York.

Erwin, D.H. (2007) Disparity: morphological pattern and developmental context. *Palaeontology* **50**, 57–73.

Erwin, D.H. and Davidson, E.H. (2002) The last common bilaterian ancestor. *Development* **129**, 3021–3032.

Ettensohn, C.A., Illies, M.R., Oliveri, P., and De Jong, D.L. (2003) Alx1, a member of the Cart1/Alx3/Alx4 subfamily of paired-class homeodomain proteins, is an essential component of the gene network controlling skeletogenic fate specification in the sea urchin embryo. *Development* **130**, 2917–2928.

Evans, N.M., Lindner, A., Raikova, E.V., Collins, A.G., and Cartwright, P. (2008) Phylogenetic placement of the enigmatic parasite, *Polypodium hydriforme*, within the Phylum Cnidaria. *BMC Evolutionary Biology* **8**, 139.

Evans, T.M. and Marcus, J.M. (2006) A simulation study of the genetic regulatory hierarchy for butterfly eyespot focus determination. *Evolution & Development* **8**, 273–283.

Eyles, N. and Eyles, C.H. (1989) Glacially-influenced deep-marine sedimentation of the late Precambrian Gaskiers Formation, Newfoundland, Canada. *Sedimentology* **36**, 601–620.

Fan, W., Huang, X., Chen, C., Gray, J., and Huang, T. (2004) *TBX3* and its isoform *TBX3+2a* are functionally distinctive in inhibition of senescence and are overexpressed in a subset of breast cancer cell lines. *Cancer Research* **64**, 5132–5139.

Farzana, L. and Brown, S.J. (2008) Hedgehog signalling pathway function conserved in *Tribolium* segmentation. *Development, Genes and Evolution* **218**, 181–192.

Fedonkin, M.A. (2003) The origin of the Metazoa in the light of the Proterozoic fossil record. *Paleontological Research* **7**, 9–41.

Fedonkin, M.A. and Waggoner, B.M. (1997) The Late Precambrian fossil *Kimberella* is a mollusc-like bilaterian organism. *Nature* **388**, 868–871.

Felsenstein, J. (1978) Cases in which parsimony or compatibility methods will be positively misleading. *Systematic Zoology* **27**, 401–410.

Felsenstein, J. (2004) *Inferring phylogenies*. Sinauer Associates, Sunderland, MA.

Fenchel, T. and Finlay, B.J. (1995) *Ecology and evolution in anoxic worlds*. Oxford University Press, Oxford.

Feng, D.F., Cho, G., and Doolittle, R.F. (1997) Determining divergence times with a protein clock: update and reevaluation. *Proceedings of the National Academy of Sciences USA* **94**, 13028–13033.

Ferkowicz, M.J. and Raff, R.A. (2001) *Wnt* gene expression in sea urchin development: heterochronies associated with the evolution of developmental mode. *Evolution & Development* **3**, 24–33.

Field, K.G., Olsen, G.J., Lane, D.J., Giovannoni, S.J., Ghiselin, M.T., Raff, E.C. *et al.* (1988) Molecular phylogeny of the animal kingdom. *Science* **239**, 748–753.

Fields, C., Adams, M.D., White, O., and Venter, J.C. (1994) How many genes in the human genome? *Nature Genetics* **7**, 345–346.

Filipowicz, W., Bhattacharyya, S.N., and Sonenberg, N. (2008) Mechanisms of post-transcriptional regulation by microRNAs: are the answers in sight? *Nature Reviews Genetics* **9**, 102–114.

Finkelstein, R. and Boncinelli, E. (1994) From fly head to mammalian forebrain: the story of *otd* and *Otx*. *Trends in Genetics* **10**, 310–315.

Finn, R.D., Mistry, J., Schuster-Bockler, B., Griffiths-Jones, S., Hollich, V., Lassmann, T. *et al.* (2006) Pfam: clans, web tools and services. *Nucleic Acids Research* **34**, D247–D251.

Fioroni, P. (1979) Phylogenetische Abänderungen der Gastrula bei Mollusken. *Zeitschrift fur Zoologische Systematik und Evolutionsforschung* **1**, 82–100.

Fisher, R.A. (1930) *The genetical theory of natural selection*. Oxford University Press, Oxford.

Fisher, S., Grice, E.A., Vinton, R.M., Bessling, S.L., and McCallion, A.S. (2006) Conservation of RET regulatory function from human to zebrafish without sequence similarity. *Science* **312**, 276–279.

Fitz-Gibbon, S.T. and House, C.H. (1999) Whole genome-based phylogenetic analysis of free-living microorganisms. *Nucleic Acids Research* **27**, 4218–4222.

Fitzhugh, K. (2008) Clarifying the role of character loss in phylogenetic inference. *Zoologica Scripta* **37**, 561–569.

Fortey, R.A., Briggs, D.E.G. and Wills, M.A. (1996) The Cambrian evolutionary 'explosion': Decoupling cladogenesis from morphological disparity. *Biological Journal of the Linnean Society* **57**, 13–33.

Foster, P.G. (2004) Modeling compositional heterogeneity. *Systematic Biology* **53**, 485–495.

Foster, P.G. and Hickey, D.A. (1999) Compositional bias may affect both DNA-based and protein-based phylogenetic reconstructions. *Journal of Molecular Evolution* **48**, 284–290.

French, V. and Brakefield, P.M. (1992) The development of eyespot patterns on butterfly wings: morphogen sources or sinks? *Development* **116**, 103–109.

French, V. and Brakefield, P.M. (1995) Eyespot development on butterfly wings: the focal signal. *Developmental Biology* **168**, 112–123.

Frieder, D., Larijani, M., Tang, E., Parsa, J.Y., Basit, W., and Martin, A. (2006) Antibody diversification: mutational mechanisms and oncogenesis. *Immunological Research* **35**, 75–88.

Friedrich, H. (1979) Nemertini. In: F. Seidel (ed.) *Morphogenese der Tiere*, Vol. D5–I. Gustav Fischer Verlag, Stuttgart.

Friedrich, M. and Tautz, D. (1995) rDNA phylogeny of the major extant arthropod classes and the evolution of myriapods. *Nature* **376**, 165–167.

Fritzsch, G., Schlegel, M., and Stadler, P.F. (2006) Alignments of mitochondrial genome arrangements: applications to metazoan phylogeny. *Journal of Theoretical Biology* **240**, 243–254.

Fryer, G. (1997) A defence of arthropod polyphyly. In: R.A. Fortey and R.H. Thomas (eds) *Arthropod relationships*, Vol. 55, pp. 23–33. Chapman and Hall, London.

Fukuda, M., Fukuda, M., Wakasugi, S., Tsuzuki, T., Nomiyama, H., Shimada, K. *et al.* (1985) Mitochondrial DNA-like sequences in the human nuclear genome: characterization and implications in the evolution of mitochondrial DNA. *Journal of Molecular Biology* **186**, 257–266.

Funch, P. (1996) The chordoid larva of *Symbion pandora* (Cycliophora) is a modified trochophore. *Journal of Morphology* **230**, 231–263.

Furlong, R.F. and Holland, P.W.H. (2002) Bayesian phylogenetic analysis supports monophyly of Ambulacraria and of cyclostomes. *Zoological Science* **19**, 593–599.

Gabriel, W.N. and Goldstein, B. (2007) Segmental expression of *Pax3/7* and *engrailed* homologs in tardigrade development. *Development Genes and Evolution* **217**, 421–433.

Galant, R., Skeath, J.B., Paddock, S., Lewis, D.L., and Carroll, S.B. (1998) Expression of an *achaete-scute* homolog during butterfly scale development reveals the homology of insect scales and sensory bristles. *Current Biology* **8**, 807–813.

Galis, F., van Dooren, T.J., and Metz, J.A. (2002) Conservation of the segmented germband stage: robustness or pleiotropy? *Trends in Genetics* **18**, 504–509.

Galko, M.J. and Krasnow, M.A. (2004) Cellular and genetic analysis of wound healing in *Drosophila*. *PLoS Biology* **2**, e239.

Garstang, W. (1894) Preliminary note on a new theory of the phylogeny of the Chordata. *Zoologischer Anzeiger* **22**, 122–125.

Garstang, W. (1928) The morphology of the Tunicata. *Quarterly Journal of Microscopical Science* **72**, 51–189.

Gee, H. (1996) *Before the backbone. Views on the origin of the vertebrates*. Chapman and Hall, London.

Gehling, J.G. (1999) Microbial mats in terminal Proterozoic siliciclastics: Ediacaran death masks. *Palaios* **14**, 40–57.

Gehling, J.G. and Rigby, J.K. (1996) Long expected sponges from the Neoproterozoic Ediacara fauna of South Australia. *Journal of Paleontology* **70**, 185–195.

Gehling, J.G., Droser, M.L., Jensen, S.R., and Runnegar, B.N. (2005) Ediacara organisms: relating for to function. In: D.E.G. Briggs (ed.) *Evolving form and function: fossils and development: proceedings of a symposium honoring Adolf Seilacher for his contributions to paleontology, in celebration of his 80th birthday*, pp. 43–66. Yale Peabody Museum Publications, New Haven, CT.

Gerhart, J. (1998) Warkany lecture: signaling pathways in development. *Teratology* **60**, 226–239.

Gerhart, J. and Kirschner, M. (1997) *Cells, embryos, and evolution : toward a cellular and developmental understanding of phenotypic variation and evolutionary adaptability.* Blackwell Science, Malden, MA.

Gerould, J.H. (1906) The development of *Phascolosoma*. *Zoologische Jahrbücher, Abteilung für Anatomie und Ontogenie der Tiere* **23**, 77–162.

Geuten, K., Massingham, T., Smets, E., and Goldman, N. (2007) Experimental design criteria in phylogenetics: where to add taxa? *Systematic Biology* **56**, 609–622.

Ghiselin, M.T. (1988) The origin of molluscs in the light of molecular evidence. *Oxford Surveys in Evolutionary Biology.* **5**, 66–95.

Gilbert, S.F., Loredo, G.A., Bruckman, A., and Burke, A.C. (2001) Morphogenesis of the turtle shell: the development of a novel structure in tetrapod evolution. *Evolution & Development* **3**, 47–58.

Giribet, G. (2002) Current advances in the phylogenetic reconstruction of metazoan evolution. A new paradigm for the Cambrian explosion? *Molecular Phylogenetics and Evolution* **24**, 345–357.

Giribet, G. (2003) Molecules, development and fossils in the study of metazoan evolution; Articulata versus Ecdysozoa revisited. *Zoology* **106**, 303–326.

Giribet, G. (2008) Assembling the lophotrochozoan (=spiralian) tree of life. *Philosophical Transactions of the Royal Society B: Biological Sciences* **363**, 1513–1522.

Giribet, G., Distel, D.L., Polz, M., Sterrer, W., and Wheeler, W.C. (2000) Triploblastic relationships with emphasis on the acoelomates and the position of Gnathostomulida, Cycliophora, Plathelminthes, and Chaetognatha: a combined approach of 18S rDNA sequences and morphology. *Systematic Biology* **49**, 539–562.

Giribet, G., Sørensen, M.V., Funch, P., Kristensen, R.M., and Sterrer, W. (2004) Investigations into the phylogenetic position of Micrognathozoa using four molecular loci. *Cladistics* **20**, 1–13.

Giribet, G., Okusu, A., Lindgren, A.R., Huff, S.W., Schrödl, M., and Nishiguchi, M.K. (2006) Evidence for a clade composed of molluscs with serially repeated structures: monoplacophorans are related to chitons. *Proceedings of the National Academy of Sciences USA* **103**, 7723–7728.

Giribet, G., Dunn, C.W., Edgecombe, G.D., and Rouse, G.W. (2007) A modern look at the animal tree of life. In: Z.-Q. Zhang and W. A. Shear (eds) Linnaeus tercentenary: progress in invertebrate taxonomy. *Zootaxa* **1668**, 61–79.

Glazov, E.A., Pheasant, M., McGraw, E.A., Bejerano, G., and Mattick, J.S. (2005) Ultraconserved elements in insect genomes: a highly conserved intronic sequence implicated in the control of homothorax mRNA splicing. *Genome Research* **15**, 800–808.

Glenner, H., Hansen, A.J., Sørensen, M.V., Ronquist, F., Huelsenbeck, J.P., and Willerslev, E. (2004) Bayesian inference of the metazoan phylogeny: a combined molecular and morphological approach. *Current Biology* **14**, 1644–1649.

Goldstein, B. and Freeman, G. (1997) Axis specification in animal development. *BioEssays* **19**, 105–116.

Goodstadt, L. and Ponting, C.P. (2006) Phylogenetic reconstruction of orthology, paralogy, and conserved synteny for dog and human. *PLoS Computational Biology* **2**, e133.

Goodstadt, L., Heger, A., Webber, C., and Ponting, C.P. (2007) An analysis of the gene complement of a marsupial, *Monodelphis domestica*: evolution of lineage-specific genes and giant chromosomes. *Genome Research* **17**, 969–981.

Gordon, D.P. (1975) The resemblance of bryozoan gizzard teeth to 'annelid-like' setae. *Acta Zoologica* **56**, 283–289.

Gosling, N.J., Thomas, C.-W., Greenwood, J.M., Dong, X., Bengtson, S., Raff, E.C. *et al.* (2008) Experimental taphonomy of lophotrochozoan and deuterostome embryos. *Evolution & Development* **10**, 339–349.

Gould, S.J. (1989) Wonderful life: the Burgess Shale and the nature of history. Norton, New York

von Graff, L. (1891) *Die Organisation der Turbellaria Acoela*. Verlag von Wilhelm Engelmann, Leipzig.

Grande, C. and Patel, N.H. (2009) Nodal signalling is involved in left-right asymmetry in snails. *Nature* **457**, 1007–1011.

Graybeal, A. (1998) Is it better to add taxa or characters to a difficult phylogenetic problems? *Systematic Biology* **47**, 9–17.

Grazhdankin, D. and Seilacher, A. (2002) Underground Vendobionta from Namibia. *Palaeontology* **45**, 57–78.

Grbic, M. and Strand, M.R. (1998) Shifts in the life history of parasitic wasps correlate with pronounced alterations in early development. *Proceedings of the National Academy of Sciences USA* **95**, 1097–1101.

Gribaldo, S. and Philippe, H. (2002) Ancient phylogenetic relationships. *Theoretical Population Biology* **61**, 391–408.

Gridley, T. (2006) The long and short of it: somite formation in mice. *Developmental Dynamics* **235**, 2330–2336.

Griffiths-Jones, S., Grocock, R.J., van Dongen, S., Bateman, A., and Enright, A.J. (2006) miRBase: microRNA sequences, targets and gene nomenclature. *Nucleic Acids Research* **34**, D140–D144.

Grishin, N.V. (2001) Mh1 domain of Smad is a degraded homing endonuclease. *Journal of Molecular Biology* **307**, 31–37.

Grobben, K. (1908) Die systematische Einteilung des Tierreichs. *Verhandlungen der k. k. Zoologisch-Botanischen Gesellschaft in Wien* **58**, 491–511.

Grotzinger, J.P., Bowring, S.A., Saylor, B.Z., and Kaufman, A.J. (1995) Biostratigraphic and geochronologic constraints on early animal evolution. *Science* **270**, 598–604.

Gu, X. and Li, W.H. (1998) Estimation of evolutionary distances under stationary and nonstationary models of nucleotide substitution. *Proceedings of the National Academy of Sciences USA* **95**, 5899–5905.

Gurtner, G.C., Werner, S., Barrandon, Y., and Longaker, M.T. (2008) Wound repair and regeneration. *Nature* **453**, 314–321.

Haag, E.S. (2005) Echinoderm rudiments, rudimentary bilaterians, and the origin of the chordate CNS. *Evolution amd Development* **7**, 280–281.

Haag, E.S. (2006) Reply to Nielsen. *Evolution & Development* **8**, 3–5.

Haag, E.S. and True, J.R. (2001) From mutants to mechanisms? Assessing the candidate gene paradigm in evolutionary biology. *Evolution* **55**, 1077–1084.

Haase, A., Stern, M., Wächtler, K., and Bicker, G. (2001) A tissue-specific marker of Ecdysozoa. *Development Genes and Evolution* **211**, 428–433.

Hadfield, M.G. (2000) Why and how marine-invertebrate larvae metamorphose so fast. *Seminars in Cell and Developmental Biology* **11**, 437–443.

Haeckel, E. (1872) *Monographie der Kalkschwämme*. Georg Reimer Verlag, Berlin.

Haeckel, E. (1874) Die Gastraea-theorie, die phylogenetische Klassifikation des Tierreichs und die Homologie der Keimblätter. *Jenaische Zeitschrift für Naturwissenschaften* **8**, 1–55.

Hagadorn, J.W., Dott, R.H.J., and Damrow, D. (2002) Stranded on a Late Cambrian shoreline: medusae from central Wisconsin. *Geology* **30**, 147–150.

Hagadorn, J.W., Xiao, S.H., Donoghue, P.C.J., Bengtson, S., Gostling, N.J., Pawlowska, M. *et al.* (2006) Cellular and subcellular structure of Neoproterozoic animal embryos. *Science* **314**, 291–294.

Hahn, M.W. and Wray, G.A. (2002) The g-value paradox. *Evolution & Development* **4**, 73–75.

Halanych, K.M. (1995) The phylogenetic position of the pterobranch hemichordates based on 18S rDNA sequence data. *Molecular Phylogenetics and Evolution* **4**, 72–76.

Halanych, K.M. (2004) The new view of animal phylogeny. *Annual Review of Ecology, Evolution and Systematics* **35**, 229–256.

Halanych, K.M., Bacheller, J.D., Aguinaldo, A.M.A., Liva, S.M., Hillis, D.M., and Lake, J.A. (1995) Evidence from 18S ribosomal DNA that the lophophorates are protostome animals. *Science* **267**, 1641–1643.

Hall, K.A., Hutchings, P.A., and Colgan, D.J. (2004) Further phylogenetic studies of the Polychaeta using

18S rDNA sequence data. *Journal of the Marine Biological Association UK* **84**, 949–960.

Hallmann, A. (2006) The pherophorins: common, versatile building blocks in the evolution of extracellular matrix architecture in Volvocales. *The Plant Journal* **45**, 292–307.

Hallström, B.M. and Janke, A. (2008) Resolution among major placental mammal interordinal relationships with genome data imply that speciation influenced their earliest radiations. *BMC Evolutionary Biology* **8**, 162.

Handel, K., Grunfelder, C.G., Roth, S., and Sander, K. (2000) *Tribolium* embryogenesis: a SEM study of cell shapes and movements from blastoderm to serosal closure. *Development Genes and Evolution* **210**, 167–179.

Handel, K., Basal, A., Fan, X., and Roth, S. (2005) *Tribolium castaneum twist*: gastrulation and mesoderm formation in a short-germ beetle. *Development Genes and Evolution* **215**, 13–31.

Hara, Y., Yamaguchi, M., Akasaka, K., Nakano, H., Nonaka, M., and Amemiya, S. (2006) Expression patterns of *Hox* genes in larvae of the sea lily *Metacrinus rotundus*. *Development Genes and Evolution* **216**, 797–809.

Harada, Y., Okai, N., Taguchi, S., Tagawa, K., Humphreys, T., and Satoh, N. (2000) Developmental expression of the hemichordate *otx* ortholog. *Mechanisms of Development* **91**, 337–339.

Harada, Y., Shoguchi, E., Taguchi, S., Okai, N., Humphreys, T., Tagawa, K. *et al.* (2002) Conserved expression pattern of *BMP-2/4* in hemichordate acorn worm and echinoderm sea cucumber embryos. *Zoological Science* **19**, 1113–1121.

Harland, R. and Gerhart, J. (1997) Formation and function of Spemann's organizer. *Annual Review of Cell and Developmental Biology* **13**, 611–667.

Hartmann, S. and Vision, T.J. (2008) Using ESTs for phylogenomics: can one accurately infer a phylogenetic tree from a gappy alignment? *BMC Evolutionary Biology* **8**, 95.

Harwood, A. and Coates, J.C. (2004) A prehistory of cell adhesion. *Current Opinion in Cell Biology* **16**, 470–476.

Harzsch, S. (2002) The phylogenetic significance of crustacean optic neuropils and chiasmata: a re-examination. *Journal of Comparative Neurology* **453**, 10–21.

Harzsch, S. (2004) Phylogenetic comparison of serotonin-immunoreactive neurons in representatives of the Chilopoda, Diplopoda, and Chelicerata: implications for arthropod relationships. *Journal of Morphology* **259**, 198–213.

Harzsch, S. and Müller, C.H.G. (2007) A new look at the ventral nerve centre of *Sagitta*: implications for the phylogenetic position of Chaetognatha (arrow worms) and the evolution of the bilaterian nervous system. *Frontiers in Zoology* **4**, 14.

Harzsch, S., Müller, C.H., and Wolf, H. (2005) From variable to constant cell numbers: cellular characteristics of the arthropod nervous system argue against a sister-group relationship of Chelicerata and 'Myriapoda' but favour the Mandibulata concept. *Development Genes and Evolution* **215**, 53–68.

Haszprunar, G. (1996a) Plathelminthes and Plathelminthomorpha—paraphyletic taxa. *Journal of Zoological Systematics and Evolutionary Research* **34**, 41–48.

Haszprunar, G. (1996b) The Mollusca: coelomate turbellarians or mesenchymate annelids? In: J.D. Taylor (ed.) *Origin and evolutionary radiation of the Mollusca* , pp. 1–28. Oxford University Press, Oxford.

Haszprunar, G. (2000) Is the Aplacophora monophyletic? A cladistic point of view. *American Malacological Bulletin* **15**, 115–130.

Haszprunar, G. and Wanninger, A. (2008) On the fine structure of the creeping larva of *Loxosomella murmanica*: additional evidence for a clade of Kamptozoa (Entoprocta) and Mollusca. *Acta Zoologica* **89**, 137–148.

Haszprunar, G., Rieger, R.M., and Schuchert, P. (1991) Extant 'Problematica' within or near the Metazoa. In: A.M. Simonetta and S. Conway Morris (eds) *The early evolution of Metazoa and the significance of problematic taxa*, pp. 99–105. Oxford University Press, Oxford.

Hausdorf, B., Helmkampf, M., Meyer, A., Witek, A., Herlyn, H., Bruchhaus, I. *et al.* (2007) Spiralian phylogenomics supports the resurrection of Bryozoa comprising Ectoprocta and Entoprocta. *Molecular Biology and Evolution*. **24**, 2723–2729.

Hausen, H. (2005) Chaetae and chaetogenesis in polychaetes (Annelida). *Hydrobiologia* **535/536**, 37–52.

Hedges, S.B. and Kumar, S. (2004) Precision of molecular time estimates. *Trends in Genetics* **20**, 242–247.

Heider, K. (1909) Entwicklung von *Balanoglossus clavigerus* Delle Chiaje. *Zoologischer Anzeiger* **34**, 695–704.

Heimberg, A.M., Sempere, L.F., Moy, V.N., Donoghue, P.C., and Peterson, K.J. (2008) MicroRNAs and the advent of vertebrate morphological complexity. *Proceedings of the National Academy of Sciences USA* **105**, 2946–2950.

Hejnol, A. and Martindale, M.Q. (2008a) Acoel development supports a simple planula-like urbilaterian. *Philosophical Transactions of the Royal Society B: Biological Sciences* **363**, 1493–1501.

Hejnol, A. and Martindale, M.Q. (2008b) Acoel development indicates the independent evolution of the bilaterian mouth and anus. *Nature* **456**, 382–386.

Hejnol, A. and Schnabel, R. (2005) The eutardigrade *Thulinia stephaniae* has an indeterminate development and the potential to regulate early blastomere ablations. *Development* **132**, 1349–1361.

Hejnol, A., Martindale, M.Q., and Henry, J.Q. (2007) High-resolution fate map of the snail *Crepidula fornicata*: the origins of ciliary bands, nervous system, and muscular elements. *Developmental Biology* **305**, 63–76.

Helaers, R., Tzika, A.C., van de Peer, Y., and Milinkovitch, M. (2008) MANTIS, a phylogenetic framework for multi-species genome comparisons. *Bioinformatics* **24**, 151–157.

Helfenbein, K.G. and Boore, J.L. (2004) The mitochondrial genome of *Phoronis architecta*—comparisons demonstrate that phoronids are lophotrochozoan protostomes. *Molecular Biology and Evolution* **21**, 153–157.

Helmkampf, M., Bruchhaus, I., and Hausdorf, B. (2008a) Multigene analysis of lophophorate and chaetognath phylogenetic relationships. *Molecular Phylogenetics and Evolution* **46**, 206–214.

Helmkampf, M., Bruchhaus, I., and Hausdorf, B. (2008b) Phylogenomic analyses of lophophorates (brachiopods, phoronids and bryozoans) confirm the Lophotrochozoa concept. *Proceedings of the Royal Society B: Biological Sciences* **275**, 1927–1933.

Hendy, M.D. and Penny, D. (1989) A framework for the quantitative study of evolutionary trees. *Systematic Zoology* **38**, 297–309.

Henry, J.Q., Martindale, M.Q., and Boyer, B.C. (2000) The unique developmental program of the acoel flatworm, *Neochildia fusca*. *Developmental Biology* **220**, 285–295.

Henry, J.Q., Hejnol, A., Perry, K.J., and Martindale, M.Q. (2007) Homology of ciliary bands in spiralian Trochophores. *Integrative and Comparative Biology* **47**, 865–871.

Hertwig, O. (1880) Über die Entwicklungsgeschichte der Sagitten. *Jenaische Zeitschrift für Medizin und Naturwissenschaft* **14**, 196–303.

Hessling, R. and Westheide, W. (2002) Are Echiura derived from a segmented ancestor? Immunohistochemical analysis of the nervous system in developmental stages of *Bonellia viridis*. *Journal of Morphology* **252**, 100–113.

Hibberd, D.J. (1975) Observations on the ultrastructure of the choanoflagellates *Codosiga botrytis* Saville-Kent with special reference to the flagellar apparatus. *Journal of Cell Science* **17**, 191–219.

Hillier, L.W., Coulson, A., Murray, J.I., Bao, Z., Sulston, J.E., and Waterston, R.H. (2005) Genomics in *C. elegans*: so many genes, such a little worm. *Genome Research* **15**, 1651–1660.

Hillis, D.M., Pollock, D.D., McGuire, J.A., and Zwickl, D.J. (2003) Is sparse taxon sampling a problem for phylogenetic inference? *Systematic Biology* **52**, 124–126.

Hirth, F., Kammermeier, L., Frei, E., Walldorf, U., Noll, M., and Reichert, H. (2003) An urbilaterian origin of the tripartite brain: developmental genetic insights from *Drosophila*. *Development* **130**, 2365–2373.

Hittinger, C.T., Stern, D.L., and Carroll, S.B. (2005) Pleiotropic functions of a conserved insect-specific Hox peptide motif. *Development* **132**, 5261–5270.

Ho, K., Dunin-Borkowski, O.M., and Akam, M. (1997) Cellularization in locust embryos occurs before blastoderm formation. *Development* **124**, 2761–2768.

Ho, S.Y., Phillips, M.J., Drummond, A.J., and Cooper, A. (2005) Accuracy of rate estimation using relaxed-clock models with a critical focus on the early metazoan radiation. *Molecular Biology and Evolution* **22**, 1355–1363.

Hobert, O. (2008) Gene regulation by transcription factors and microRNAs. *Science* **319**, 1785–1786.

Hoekstra, H.E. and Coyne, J.A. (2007) The locus of evolution: evo devo and the genetics of adaptation. *Evolution & Development* **61**, 995–1016.

Holland, H.D. (2006) The oxygenation of the atmosphere and oceans. *Philosophical Transactions of the Royal Society B: Biological Sciences* **361**, 903–915.

Holland, N.D. (2003) Early central nervous system evolution: an era of skin brains? *Nature Reviews Neuroscience* **4**, 617–627.

Holley, S.A. and Ferguson, E.L. (1997) Fish are like flies are like frogs: conservation of dorsal-ventral patterning mechanisms. *BioEssays* **19**, 281–284.

Holmer, L.E., Skovsted, C.B., and Williams, A. (2002) A stem group brachiopod from the Lower Cambrian: support for a *Micrina* (halkieriid) ancestry. *Palaeontology* **45**, 875–882.

Holmer, L.E., Skovsted, C.B., Brock, G.A., Valentine, J.L., and Paterson, J.R. (2008) The Early Cambrian tommotiid *Micrina*, a sessile bivalved stem group brachiopod. *Biology Letters* **4**, 724–728.

Hordijk, W. and Gascuel, O. (2005) Improving the efficiency of SPR moves in phylogenetic tree search methods based on maximum likelihood. *Bioinformatics* **21**, 4338–4347.

Hou, X.-G., Aldridge, R.J., Bergström, J., Siveter, D.J., and Feng, X.-H. (2004) *The Cambrian fossils of Chengjiang, China*. Blackwell Science, Oxford.

House, C.H. and Fitz-Gibbon, S.T. (2002) Using homolog groups to create a whole-genomic tree of free-living organisms: an update. *Journal of Molecular Evolution* **54**, 539–547.

Huang, D.-Y., Chen, J.-Y., Vannier, J., and Saiz Salinas, J.I. (2004) Early Cambrian sipunculan worms from southwest China. *Proceedings of the Royal Society B: Biological Sciences* **271**, 1671–1676.

Hubrecht, A.A.W. (1887) The relation of the Nemertea to the Vertebrata. *Quarterly Journal of Microscopical Science* **27**, 605–644.

Hug, L.A. and Roger, A.J. (2007) The impact of fossils and taxon sampling on ancient molecular dating analyses. *Molecular Biology and Evolution.* **24**, 1889–1897.

Hughes, A.L. and Friedman, R. (2005) Loss of ancestral genes in the genomic evolution of *Ciona intestinalis. Evolution & Development* **7**, 196–200.

Hughes, C.L. and Kaufman, T.C. (2000) RNAi analysis of *Deformed, proboscipedia* and *Sex combs reduced* in the milkweed bug *Oncopeltus fasciatus*: novel roles for Hox genes in the hemipteran head. *Development* **127**, 3683–3694.

Hunter, T. (2000) Signaling—2000 and beyond. *Cell* **100**, 113–127.

Huntley, S., Baggott, D.M., Hamilton, A.T., Tran-Gyamfi, M., Yang, S., Kim, J. *et al.* (2006) A comprehensive catalog of human KRAB-associated zinc finger genes: insights into the evolutionary history of a large family of transcriptional repressors. *Genome Research* **16**, 669–677.

Huson, D.H. and Steel, M. (2004) Phylogenetic trees based on gene content. *Bioinformatics* **20**, 2044–2049.

Hwang, U.W., Friedrich, M., Tautz, D., Park, C.J., and Kim, W. (2001) Mitochondrial protein phylogeny joins myriapods with chelicerates. *Nature* **413**, 154–157.

Hyman, L.H. (1940) *The invertebrates*, 1st edn. McGraw-Hill, New York.

Hyman, L.H. (1951) *The invertebrates. Vol. 2. Platyhelminthes and Rhynchocoela.* McGraw-Hill, New York.

Hyman, L.H. (1959) *The invertebrates. Vol. 5. Smaller coelomate groups.* McGraw-Hill, New York.

In der Rieden, P.M., Mainguy, G., Woltering, J.M., and Durston, A.J. (2004) Homeodomain to hexapeptide or PBC-interaction-domain distance: size apparently matters. *Trends in Genetics* **20**, 76–79.

Inoue, I. (1958) Studies on the life history of *Chordodes japonensis*, a species of Gordiacea. I. The development and structure of the larva. *Japanese Journal of Zoology* **12**, 203–218.

Irimia, M., Maeso, I., Penny, D., Garcia-Fernàndez, J., and Roy, S. (2007) Rare coding sequence changes are consistent with Ecdysozoa, not Coelomata. *Molecular Biology and Evolution* **24**, 1604–1607.

Irish, V.F., Martinez-Arias, A., and Akam, M. (1989) Spatial regulation of the *Antennapedia* and *Ultrabithorax* homeotic genes during *Drosophila* early development. *EMBO Journal* **8**, 1527–1537.

Israelsson, O. and Budd, G.E. (2005) Eggs and embryos in *Xenoturbella* (phylum uncertain) are not ingested prey. *Development Genes and Evolution* **215**, 358–363.

Iwata, F. (1985) Foregut formation of the nemerteans and its role in nemertean systematics. *American Zoologist* **25**, 23–36.

Jackson, M., Watt, A.J., Gautier, P., Gilchrist, D., Driehaus, J., Graham, G.J. *et al.* (2006) A murine specific expansion of the *Rhox* cluster involved in embryonic stem cell biology is under natural selection. *BMC Genomics* **7**, 212.

Jager, M., Murienne, J., Clabaut, C., Deutsch, J., Le Guyader, H., and Manuel, M. (2006) Homology of arthropod anterior appendages revealed by Hox gene expression in a sea spider. *Nature* **441**, 506–508.

Jägersten, G. (1940) Zur Kenntnis der Morphologie, Entwicklung und Taxonomie der Myzostomida. *Nova Acta Regiase Societatis Scientiarum Upsaliensis Ser. 4* **11**, 1–84.

Jägersten, G. (1955) On the early phylogeny of the Metazoa: the bilatero-gastrea theory. *Zoologiska Bidrag från Uppsala* **30**, 321–354.

Jågersten, G. (1972) *Evolution of the metazoan life cycle.* Academic Press, London.

James-Clark, H. (1866) Note on the infusoria flagellata and the spongiae ciliatae. *American Journal of Science* **1**, 113–114.

James-Clark, H. (1868) On the spongiae ciliatae as infusoria flagellata, or observations on the structure, animality and relationship of *Leucosolenia botryoides* Bowerbank. *Annals and Magazine of Natural History* **4**, 133–142, 188–215, 250–264.

Jaynes, J.B. and Fujioka, M. (2004) Drawing lines in the sand: *even skipped* et al. and parasegment boundaries. *Developmental Biology* **269**, 609–622.

Jenner, R.A. (2000) Evolution of animal body plans: the role of metazoan phylogeny at the interface between pattern and process. *Evolution & Development* **2**, 208–221.

Jenner, R.A. (2001) Bilaterian phylogeny and uncritical recycling of morphological data sets. *Systematic Biology* **50**, 730–742.

Jenner, R.A. (2002) Boolean logic and character state identity: pitfalls of character coding in metazoan cladistics. *Contributions to Zoology* **71**, 67–91.

Jenner, R.A. (2003) Unleashing the force of cladistics? Metazoan phylogenetics and hypothesis testing. *Integrative and Comparative Biology* **43**, 207–218.

Jenner, R.A. (2004a) The scientific status of metazoan cladistics: why current research practice must change. *Zoologica Scripta* **33**, 293–310.

Jenner, R.A. (2004b) Towards a phylogeny of the Metazoa: evaluating alternative phylogenetic positions of Platyhelminthes, Nemertea, and Gnathostomulida, with a critical reappraisal of cladistic characters. *Contributions to Zoology* **73**, 3–163.

Jenner, R.A. (2004c) When molecules and morphology clash: reconciling conflicting phylogenies of the Metazoa by considering secondary character loss. *Evolution & Development* **6**, 372–378.

Jenner, R.A. (2006a) Unburdening evo-devo: ancestral attractions, model organisms, and basal baloney. *Evolution & Development* **216**, 385–394.

Jenner, R.A. (2006b) Challenging received wisdoms: some contributions of the new microscopy to the new animal phylogeny. *Integrative and Comparative Biology* **46**, 93–103.

Jenner, R.A. and Littlewood, D.T.J. (2008) Problematica old and new. *Philosophical Transactions of the Royal Society B: Biological Sciences* **363**, 1503–1512.

Jenner, R.A. and Scholtz, G. (2005) Playing another round of metazoan phylogenetics: historical epistemology, sensitivity analysis, and the position of Arthropoda within Metazoa on the basis of morphology. *Crustacean Issues* **16**, 355–385.

Jenner, R.A. and Wills, M.A. (2007) The choice of model organisms in evo-devo. *Nature Reviews Genetics* **8**, 311–319.

Jensen, S., Droser, M.L., and Gehling, J.G. (2005) Trace fossil preservation and the early evolution of animals. *Palaeogeography, Palaeoclimatology, Palaeoecology* **220**, 19–29.

Jimenez-Guri, E., Philippe, H., Okamura, B., and Holland, P.W.H. (2007) *Buddenbrockia* is a cnidarian worm. *Science* **317**, 116–118.

Jobb, G., von Haeseler, A., and Strimmer, K. (2004) TREEFINDER: a powerful graphical analysis environment for molecular phylogenetics. *BMC Evolutionary Biology* **4**, 18.

Johnston, R.J. and Hobert, O. (2003) A microRNA controlling left/right neuronal asymmetry in *Caenorhabditis elegans*. *Nature* **426**, 845–849.

Jondelius, U., Ruiz-Trillo, I., Baguñà, J., and Riutort, M. (2002) The Nemertodermatida are basal bilaterians and not members of the Platyhelminthes. *Zoologica Scripta* **31**, 201–215.

Kabat, J.L., Barberan-Soler, S., McKenna, P., Clawson, H., Farrer, T., and Zahler, A.M. (2006) Intronic alternative splicing regulators identified by comparative genomics in nematodes. *PLoS Computational Biology* **2**, e86.

Kadner, D. and Stollewerk, A. (2004) Neurogenesis in the chilopod *Lithobius forficatus* suggests more similarities to chelicerates than to insects. *Development Genes and Evolution* **214**, 367–379.

Karaulanov, E., Knochel, W., and Niehrs, C. (2004) Transcriptional regulation of BMP4 synexpression in transgenic *Xenopus*. *EMBO Journal* **23**, 844–856.

Karpov, S.A. and Coupe, S.J. (1998) A revision of choanoflagellates genera *Kentrosiga*, Schiller, 1953 and *Desmarella*, Kent, 1880. *Acta Protozoologica* **37**, 23–27.

Kendall, B., Creaser, R.A., and Selby, D. (2006) Re-Os geochronology of postglacial black shales in Australia: constraints on the timing of 'Sturtian' glaciation. *Geology* **34**, 729–732.

Kennedy, M.J., Runnegar, B., Prave, A.R., Hoffmann, K.H., and Arthur, M.A. (1998) Two or four Neoproterozoic glaciations? *Geology* **26**, 1059–1063.

Kennel, J. (1885) Entwicklungsgeschichte von *Peripatus edwardsii* Blanch. und *Peripatus torquatus* n. sp. I. Theil. *Arbeiten aus dem Zoologisch-Zootomischen Institut in Würzburg* **7**, 95–229.

Keys, D.N., Lewis, D.L., Selegue, J.E., Pearson, B.J., Goodrich, L.V., Johnson, R.L. *et al.* (1999) Recruitment of a *hedgehog* regulatory circuit in butterfly eyespot evolution. *Science* **283**, 532–534.

Kim, E., Magen, A., and Ast, G. (2007) Different levels of alternative splicing among eukaryotes. *Nucleic Acids Research* **35**, 125–131.

Kim, V.N. (2005) MicroRNA biogenesis: coordinated cropping and dicing. *Nature Reviews Molecular Cell Biology* **6**, 376–385.

King, N. (2004) The unicellular ancestry of animal development. *Developmental Cell* **7**, 313–325.

King, N., Westbrook, M.J., Young, S.L., Kuo, A., Abedin, M., Chapman, J. *et al.* (2008) The genome of the choanoflagellate *Monosiga brevicollis* and the origin of metazoans. *Nature* **451**, 783–788.

Knauss, E. (1979) Indication of an anal pore in Gnathostomulida. *Zoologica Scripta* **8**, 181–186.

Knauth, L.P. (2005) Temperature and salinity history of the Precambrian ocean: implications for the course of microbial evolution. *Palaeogeography, Palaeoclimatology, Palaeoecology* **219**, 53–69.

Knight-Jones, E. (1952) On the nervous system of *Saccoglossus cambriensis* (Enteropneusta). *Philosophical Transactions of the Royal Society of London B* **236**, 315–354.

Knoll, A.H. (1992) The early evolution of eukaryotes: a geological perspective. *Science* **256**, 622–627.

Knoll, A.H. (2003) Life on a young planet: the first three billion years of evolution on Earth. Princeton University Press, Princeton, NJ.

Knoll, A.H. and Carroll, S.B. (1999) Early animal evolution: emerging views from comparative biology and geology. *Science* **284**, 2129–2137.

Knoll, A.H., Walter, M.R., Narbonne, G.M., and Christie-Blick, N. (2004) A new period for the geologic time scale. *Science* **305**, 621–622.

Knoll, A.H., Walter, M.R., Narbonne, G.M., and Christie-Blick, N. (2006) The Ediacaran Period: a new addition to the geologic time scale. *Lethaia* **39**, 13–30.

Koch, P.B. and Nijhout, H.F. (2002) The role of wing veins in colour pattern development in the butterfly *Papilio xuthus*. *European Journal of Entomology* **99**, 67–72.

Konuma, J. and Chiba, S. (2007) Trade-offs between force and fit: extreme morphologies associated with feeding behavior in carabid beetles. *The American Naturalist* **170**, 90–100.

Kraft, R. and Jackle, H. (1994) *Drosophila* mode of metamerization in the embryogenesis of the lepidopteran insect *Manduca sexta*. *Proceedings of the National Academy of Sciences USA* **91**, 6634–6638.

Kristensen, R.M. and Funch, P. (2000) Micrognathozoa: a new class with complicated jaws like those of Rotifera and Gnathostomulida. *Journal of Morphology* **246**, 1–49.

Krumlauf, R., Marshall, H., Studer, M., Nonchev, S., Sham, M.H., and Lumsden, A. (1993) *Hox* homeobox genes and regionalisation of the nervous system. *Journal of Neurobiology* **24**, 1328–1340.

Kühn, A. and Von Englehardt, A. (1933) Uber die Determination des symmetrie Systems auf dem vorderflugel von Phylogeny. *Evolution* **42**, 862–884.

Kumar, S. and Hedges, S.B. (1998) A molecular timescale for vertebrate evolution. *Nature* **392**, 917–920.

Kuraku, S., Usuda, R., and Kuratani, S. (2005) Comprehensive survey of carapacial ridge-specific genes in turtle implies co-option of some regulatory genes in carapace evolution. *Evolution & Development* **7**, 3–17.

Kuzniar, A., van Ham, R.C.H.J., Pongor, S., and Leunissen, J.A.M. (2008) The quest for orthologs: finding the corresponding gene across genomes. *Trends in Genetics* **24**, 539–551.

Lacalli, T.C. (2002) Vetulicolians—are they deuterostomes? chordates? *BioEssays* **24**, 208–211.

Ladurner, P. and Rieger, R. (2000) Embryonic muscle development of *Convoluta pulchra* (Turbellaria-Acoelomorpha, Platyhelminthes). *Developmental Biology* **222**, 359–375.

Lagos-Quintana, M., Rauhut, R., Lendeckel, W., and Tuschl, T. (2001) Identification of novel genes coding for small expressed RNAs. *Science* **294**, 853–858.

Lake, J.A. (1990) Origin of the Metazoa. *Proceedings of the National Academy of Sciences USA* **87**, 763–766.

Lambert, C.C. (2005) Historical introduction, overview, and reproductive biology of protochordates. *Canadian Journal of Zoology* **83**, 1–7.

Lambert, I.B. and Donnelly, T.H. (1991) Atmospheric oxygen levels in the Precambrian—a review of isotopic and geological evidence. *Global Planetary Change* **97**, 83–91.

Lander, E.S., Linton, L.M., Birren, B., Nusbaum, C., Zody, M.C., Baldwin, J. *et al.* (2001) Initial sequencing and analysis of the human genome. *Nature* **409**, 860–921.

Lanterbecq, D., Bleidorn, C., Michel, S., and Eeckhaut, I. (2008) Locomotion and fine structure of parapodia in *Myzostoma cirriferum* (Myzostomida). *Zoomorphology* **127**, 59–68.

Lareau, L.F., Inada, M., Green, R.E., Wengrod, J.C., and Brenner, S.E. (2007) Unproductive splicing of SR genes associated with highly conserved and ultraconserved DNA elements. *Nature* **446**, 926–929.

LaRonde-LeBlanc, N.A. and Wolberger, C. (2003) Structure of HoxA9 and Pbx1 bound to DNA: Hox hexapeptide and DNA recognition anterior to posterior. *Genes and Development* **17**, 2060–2072.

Larroux, C., Luke, G.N., Koopman, P., Rokhsar, D.S., Shimeld, S.M., and Degnan, B.M. (2008) Genesis and expansion of metazoan transcription factor gene classes. *Molecular Biology and Evolution* **25**, 980–996.

Lartillot, N. and Philippe, H. (2004) A Bayesian mixture model for across-site heterogeneities in the amino-acid replacement process. *Molecular Biology and Evolution* **21**, 1095–1109.

Lartillot, N. and Philippe, H. (2008) Improvement of molecular phylogenetic inference and the phylogeny of Bilateria. *Philosophical Transactions of the Royal Society B: Biological Sciences* **363**, 1463–1472.

Lartillot, N., Brinkmann, H., and Philippe, H. (2007) Suppression of long-branch attraction artefacts in the animal phylogeny using a site-heterogeneous model. *BMC Evolutionary Biology* **7**(Suppl. 1), S4.

Lau, N.C., Lim, L.P., Weinstein, E.G., and Bartel, D.P. (2001) An abundant class of tiny RNAs with probable regulatory roles in *Caenorhabditis elegans*. *Science* **294**, 858–862.

Lavrov, D.V. and Lang, B.F. (2005) Poriferan mtDNA and animal phylogeny based on mitochondrial gene arrangements. *Systematic Biology* **54**, 651–659.

Lavrov, D., Brown, W.M., and Boore, J.L. (2004) Phylogenetic position of the Pentastomida and (pan) crustacean relationships. *Proceedings of the Royal Society B: Biological Sciences* **271**, 537–544.

Lavrov, D.V., Forget, L., Kelly, M., and Lang, B.F. (2005) Mitochondrial genomes of two demosponges provide insights into an early stage of animal evolution. *Molecular Biology and Evolution* **22**, 1231–1239.

Lavrov, D.V., Wang, X., and Kelly, M. (2008) Reconstructing ordinal relationships in the Demospongiae using mitochondrial genomic data. *Molecular Phylogenetics and Evolution* **49**, 111–124.

Lawrence, P.A. (1992) *The making of a fly: the genetics of animal design*. Blackwell Scientific, Oxford.

Leadbeater, B. (1983) Life-history and ultrastructure of a new marine species of *Proterospongia* (Choanoflagellida).

Journal of the Marine Biological Association UK **63**, 135–160.

Lechner, M. (1966) Untersuchungen zur Embryonalentwicklung des Rädertieres *Asplanchna girodi* de Guerne. *Wilhelm Roux' Archiv für Entwicklungsmechanik der Organismen* **157**, 117–173.

Lecointre, G. and Deleporte, P. (2004) Total evidence requires exclusion of phylogenetically misleading data. *Zoologica Scripta* **34**, 101–117.

Lee, D.-C. and Bryant, H.N. (1999) A reconsideration of the coding of inapplicable characters: assumptions and problems. *Cladistics* **15**, 373–378.

Lee, R.C. and Ambros, V. (2001) An extensive class of small RNAs in *Caenorhabditis elegans*. *Science* **294**, 862–864.

Lee, R.C., Feinbaum, R.L., and Ambros, V. (1993) The *C. elegans* heterochronic gene *lin-4* encodes small RNAs with antisense complementarity to *lin-14*. *Cell* **75**, 843–854.

Leebens-Mack, J., Vision, T., Brenner, E., Bowers, J.E., Cannon, S., Clement, M.J. *et al.* (2006) Taking the first steps towards a standard for reporting on phylogenies: minimum information about a phylogenetic analysis (MIAPA). *OMICS: a Journal of Integrative Biology* **10**, 231–237.

Leise, E.M. and Cloney, R.A. (1982) Chiton integument: ultrastructure of the sensory hairs of *Mopalia muscosa* (Mollusca: Polyplacophora). *Cell and Tissue Research* **223**, 43–59.

Lemons, D. and McGinnis, W. (2006) Genomic evolution of *Hox* gene clusters. *Science* **313**, 1918–1922.

Letunic, I., Copley, R.R., Pils, B., Pinkert, S., Schultz, J., and Bork, P. (2006) SMART 5: domains in the context of genomes and networks. *Nucleic Acids Research* **34**, D257–D260.

Leuckart, F.S. (1827) *Versuch einer naturgemässen Eintheilung der Helminthen*. Neue Akademische Buchhandlung von Karl Gross, Heidelberg, Leipzig.

Levine, A.J. and Brivanlou, A.H. (2007) Proposal of a model of mammalian neural induction. *Developmental Biology* **308**, 247–256.

Lewis, P.O. (2001) Phylogenetic systematics turns over a new leaf. *Trends in Ecology and Evolution* **16**, 30–37.

Leys S.P. and Eerkes-Medrano, D.I. (2006) Feeding in a calcareous sponge: particle uptake by pseudopodia. *Biological Bulletin* **211**, 157–171.

Li, C.W., Chen, J.Y., and Hua, T.E. (1998) Precambrian sponges with cellular structures. *Science* **279**, 879–882.

Li, W., Young, S.L., King, N. and Miller, W.T. 2008. Signaling properties of a non-metazoan Src kinase and the evolutionary history of Src negative regulation. *Journal of Biological Chemistry* **283**, 15491–15501.

Lichtneckert, R. and Reichert, H. (2005) Insights into the urbilaterian brain: conserved genetic patterning mechanisms in insect and vertebrate brain development. *Heredity* **94**, 465–477.

Lim, L.P., Lau, N.C., Garrett-Engele, P., Grimson, A., Schelter, J.M., Castle, J. *et al.* (2005) Microarray analysis shows that some microRNAs downregulate large numbers of target mRNAs. *Nature* **433**, 769–773.

Lin, J.-P., Gon, S.M.I., Gehling, J.G., Babcock, L.E., Zhao, Y.-L., Zhang, X.-L. *et al.* (2006) A *Parvancorina*-like arthropod from the Cambrian of South China. *Historical Biology* **18**, 33–45.

Linder, H., Hardy, C.R., and Rutschmann, F. (2005) Taxon sampling effects in molecular clock dating: an example from the African Restionaceae. *Molecular Phylogenetics and Evolution.* **35**, 569–582.

Littlewood, D.T.J., Rohde, K., and Clough, K.A. (1999) Interrelationships of all major groups of Platyhelminthes – phylogenetic evidence from morphology and molecules. *Biological Journal of the Linnean Society.* **66**, 75–114.

Liu, J., Perumal, N.B., Oldfield, C.J., Su, E.W., Uversky, V.N., and Dunker, A.K. (2006) Intrinsic disorder in transcription factors. *Biochemistry* **45**, 6873–8688.

Liu, P.J., Yin, C.Y., and Tang, F. (2006) Discovery of the budding phosphatized globular fossils from the Neoproterozoic Doushantuo Formation at Weng'an, Guizhou Province, China. *Progress in Natural Science* **16**, 1079–1083.

Liu, P.J., Xiao, S.H., Yin, C.Y., Zhou, C.M., Gao, L.Z., and Tang, F. (2008) Systematic description and phylogenetic affinity of tubular microfossils from the Ediacaran Doushantuo formation at Weng'an, South China. *Palaeontology* **51**, 339–366.

Liu, P.Z. and Kaufman, T.C. (2005) Short and long germ segmentation: unanswered questions in the evolution of a developmental mode. *Evolution & Development* **7**, 629–646.

Locke, D.P., Segraves, R., Carbone, L., Archidiacono, N., Albertson, D.G., Pinkel, D. *et al.* (2003) Large-scale variation among human and great ape genomes determined by array comparative genomic hybridization. *Genome Research* **13**, 347–357.

Logan, C.Y. and Nusse, R. (2004) The Wnt signaling pathway in development and disease. *Annual Reviews of Cell and Developmental Biology* **20**, 781–810.

Lohr, U. and Pick, L. (2005) Cofactor-interaction motifs and the cooption of a homeotic Hox protein into the segmentation pathway of *Drosophila melanogaster*. *Current Biology* **15**, 643–649.

Long, S., Martinez, P., Chen, W.C., Thorndyke, M., and Byrne, M. (2003) Evolution of echinoderms may not have required modification of the ancestral deuterostome HOX gene cluster: first report of PG4 and PG5 Hox orthologues in echinoderms. *Development Genes and Evolution* **213**, 573–576.

Love, A., Andrews, M., and Raff, R.A. (2007) Pluteus larval arm morphogenesis and evolution: gene expression patterns in a novel animal appendage and their transformation in the origin of direct development. *Evolution & Development* **9**, 51–68.

Love, A., Lee, A., and Raff, R.A. (2008) Evolutionary patterns of gene expression in the sea urchin larval gut: co-option and dissociation in larval origins and evolution. *Evolution & Development* **10**, 74–88.

Love, G.D., Fike, D. A., Grosjean, E., Stalvies, C., Grotzinger, J., Bradley, A.S. *et al.* (2006) Constraining the timing of basal metazoan radiation using molecular biomarkers and U-Pb isotope dating. *Geochimica et Cosmochimica Acta* **70**, A371.

Lowe C.J. (2008) Molecular genetic insights into deuterostome evolution from the direct-developing hemichordate *Saccoglossus kowalevskii*. *Philosophical Transactions of the Royal Society B: Biological Sciences* **363**, 1569–1578.

Lowe, C.J. and Wray, G.A. (1997) Radical alterations in the roles of homeobox genes during echinoderm evolution. *Nature* **389**, 718–721.

Lowe, C.J., Issel-Tarver, L., and Wray, G.A. (2002) Gene expression and larval evolution: changing roles of *distal-less* and *orthodenticle* in echinoderm larvae. *Evolution & Development* **4**, 111–123.

Lowe, C.J., Wu, M., Salic, A., Evans, L., Lander, E., Stange-Thomann, N. *et al.* (2003) Anteroposterior patterning in hemichordates and the origins of the chordate nervous system. *Cell* **113**, 853–865.

Lowe, C.J., Tagawa, K., Humphreys, T., Kirschner, M., and Gerhart, J. (2004) Hemichordate embryos: procurement, culture, and basic methods. *Methods in Cell Biology* **74**, 171–194.

Lowe, C.J., Terasaki, M., Wu, M., Freeman, R.M., Jr, Runft, L., Kwan, K. *et al.* (2006) Dorsoventral patterning in hemichordates: insights into early chordate evolution. *PLoS Biology* **4**, e291.

Ludwig, M.Z., Bergman, C., Patel, N.H., and Kreitman, M. (2000) Evidence for stabilizing selection in a eukaryotic enhancer element. *Nature* **403**, 564–567.

Lumsden, A. and Krumlauf, R. (1996) Patterning the vertebrate neuraxis. *Science* **274**, 1109–1115.

Luo, Z. (2000) In search of the whales' sisters. *Nature* **404**, 235–239.

Lüter, C. (2000) Ultrastructure of larval and adult setae of Brachiopoda. *Zoologischer Anzeiger* **239**, 75–90.

Lynch, J.A., Brent, A.E., Leaf, D.S., Pultz, M.A., and Desplan, C. (2006a) Localized maternal *orthodenticle*

patterns anterior and posterior in the long germ wasp *Nasonia*. *Nature* **439**, 728–732.

Lynch, J.A., Olesnicky, E.C., and Desplan, C. (2006b) Regulation and function of *tailless* in the long germ wasp *Nasonia vitripennis*. *Development Genes and Evolution* **216**, 493–498.

Lynch, M. (2007a) The frailty of adaptive hypotheses for the origins of organismal complexity. *Proceedings of the National Academy of Sciences USA* **104**(Suppl. 1), 8597–8604.

Lynch, M. (2007b) *The origins of genome architecture.* Sinauer Associates, Sunderland, MA.

Maas, A., Huang, D., Chen, J., Waloszek, D., and Braun, A. (2007) Maotianshan-shale nemathelminths—morphology, biology, and the phylogeny of Nemathelminthes. *Palaeogeography, Palaeoclimatology, Palaeoecology* **254**, 288–306.

McCaffrey, M.A., Moldowan, J.M., Lipton, P.A., Summons, R.E., Peters, K.E., Jeganathan, A. *et al.* (1994) Paleoenvironmental implications of novel C30 steranes in Precambrian to Cenozoic age petroleum and bitumen. *Geochimica et Cosmochimica Acta* **58**, 529–532.

Mace, K.A., Pearson, J.C., and McGinnis, W. (2005) An epidermal barrier wound repair pathway in *Drosophila* is mediated by *grainy head*. *Science* **308**, 381–385.

McEdward, L.R. (2000) Adaptive evolution of larvae and life cycles. *Seminars in Cell and Developmental Biology* **11**, 403–409.

McGinnis, W., Garber, R.L., Wirz, J., Kuroiwa, A., and Gehring, W.J. (1984) A homologous protein-coding sequence in *Drosophila* homeotic genes and its conservation in other metazoans. *Cell* **37**, 403–408.

McGregor, A.P., Shaw, P.J., Hancock, J.M., Bopp, D., Hediger, M., Wratten, N.S. *et al.* (2001) Rapid restructuring of bicoid-dependent hunchback promoters within and between dipteran species: implications for molecular coevolution. *Evolution & Development* **3**, 397–407.

McHugh, D. (1997) Molecular evidence that echiurans and pogonophorans are derived annelids. *Proceedings of the National Academy of Sciences USA* **94**, 8006–8009.

Mackie, G.O. (2004) Central neural circuitry in the jellyfish *Aglantha*: a model 'simple nervous system'. *Neurosignals* **13**, 5–19.

Maclean, J.A., 2nd, Chen, M.A., Wayne, C.M., Bruce, S.R., Rao, M., Meistrich, M.L. *et al.* (2005) *Rhox*: a new homeobox gene cluster. *Cell* **120**, 369–382.

McMillan, W.O., Monteiro, A., and Kapan, D.D. (2002) Development and evolution on the wing. *Trends in Ecology and Evolution* **17**, 125–133.

McMullin, E.R., Bergquist, D.C., and Fisher, C.R. (2000) Metazoans in extreme environments: adaptations of hydrothermal vent and hydrocarbon seep fauna. *Gravitational and Space Biology Bulletin* **13**, 13–23.

Malakhov, V.V. (1994) *Nematodes: structure, development, classification and phylogeny*. Smithsonian Institution, Washington, DC.

Malakhov, V.V. (2004) [Origin of bilateral-symmetrical animals (Bilateria)]. *Zhurnal Obshchei Biologii* **65**, 371–388 (in Russian).

Maldonado, M. (2004) Choanoflagellates, choanocytes, and animal multicellularity. *Invertebrate Biology* **123**, 1–22.

Maletz, J., Steiner, M., and Fatka, O. (2005) Middle cambrian pterobranchs and the question: what is a graptolite. *Lethaia* **38**, 73–85.

Mallatt, J. and Chen, J.Y. (2003) Fossil sister group of craniates: predicted and found. *Journal of Morphology* **258**, 1–31.

Mallatt, J. and Giribet, G. (2006) Further use of nearly complete 28S and 18S rRNA genes to classify Ecdysozoa: 37 more arthropods and a kinorhynch. *Molecular Phylogenetics and Evolution* **40**, 772–794.

Mallatt, J. and Sullivan, J. (1998) 28S and 18S rDNA sequences support the monophyly of lampreys and hagfishes. *Molecular Biology and Evolution* **15**, 1706–1718.

Mallatt, J. and Winchell, C.J. (2002) Testing the new animal phylogeny: first use of combined large-subunit and small-subunit rRNA gene sequences to classify the protostomes. *Molecular Biology and Evolution* **19**, 289–301.

Mallatt, J., Chen, J., and Holland, N.D. (2003) Comment on 'A new species of yunnanozoan with implications for deuterostome evolution'. *Science* **300**, 1372; author reply 1372.

Mallatt, J.M., Garey, J.R., and Shultz, J.W. (2004) Ecdysozoan phylogeny and bayesian inference: first use of nearly complete 28S and 18S rRNA gene sequences to classify the arthropods and their kin. *Molecular Phylogenetics and Evolution* **31**, 178–191.

Manning, G., Young, S.L., Miller, W.T., and Zhai, Y. (2008) The protist, *Monosiga brevicollis*, has a tyrosine kinase signaling network more elaborate and diverse than found in any known metazoan. *Proceedings of the National Academy of Sciences USA* **105**, 9674–9679.

Manton, S.M. (1949) Studies on the Onychophora VII. The early embryonic stages of *Peripatopsis* and some general considerations concerning the morphology and phylogeny of the Arthropoda. *Philosophical Transactions of the Royal Society of London* **233**, 483–580.

Manuel, M., Kruse, M., Müller, W.E., and Le Parco, Y. (2000) The comparison of beta-thymosin homologues among Metazoa supports an arthropod-nematode clade. *Journal of Molecular Evolution* **51**, 378–381.

Marcellini, S. (2006) When Brachyury meets Smad1: the evolution of bilateral symmetry during gastrulation. *BioEssays* **28**, 413–420.

Marcus, E. (1929) Zur Embryologie der Tardigraden. *Zoologische Jahrbücher, Abteilung für Anatomie und Ontogenie der Tiere* **50**, 333–384.

Marcus, E. (1939) Bryozoarios marinhos brasileiros III. *Boletim da Faculdade de Fiosofia, Ciêcias e Letras, Universidade di São Paolo, Zoologia* **3**, 111–354.

Marcus, J.M. (2001) The development and evolution of crossveins in insect wings. *Journal of Anatomy* **199**, 211–216.

Markstein, M., Markstein, P., Markstein, V., and Levine, M.S. (2002) Genome-wide analysis of clustered Dorsal binding sites identifies putative target genes in the *Drosophila* embryo. *Proceedings of the National Academy of Sciences USA* **99**, 763–768.

Markstein, M., Zinzen, R., Markstein, P., Yee, K.P., Erives, A., Stathopoulos, A. *et al.* (2004) A regulatory code for neurogenic gene expression in the *Drosophila* embryo. *Development* **131**, 2387–2394.

Marlétaz, F., Martin, E., Perez, Y., Papillon, D., Caubit, X., Lowe, C.J. *et al.* (2006) Chaetognath phylogenomics: a protostome with deuterostome-like development. *Current Biology* **16**, R577–R578.

Marlétaz, F., Gilles, A., Caubit, X., Perez, Y., Dossat, C., Samain, S. *et al.* (2008) Chaetognath transcriptome reveals ancestral and unique features among bilaterians. *Genome Biology* **9**, R94.

Marshall, C.R. (2006) Explaining the Cambrian 'explosion' of animals. *Annual Reviews in Earth and Planetary Sciences* **34**, 355–384.

Marshall, C.R. (2008) A simple method for bracketing absolute divergence times on molecular phylogenies using multiple fossil calibration points. *The American Naturalist* **171**, 726–742.

Martin, M.W., Grazhdankin, D.V., Bowring, S.A., Evans, D.A.D., Fedonkin, M.A., and Kirschvink, J.L. (2000) Age of Neoproterozoic bilaterian body and trace fossils, White Sea, Russia: implications for metazoan evolution. *Science* **288**, 841–845.

Martin, P and Parkhurst, S.M. (2004) Parallels between tissue repair and embryo morphogenesis. *Development* **131**, 3021–3034.

Martin, W., Rotte, C., Hoffmeister, M., Theissen, U., Gelius-Dietrich, G., Ahr, S. *et al.* (2003) Early cell evolution, eukaryotes, anoxia, sulfide, oxygen, fungi first (?), and a tree of genomes revisited. *IUBMB Life* **55**, 193–204.

Martindale, M.Q., Pang, K., and Finnerty, J.R. (2004) Investigating the origins of triploblasty: 'mesodermal' gene expression in a diploblastic animal, the sea anemone *Nematostella vectensis* (phylum, Cnidaria; class, Anthozoa). *Development* **131**, 2463–2474.

Masamizu, Y., Ohtsuka, T., Takashima, Y., Nagahara, H., Takenaka, Y., Yoshikawa, K. *et al.* (2006) Real-time imaging of the somite segmentation clock: revelation of unstable oscillators in the individual presomitic mesoderm cells. *Proceedings of the National Academy of Sciences USA* **103**, 1313–1318.

Maslakova, S. A., Martindale, M.Q., and Norenburg, J.L. (2004a) Fundamental properties of the spiralian developmental program are displayed by the basal nemertean *Carinoma tremaphoros* (Palaeonemertea, Nemertea). *Developmental Biology* **267**, 342–360.

Maslakova, S.A., Martindale, M.Q., and Norenburg, J.L. (2004b) Vestigial prototroch in a basal nemertean, *Carinoma tremaphoros* (Nemertea; Palaeonemertea). *Evolution & Development* **6**, 219–226.

Mastick, G.S., Davis, N.M., Andrew, G.L., and Easter, S.S. (1997) Pax-6 functions in boundary formation and axon guidance in the embryonic mouse forebrain. *Development* **124**, 1985–1997.

Matsunaga, E., Araki, I., and Nakamura, H. (2001) Role of Pax3/7 in the tectum regionalization. *Development* **128**, 4069–4077.

Mattei, X. and Marchand, B. (1987) Les spermatozoïdes des Acanthocéphales et des Myzostomides: Ressemblance et conséquences phylétiques. *Comptes Rendus De L'Academie Des Sciences de Paris, Sciences de la Vie* **305**, 525–529.

Matus, D.Q., Pang, K., Marlow, H., Dunn, C.W., Thomsen, G.H., and Martindale, M.Q. (2006a) Molecular evidence for deep evolutionary roots of bilaterality in animal development. *Proceedings of the National Academy of Sciences USA* **103**, 11195–11200.

Matus, D.Q., Copley, R.R., Dunn, C.W., Hejnol, A., Eccleston, H., Halanych, K.M. *et al.* (2006b) Broad taxon and gene sampling indicate that chaetognaths are protostomes. *Current Biology* **16**, R575–R576.

Matus, D.Q., Magi, C.R., Pang, K., Martindale, M.Q., and Thomsen, G.H. (2008) The hedgehog gene family of the cnidarian, *Nematostella vectensis*, and implications for understanding metazoan Hedgehog pathway evolution. *Developmental Biology* **313**, 501–518.

Maxmen, A.B., King, B.F., Cutler, E.B., and Giribet, G. (2003) Evolutionary relationships within the protostome phylum Sipuncula: a molecular analysis of ribosomal genes and histone H3 sequence data. *Molecular Phylogenetics and Evolution* **27**, 489–503.

Maxmen, A., Browne, W.E., Martindale, M.Q., and Giribet, G. (2005) Neuroanatomy of sea spiders implies

an appendicular origin of the protocerebral segment. *Nature* **437**, 1144–1148.

Mayr, E. (1960) The emergence of evolutionary novelties. In: S. Tax (ed.) *The evolution of life*, pp. 349–380. University of Chicago Press, Chicago.

Mead, A.D. (1897) The early development of marine annelids. *Journal of Morphology* **13**, 227–326.

Medina, M., Collins, A.G., Silberman, J.D., and Sogin, M.L. (2001) Evaluating hypotheses of basal animal phylogeny using complete sequences of large and small subunit rRNA. *Proceedings of the National Academy of Sciences USA* **98**, 9707–9712.

Medina, M., Collins, A.G., Taylor, J.W., Valentine, J.W., Lipps, J.H., Amaral-Zettler, L. *et al.* (2003). Phylogeny of Opisthokonta and the evolution of multicellularity and complexity in Fungi and Metazoa. *International Journal of Astrobiology* **2**, 203–211.

Michod, R.E. (2007) Evolution of individuality during the transition from unicellular to multicellular life. *Proceedings of the National Academy of Sciences USA* **104**(Suppl. 1), 8613–8618.

Mierzejewski, P. and Urbanek, A. (2004) The morphology and fine structure of the ordovician *Cephalodiscus*-like genus *Melanostrophus*. *Acta Palaeontologica Polonica* **49**, 519–528.

Miljkovic-Licina, M., Gauchat, D., and Galliot, B. (2004) Neuronal evolution: analysis of regulatory genes in a first-evolved nervous system, the hydra nervous system. *Biosystems* **76**, 75–87.

Mineta, K., Nakazawa, M., Cebria, F., Ikeo, K., Agata, K., and Gojobori, T. (2003) Origin and evolutionary process of the CNS elucidated by comparative genomics analysis of planarian ESTs. *Proceedings of the National Academy of Sciences USA* **100**, 7666–7671.

Minezaki, Y., Homma, K., Kinjo, A.R., and Nishikawa, K. (2006) Human transcription factors contain a high fraction of intrinsically disordered regions essential for transcriptional regulation. *Journal of Molecular Biology* **359**, 1137–1149.

Minguillón, C., Gardenyes, J., Serra, E., Castro, L.F., Hill-Force, A., Holland, P.W.H. *et al.* (2005) No more than 14: the end of the amphioxus *Hox* cluster. *International Journal of Biological Sciences* **1**, 19–23.

Mistry, N., Harrington, W., Lasda, E., Wagner, E.J., and Garcia-Blanco, M.A. (2003) Of urchins and men: evolution of an alternative splicing unit in fibroblast growth factor receptor genes. *RNA* **9**, 209–217.

Miyawaki, K., Mito, T., Sarashina, I., Zhang, H., Shinmyo, Y., Ohuchi, H. *et al.* (2004) Involvement of Wingless/Armadillo signaling in the posterior sequential segmentation in the cricket, *Gryllus bimaculatus* (Orthoptera), as revealed by RNAi analysis. *Mechanisms of Development* **121**, 119–130.

Mizutani, C.M., Nie, Q., Wan, F.Y., Zhang, Y.T., Vilmos, P., Sousa-Neves, R. *et al.* (2005) Formation of the BMP activity gradient in the *Drosophila* embryo. *Developmental Cell* **8**, 915–924.

Mizutani, C.M., Meyer, N., Roelink, H., and Bier, E. (2006) Threshold-dependent BMP-mediated repression: a model for a conserved mechanism that patterns the neuroectoderm. *PLoS Biology* **4**, e313.

Moczek, A.P. (2008) On the origin of novelty in development and evolution. *BioEssays*, **5**, 432–447.

Moczek, A.P. and Nagy, L.M. (2005) Diverse developmental mechanisms contribute to different levels of diversity in horned beetles. *Evolution & Development* **7**, 175–185.

Modrek, B., Resch, A., Grasso, C., and Lee, C. (2001) Genome-wide detection of alternative splicing in expressed sequences of human genes. *Nucleic Acids Research* **29**, 2850–2859.

Monteiro, A. (2008) Alternative models for the evolution of eyespots and of serial homology on lepidopteran wings. *BioEssays* **30**, 358–366.

Monteiro, A., Glaser, G., Stockslager, S., Glansdorp, N., and Ramos, D. (2006) Comparative insights into questions of lepidopteran wing pattern homology. *BMC Developmental Biology* **6**, 52–65.

Montgomery, T.H. (1904) Development and structure of the larva of *Paragordius*. *Proceedings of the Academy of Natural Sciences Philadelphia* **56**, 738–755.

Moran, A.L. (2004) Egg size evolution in tropical American arcid bivalves: the comparative method and the fossil record. *Evolution* **58**, 2718–2733.

Moreira, D., von der Heyden, S., Bass, D., Lopez-Garcia, P., Chao, E., and Cavalier-Smith, T. (2007) Global eukaryote phylogeny: combined small- and large-subunit ribosomal DNA trees support monophyly of Rhizaria, Retaria and Excavata. *Molecular Phylogenetics and Evolution* **44**, 255–266.

Morgan, T. (1891) The growth and metamorphosis of tornaria. *Journal of Morphology* **5**, 407–458.

Morgan, T. (1894) Development of *Balanoglossus*. *Journal of Morphology* **9**, 1–86.

Morris, V.B. and Byrne, M. (2005) Involvement of two *Hox* genes and *Otx* in echinoderm body-plan morphogenesis in the sea urchin *Holopneustes purpurescens*. *Journal of Experimental Zoology (Molecular and Developmental Evolution)* **304B**, 456–467.

Morris, V.B., Zhao, J.T., Shearman, D.C., Byrne, M., and Frommer, M. (2004) Expression of an *Otx* gene in the adult rudiment and the developing central nervous system in the vestibula larva of the sea urchin *Holopneustes purpurescens*. *International Journal of Developmental Biology* **48**, 17–22.

Moses, A.M., Pollard, D.A., Nix, D.A., Iyer, V.N., Li, X.Y., Biggin, M.D. *et al.* (2006) Large-scale turnover of functional transcription factor binding sites in *Drosophila*. *PLoS Computational Biology* **2**, e130.

Müller, G.B. and Wagner, G.P. (1991) Novelty in evolution: restructuring the concept. *Annual Review of Ecology and Systematics* **22**, 229–256.

Müller, M. and Schmidt-Rhaesa, A. (2003) Reconstruction of the muscle system in *Antygomonas* sp. (Kinorhyncha, Cyclorhagida) by means of phalloidin labeling and cLSM. *Journal of Morphology* **256**, 103–110.

Murphy, W.J., Pevzner, P.A., and O'Brien, S.J. (2004) Mammalian phylogenomics comes of age. *Trends in Genetics* **20**, 631–639.

Murrell, A., Campbell, N.J., and Barker, S.C. (2003) The value of idiosyncratic markers and changes to conserved tRNA sequences from the mitochondrial genome of hard ticks (Acari: Ixodida: Ixodidae) for phylogenetic inference. *Systematic Biology* **52**, 296–310.

Murzin, A.G., Brenner, S.E., Hubbard, T., and Chothia, C. (1995) SCOP: a structural classification of proteins database for the investigation of sequences and structures. *Journal of Molecular Biology* **247**, 536–540.

Nagy, L.M. (1995) A summary of lepidopteran embryogenesis and experimental embryology. In: M.R. Goldsmith and A.S. Wilkins (eds) *Molecular model systems in the Lepidoptera*, pp.139–164. Cambridge University Press, Cambridge.

Narbonne, G.M. (2004) Modular construction of early Ediacaran complex life forms. *Science* **305**, 1141–1144.

Narbonne, G.M. (2005) The Ediacara biota: Neoproterozoic origin of animals and their ecosystems. *Annual Reviews in Earth and Planetary Sciences* **33**, 421–442.

Nardi, F., Spinsanti, G., Boore, J.L., Carapelli, A., Dallai, R., and Frati, F. (2003) Hexapod origins: monophyletic or paraphyletic? *Science* **299**, 1887–1889.

Naylor, G.J.P. and Brown W.M. (1998) Amphioxus mitochondrial DNA, chordate phylogeny, and the limits of inference based on comparisons of sequences. *Systematic Biology* **47**, 61–76.

Near, T.J., Meylan, P.A., and Shaffer, H.B. (2005) Assessing concordance of fossil calibration points in molecular clock studies: an example using turtles. *The American Naturalist* **165**, 137–146.

Neduva, V. and Russell, R.B. (2005) Linear motifs: evolutionary interaction switches. *FEBS Letters* **579**, 3342–3345.

Neduva, V., Linding, R., Su-Angrand, I., Stark, A., de Masi, F., Gibson, T.J. *et al.* (2005) Systematic discovery of new recognition peptides mediating protein inter-action networks. *PLoS Biology* **3**, e405.

Negrisolo, E., Minelli, A., and Valle, G. (2004) The mito-chondrial genome of the house centipede *Scutigera* and the monophyly versus paraphyly of myriapods. *Molecular Biology and Evolution* **21**, 770–780.

Nei, M., Xu, P., and Glazko, G. (2001) Estimation of diver-gence times from multiprotein sequences for a few mammalian species and several distantly related organisms. *Proceedings of the National Academy of Sciences USA* **98**, 2497–2502.

Ni, J.Z., Grate, L., Donohue, J.P., Preston, C., Nobida, N., O'Brien, G. *et al.* (2007) Ultraconserved elements are associated with homeostatic control of splicing regu-lators by alternative splicing and nonsense-mediated decay. *Genes and Development* **21**, 708–718.

Nichols, S.A. (2005) An evaluation of support for order-level monophyly and interrelationships within the class Demospongiae using partial data from the large subunit rDNA and cytochrome oxidase subunit I. *Molecular Phylogenetics and Evolution* **34**, 81–96.

Nichols, S.A., Dirks, W., Pearse, J.S., and King, N. (2006) Early evolution of animal cell signaling and adhesion genes. *Proceedings of the National Academy of Sciences USA* **103**, 12451–12456.

von Nickisch-Rosenegk, M., Brown, W.M., and Boore, J.L. (2001) Sequence and structure of the mitochon-drial genome of the tapeworm *Hymenolepis diminuta*: gene arrangement indicates that platyhelminths are derived eutrochozoans. *Molecular Biology and Evolution* **18**, 721–730.

Nicol, D. (1966) Cope's rule and Precambrian and Cambrian invertebrates. *Journal of Paleontology* **40**, 1397–1399.

Niehrs, C. and Pollet, N. (1999) Synexpression groups in eukaryotes. *Nature* **402**, 483–487.

Nielsen, C. (1995) *Animal evolution: interrelationships of the living phyla*, 1st edn. Oxford University Press, Oxford.

Nielsen, C. (1999) Origin of the chordate central nervous system—and the origin of chordates. *Development Genes and Evolution* **209**, 198–205.

Nielsen, C. (2001) *Animal evolution: interrelationships of the living phyla*, 2nd edn. Oxford University Press, Oxford.

Nielsen, C. (2008) Six major steps in animal evolution: are we derived sponge larvae? *Evolution & Development* **10**, 241–257.

Nielsen, C. and Nørrevang, A (1985) The trochaea the-ory: an example of life cycle phylogeny. In: S. Conway-Morris, J.D. George, R. Gibson, and H.M. Platt (eds) *The origins and relationships of lower invertebrates*, pp. 297–309. Clarendon Press, Oxford.

Nielsen, C., Scharff, N., and Eibye-Jacobsen, D. (1996) Cladistic analyses of the animal kingdom. *Biological Journal of the Linnean Society* **57**, 385–410.

Nieuwenhuys, R. (2002) Deuterostome brains: syn-opsis and commentary. *Brain Research Bulletin* **57**, 257–270.

Nijhout, H.F. (1980) Pattern formation on lepidopteran wings: determination of an eyespot. *Developmental Biology* **80**, 267–274.

Nijhout, H.F. (1991) *The development and evolution of butterfly wing patterns*. Smithsonian Institution Press, Washington, DC.

Nijhout, H.F., Maini, P.K., Madzvamuse, A., Wathen, A.J., and Sekimura, T. (2003) Pigmentation pattern forma-tion in butterflies: experiments and models. *Comptes Rendus Biologies* **326**, 717–727.

Nikaido, M., Rooney, A.P., and Okada, N. (1999) Phylogenetic relationships among cetartiodactyls based on insertions of short and long interspersed elements: hippopotamuses are the closest extant rela-tives of whales. *Proceedings of the National Academy of Sciences USA* **96**, 10261–10266.

Nikaido, M., Matsuno, F., Hamilton, H., Brownell, R.L., Cao, Y., Ding, W. *et al.* (2001) Retroposon ana-lysis of major cetacean lineages: the monophyly of toothed whales and the paraphyly of river dolphins. *Proceedings of the National Academy of Sciences USA* **98**, 7384–7389.

Norden-Krichmaer, T.M., Holtz, J., Pasquinelli, A.E., and Gaasterland, T. (2007) Computational prediction and experimental validation of *Ciona intestinalis* microRNA genes. *BMC Genomics* **8**, 445.

Noro, B., Culi, J., McKay, D.J., Zhang, W., and Mann, R.S. (2006) Distinct functions of homeodomain-containing and homeodomain-less isoforms encoded by *homotho-rax*. *Genes and Development* **20**, 1636–1650.

Nübler-Jung, K. and Arendt, D. (1996) Enteropneusts and chordate evolution. *Current Biology* **6**, 352–353.

Nugent, J.M. and Palmer, J.D. (1991) RNA-mediated transfer of the gene *coxII* from the mitochondrion to the nucleus during flowering plant evolution. *Cell* **66**, 473–481.

Nursall, J.R. (1959) Oxygen as a prerequisite to the origin of the Metazoa. *Nature* **183**, 1170–1172.

Nützel, A., Lehnert, O., and Fryda, J. (2006) Origin of planktotrophy—evidence from early mollusks. *Evolution & Development* **8**, 325–330.

Ogasawara, M., Wada, H., Peters, H., and Satoh, N. (1999) Developmental expression of *Pax1/9* genes in urochord-ate and hemichordate gills: insight into function and evolution of the pharyngeal epithelium. *Development* **126**, 2539–5250.

O'Hara, R.J. (1992) Telling the tree: narrative representation and the study of evolutionary history. *Biology and Philosophy* **7**, 135–160.

Okada, N., Shedlock, A.M., and Nikaido, M. (2004) Retroposon mapping in molecular systematics. *Methods in Molecular Biology* **260**, 189–226.

Okai, N., Tagawa, K., Humphreys, T., Satoh, N., and Ogasawara, M. (2000) Characterization of gill-specific genes of the acorn worm *Ptychodera flava*. *Developmental Dynamics* **217**, 309–319.

Okuse, T., Chiba, T., Katsuumi, I., and Imai, K. (2005) Differential expression and localization of WNTs in an animal model of skin wound healing. *Wound Repair and Regeneration* **13**, 491–497.

Olesnicky, E.C. and Desplan, C. (2007) Distinct mechanisms for mRNA localization during embryonic axis specification in the wasp *Nasonia*. *Developmental Biology* **306**, 134–142.

Olesnicky, E.C., Brent, A.E., Tonnes, L., Walker, M., Pultz, M.A., Leaf, D. *et al.* (2006) A *caudal* mRNA gradient controls posterior development in the wasp *Nasonia*. *Development* **133**, 3973–3982.

Olsen, G. (1987) Earliest phylogenetic branching: comparing rRNA-based evolutionary trees inferred with various techniques. *Cold Spring Harbor Symposia on Quantitative Biology* **52**, 825–837.

Padgett, R.W., Wozney, J.M., and Gelbart, W.M. (1993) Human BMP sequences can confer normal dorsalventral patterning in the *Drosophila* embryo. *Proceedings of the National Academy of Sciences USA* **90**, 2905–2909.

Panfilio, K.A., Liu, P.Z., Akam, M., and Kaufman, T.C. (2006) *Oncopeltus fasciatus zen* is essential for serosal tissue function in katatrepsis. *Developmental Biology* **292**, 226–243.

Panne, D., Maniatis, T., and Harrison, S.C. (2007) An atomic model of the interferon-β enhanceosome. *Cell* **129**, 1111–1123.

Papatsenko, D. and Levine, M.S. (2008) Dual regulation by the Hunchback gradient in the *Drosophila* embryo. *Proceedings of the National Academy of Sciences USA* **105**, 2901–2906.

Papillon, D., Perez, Y., Caubit, X., and Le Parco, Y. (2004) Identification of chaetognaths as protostomes is supported by the analysis of their mitochondrial genome. *Molecular Biology and Evolution* **21**, 2122–2129.

Park, J.-K., Rho, H.S., Kristensen, R.M., Kim, W., and Giribet, G. (2006) First molecular data on the phylum Loricifera – an investigation into the phylogeny of Ecdysozoa with emphasis on the positions of Loricifera and Priapulida. *Zoological Science* **23**, 943–954.

Pasquinelli, A.E., Reinhart, B.J., Slack, F., Martindale, M.Q., Kuroda, M.I., Maller, B. *et al.* (2000) Conservation of the sequence and temporal expression of *let-7* heterochronic regulatory RNA. *Nature* **408**, 86–89.

Pasquinelli, A.E., McCoy, A., Jiménez, E., Saló, E., Ruvkun, G., Martindale, M.Q. *et al.* (2003) Expression of the 22 nucleotide *let-7* heterochronic RNA throughout the Metazoa: a role in life history evolution? *Evolution & Development* **5**, 372–378.

Passamaneck, Y. and Halanych, K.M. (2006) Lophotrochozoan phylogeny assessed with LSU and SSU data: evidence of lophophorate polyphyly. *Molecular Phylogenetics and Evolution* **40**, 20–28.

Passamaneck, Y.J., Schander, C., and Halanych, K.M. (2004) Investigation of molluscan phylogeny using large-subunit and small-subunit nuclear rRNA sequences. *Molecular Phylogenetics and Evolution* **32**, 25–38.

Patel, N.H., Martin-Blanco, E., Coleman, K.G., Poole, S.J., Ellis, M.C., Kornberg, T.B. *et al.* (1989) Expression of *engrailed* proteins in arthropods, annelids, and chordates. *Cell* **58**, 955–968.

Patel, N.H., Condron, B.G., and Zinn, K. (1994) Pair-rule expression patterns of *even-skipped* are found in both short- and long-germ beetles. *Nature* **367**, 429–434.

Patthy, L. (1999) Genome evolution and the evolution of exon-shuffling – a review. *Gene* **238**, 103–114.

Pattyn, A., Vallstedt, A., Dias, J.M., Sander, M., and Ericson, J. (2003) Complementary roles for Nkx6 and Nkx2 class proteins in the establishment of motoneuron identity in the hindbrain. *Development* **130**, 4149–4159.

Pearson, J.C., Lemons, D., and McGinnis, W. (2005) Modulating *Hox* gene functions during animal body patterning. *Nature Reviews Genetics* **6**, 893–904.

Pedersen, K.J. and Pedersen, L.R. (1986) Ultrastructural observations on the epidermis of *Xenoturbella bocki* Westblad, 1949. *Acta Zoologica* **67**, 103–113.

Peel, A. and Akam, M. (2003) Evolution of segmentation: rolling back the clock. *Current Biology* **13**, R708–R710.

Peel, A.D. (2008) The evolution of developmental gene networks: lessons from comparative studies on holometabolous insects. *Philosophical Transactions of the Royal Society B: Biological Sciences* **363**, 1539–1547.

Peel, A.D., Chipman, A.D., and Akam, M. (2005) Arthropod segmentation: beyond the *Drosophila* paradigm. *Nature Reviews Genetics* **6**, 905–916.

Pennacchio, L.A., Ahituv, N., Moses, A.M., Prabhakar, S., Nobrega, M.A., Shoukry, M. *et al.* (2006) *In vivo* enhancer analysis of human conserved non-coding sequences. *Nature* **444**, 499–502.

Pennisi, E. (2007) Working the (gene count) numbers: finally, a firm answer? *Science* **316**, 1113.

Perina, D., Cetkovic, H., Harcet, M., Premzl, M., Lukic-Bilela, L., Müller, W.E.G. *et al.* (2006) The complete set of ribosomal proteins from the marine sponge *Suberites domuncula*. *Gene* **366**, 275–284.

Perler, F.B. (1998) Protein splicing of inteins and hedgehog autoproteolysis: structure, function, and evolution. *Cell* **92**, 1–4.

Perseke, M., Hankeln, T., Weich, B., Fritzsch, G., Stadler, P.F., Israelsson, O. *et al.* (2007) The mitochondrial DNA of *Xenoturbella bocki*: genomic architecture and phylogenetic analysis. *Theory in Biosciences* **126**, 35–42.

Peterson, K.J. (2004) Isolation of *Hox* and *Parahox* genes in the hemichordate *Ptychodera flava* and the evolution of deuterostome *Hox* genes. *Molecular Phylogenetics and Evolution* **31**, 1208–1215.

Peterson, K.J. (2005) Macroevolutionary interplay between planktonic larvae and benthic predators. *Geology* **33**, 929–932.

Peterson, K.J. and Butterfield, N.J. (2005) Origin of the Eumetazoa: testing ecological predictions of molecular clocks against the Proterozoic fossil record. *Proceedings of the National Academy of Sciences USA* **102**, 9547–9552.

Peterson, K.J. and Eernisse, D.J. (2001) Animal phylogeny and the ancestry of bilaterians: inferences from morphology and 18S rDNA gene sequences. *Evolution & Development* **3**, 170–205.

Peterson, K.J., Cameron, R.A., Tagawa, K., Satoh, N., and Davidson, E.H. (1999) A comparative molecular approach to mesodermal patterning in basal deuterostomes: the expression pattern of *Brachyury* in the enteropneust hemichordate *Ptychodera flava*. *Development* **126**, 85–95.

Peterson, K.J., Arenas-Mena, C., and Davidson, E.H. (2000) The A/P axis in echinoderm ontogeny and evolution: evidence from fossils and molecules. *Evolution & Development* **2**, 93–101.

Peterson, K.J., Lyons, J.B., Nowak, K.S., Takacs, C.M., Wargo, M.J., and McPeek, M.A. (2004) Estimating metazoan divergence times with a molecular clock. *Proceedings of the National Academy of Sciences USA* **101**, 6536–6541.

Peterson, K.J., McPeek, M.A., and Evans, D.A.D. (2005) Tempo and mode of early animal evolution: inferences from rocks, Hox, and molecular clocks. *Paleobiology* **31**(Suppl.), 36–55.

Peterson, K.J., Summons, R.E., and Donoghue, P.C.J. (2007) Molecular paleobiology. *Palaeontology* **50**, 775–809.

Peterson, K.J., Cotton, J.A., Gehling, G.A., and Pisani, D. (2008) The Ediacaran emergence of bilaterians: congruence between the genetic and the geological fossil records. *Philosophical Transactions of the Royal Society B: Biological Sciences* **363**,1435–1443.

Pfeiffer, T. and Bonhoeffer, S. (2003) An evolutionary scenario for the transition to undifferentiated multicellularity. *Proceedings of the National Academy of Sciences USA* **100**, 1095–1098.

Philip, G.K., Creevey, C.J., and McInerney, J.O. (2005) The Opisthokonta and the Ecdysozoa may not be clades: stronger support for the grouping of plant and animal than for animal and fungi and stronger support for the Coelomata than Ecdysozoa. *Molecular Biology and Evolution* **22**, 1175–1184.

Philippe, H. and Laurent, J. (1998) How good are deep phylogenetic trees? *Current Opinions in Development and Evolution* **8**, 616–623.

Philippe, H. and Telford, M.J. (2006) Large-scale sequencing and the new animal phylogeny. *Trends in Ecology and Evolution* **21**, 614–620.

Philippe, H., Chenuil, A., and Adoutte, A. (1994) Can the Cambrian explosion be inferred through molecular phylogeny? *Development* **120**, S15–S25.

Philippe, H., Lartillot, N., and Brinkmann, H. (2005a) Multigene analyses of bilaterian animals corroborate the monophyly of Ecdysozoa, Lophotrochozoa and Protostomia. *Molecular Biology and Evolution* **22**, 1246–1253.

Philippe, H., Delsuc, F., Brinkmann, H., and Lartillot, N. (2005b) Phylogenomics. *Annual Review of Ecology, Evolution and Systematics* **36**, 541–562.

Philippe, H., Brinkmann, H., Martinez, P., Riutort, M., and Baguñà, J. (2007) Acoel flatworms are not Platyhelminthes: evidence from phylogenomics. *PLoS One* **2**, e717.

Phillips, M.J., Delsuc, F., and Penny, D. (2004) Genome-scale phylogeny and the detection of systematic biases. *Molecular Biology and Evolution* **21**, 1455–1458.

Pickens, P.E. (1970) Conduction along the ventral nerve cord of a hemichordate worm. *Journal of Experimental Biology* **53**, 515–528.

Pickens, P.E. (1973) Bioelectric activity during the startle response of a hemichordate worm. *Journal of Experimental Biology* **58**, 295–304.

Pierce, M.L., Weston, M.D., Fritzsch, B., Gabel, H.W., Ruvkun, G., and Soukup, G.A. (2008) MicroRNA-183 family conservation and ciliated neurosensory organ expression. *Evolution & Development* **10**, 106–113.

Pigliucci, M. (2008). What, if anything, is an evolutionary novelty? *Philosophy of Science* **75**, 887–898.

Pincus, D., Letunic, I., Bork, P., and Lim, W.A. (2008) Evolution of the phospho-tyrosine signaling machinery in pre-metazoan lineages. *Proceedings of the National Academy of Sciences USA* **105**, 9680–9684.

Pires-daSilva, A. and Sommer, R.J. (2003) The evolution of signaling pathways in animal development. *Nature Reviews Genetics* **4**, 39–49.

Pisani, D. (2004) Identifying and removing fast-evolving sites using compatibility analysis: an example from the Arthropoda. *Systematic Biology* **53**, 978–989.

Pisani, D., Poling, L.L., Lyons-Weiler, M., and Hedges, S.B. (2004) The colonization of land by animals: molecular phylogeny and divergence times among arthropods. *BMC Evolutionary Biology* **2**, 1–10.

Pisani, D., Benton, M.J., and Wilkinson, M. (2007) The congruence of molecular and morphological phylogenies. *Acta Biotheoretica* **55**, 269–281.

Pisera, A. (2006) Palaeontology of sponges—a review. *Canadian Journal of Zoology* **84**, 242–261.

Pleijel, F. (1995) On character coding for phylogeny reconstruction. *Cladistics* **11**, 309–315.

Pollock, D.D., Zwickl, D.J., McGuire, J.A., and Hillis, D.M. (2002) Increased taxon sampling is advantageous for phylogenetic inference. *Systematic Biology* **51**, 664–671.

Poole, R.J. and Hobert, O. (2006) Early embryonic programming of neuronal left/right asymmetry in *C. elegans*. *Current Biology* **16**, 2279–2292.

Pop, M. and Salzberg, S.L. (2008) Bioinformatics challenges of new sequencing technology. *Trends in Genetics* **24**, 142–149.

Pörtner, H.O. (2002) Environmental and functional limits to muscular exercise and body size in marine invertebrate athletes. *Comparative Biochemistry and Physiology A: Molecular and Integrative Physiology* **133**, 303–321.

Pourquié, O. (2004) The chick embryo: a leading model in somitogenesis studies. *Mechanisms of Development* **121**, 1069–1079.

Poustka, A.J., Kuhn, A., Radosavljevic, V., Wellenreuther, R., Lehrach, H., and Panopoulou, G. (2004) On the origin of the chordate central nervous system: expression of onecut in the sea urchin embryo. *Evolution & Development* **6**, 227–236.

Prendini, L. (2001) Species or supraspecific taxa as terminals in cladistic analysis? Groundplans versus exemplars revisited. *Systematic Biology* **50**, 290–300.

Prud'homme, B., de Rosa, R., Arendt, D., Julien, J.F., Pajaziti, R., Dorresteijn, A.W. *et al.* (2003) Arthropod-like expression patterns of *engrailed* and *wingless* in the annelid *Platynereis dumerilii* suggest a role in segment formation. *Current Biology* **13**, 1876–1881.

Prum, R.O. (1999) Development and evolutionary origin of feathers. *Journal of Experimental Zoology* **285**, 291–306.

Puelles, E., Acampora, D., Lacroix, E., Signore, M., Annino, A., Tuorto, F. *et al.* (2003) Otx dose-dependent integrated control of antero-posterior and dorso-ventral patterning of midbrain. *Nature Neuroscience* **6**, 453–460.

Pultz, M.A., Westendorf, L., Gales, S.D., Hawkins, K., Lynch, J., Pitt, J.N. *et al.* (2005) A major role for zygotic *hunchback* in patterning the *Nasonia* embryo. *Development* **132**, 3705–3715.

Putnam, N.H., Srivastava, M., Hellsten, U., Dirks, B., Chapman, J., Salamov, A. *et al.* (2007) Sea anemone genome reveals ancestral eumetazoan gene repertoire and genomic organization. *Science* **317**, 86–94.

Putnam, N.H., Butts, T., Ferrier, D.E.K., Furlong, R.F., Hellsten, U., Kawashima, T. *et al.* (2008) The amphioxus genome and the evolution of the chordate karyotype. *Nature* **453**, 1064–1071.

Qiu, Y.-L., Cho, Y., Cox, J.C., and Palmer, J.D. (1998) The gain of three mitochondrial introns identifies liverworts as the earliest land plants. *Nature* **394**, 671–674.

Raff, R.A. (2007) Written in stone: fossils, genes and evo-devo. *Nature Reviews Genetics* **8**, 911–920.

Raff, R.A. (2008) Origins of the other metazoan body plans: the evolution of larval forms. *Philosophical Transactions of the Royal Society B: Biological Sciences* **363**, 1473–1479.

Raff, R.A. and Byrne, M. (2006) The active evolutionary lives of echinoderm larvae. *Heredity* **97**, 244–252.

Raff, E.C., Villinski, J.A., Turner, F.R., Donoghue, P.C.J., and Raff, R.A. (2006) Experimental taphonomy: feasibility of fossil embryos. *Proceedings of the National Academy of Sciences USA* **103**, 5846–5851.

Raff, E.C., Schollaert, K.L. Nelson, D.E., Donoghue, P.C., Thomas, C.W., Turner, F.R. *et al.* (2008) Embryo fossilization is a biological process mediated by microbial biofilms. *Proceedings of the National Academy of Sciences USA* **150**, 19360–19365.

Raible, F., Tessmar-Raible, K., Osoegawa, K., Wincker, P., Jubin, C., Balavoine, G. *et al.* (2005) Vertebrate-type intron-rich genes in the marine annelid *Platynereis dumerilii*. *Science* **310**, 1325–1326.

Raikova, O.I., Reuter, M., Gustafsson, M.K.S., Maule, A.G., Halton, D.W., and Jondelius, U. (2004a) Basiepidermal nervous system in *Nemertoderma westbladi* (Nemertodermatida): GYIRFamide immunoreactivity. *Zoology* **107**, 75–86.

Raikova, O.I., Reuter, M., Gustafsson, M.K.S., Maule, A.G., Halton, D.W., and Jondelius, U. (2004b) Evolution of the nervous system in *Paraphanostoma* (Acoela). *Zoologica Scripta* **33**, 71–88.

Ramachandra, N.B., Gates, R.D., Ladurner, P., Jacobs, D.K., and Hartenstein, V. (2002) Embryonic development in the primitive bilaterian *Neochildia fusca*: normal morphogenesis and isolation of POU genes *Brn-1* and *Brn-3*. *Development Genes and Evolution* **212**, 55–69.

Ramírez, M.J. (2007) Homology as a parsimony problem: a dynamic homology approach for morphological data. *Cladistics* **23**, 588–612.

Ramírez, M.J., Coddington, J.A., Maddison, W.P., Midford, P.E., Prendini, L., Miller, J. *et al.* (2007) Linking of digital images to phylogenetic data matrices using a morphological ontology. *Systematic Biology* **56**, 283–294.

Ramos, D.M. and Monteiro, A. (2007) Transgenic approaches to study wing color pattern development in Lepidoptera. *Molecular Biosystems* **3**, 530–535.

Rattenbury, J.C. (1954) The embryology of *Phoronopsis viridis*. *Journal of Morphology* **95**, 289–340.

Rebeiz, M., Stone, T., and Posakony, J.W. (2005) An ancient transcriptional regulatory linkage. *Developmental Biology* **281**, 299–308.

Reed, R.D. and Gilbert, L.E. (2004) Wing venation and Distal-less expression in *Heliconius* butterfly wing pattern development. *Development, Genes and Evolution* **214**, 628–634.

Reed, R.D. and Nagy, L.M. (2005) Evolutionary redeployment of a biosynthetic module: expression of eye pigment genes *vermilion, cinnabar,* and *white* in butterfly wing development. *Evolution & Development* **7**, 301–311.

Reed, R.D. and Serfas, M.S. (2004) Butterfly wing pattern evolution is associated with changes in a *Notch/Distal-less* temporal pattern formation process. *Current Biology* **14**, 1159–1166.

Reed, R.D., McMillan, W.O., and Nagy, L.M. (2008) Gene expression underlying adaptive variation in *Heliconius* wing patterns: non-modular regulation of overlapping *cinnabar* and *vermilion* prepatterns. *Proceedings of the Royal Society B: Biological Sciences* **275**, 37–45.

Regier, J.C., Shultz, J.W., and Kambic, R.E. (2005) Pancrustacean phylogeny: hexapods are terrestrial crustaceans and maxillopods are not monophyletic. *Proceedings of the Royal Society B: Biological Sciences* **272**, 395–401.

Reinhart, B.J., Slack, F.J., Basson, M., Pasquinelli, A.E., Bettinger, J.C., Rougvie, A.E. *et al.* (2000) The 21-nucleotide *let-7* RNA regulates developmental timing in *Caenorhabditis elegans*. *Nature* **403**, 901–906.

Reisinger, E. (1970) Zur Problematik der Evolution der Coelomaten. *Zeitschrift fur zoologische Systematik und Evolutionsforschung* **8**, 81–108.

Reitner, J. and Wörheide, G. (2002) Non-lithistid fossil Demospongiae—origins of the palaeobiodiversity and highlights in history of preservation. In: J.N.A. Hooper and R.W.M. van Soest (eds) *Systema Porifera: a guide to the classification of sponges,* pp. 52–68 Kluwer Academic/Plenum Publishers, New York.

Remane, A. (1950) Die Entstehung der Metamerie der Wirbellosen. *Zoologischer Anzeiger* **14**(Suppl.), 18–23.

Rentzsch, F., Anton, R., Saina, M., Hammerschmidt, M., Holstein, T.W., and Technau, U. (2006) Asymmetric expression of the BMP antagonists chordin and gremlin in the sea anemone *Nematostella vectensis*: implications for the evolution of axial patterning. *Developmental Biology* **296**, 375–387.

Rhinn, M., Lun, K., Luz, M., Werner, M., and Brand, M. (2005) Positioning of the midbrain-hindbrain boundary organizer through global posteriorization of the neuroectoderm mediated by Wnt8 signaling. *Development* **132**, 1261–1272.

Rhoads, D.C. and Morse, J.W. (1971) Evolutionary and ecologic significance of oxygen-deficient marine basins. *Lethaia* **4**, 413–428.

Richly, E. and Leister, D. (2004) NUMTs in sequenced eukaryotic genomes. *Molecular Biology and Evolution* **21**, 1081–1084.

Rickards, B. and Durman, P.N. (eds) (2006) *Evolution of the earliest graptolites and other hemichordates*. Studies in Palaeozoic Palaeontology. National Museum of Wales Geological Series 25. National Museum of Wales, Cardiff.

Riedl, R.J. (1969) Gnathostomulida from America. *Science* **163**, 445–452.

Robinson-Rechavi, M., Maina, C.V., Gissendanner, C.R., Laudet, V., and Sluder, A. (2005) Explosive lineage-specific expansion of the orphan nuclear receptor HNF4 in nematodes. *Journal of Molecular Biology* **60**, 577–586.

Rodríguez-Ezpeleta, N., Brinkmann, H., Roure, B., Lartillot, N., Lang, B.F., and Philippe, H. (2007) Detecting and overcoming systematic errors in genome-scale phylogenies. *Systematic Biology* **56**, 389–399.

Roeding, F., Hagner-Holler, S., Ruhberg, H., Ebersberger, I., von Haeseler, A., Kube, M. *et al.* (2007) EST sequencing of Onychophora and phylogenomic analysis of Metazoa. *Molecular Phylogenetics and Evolution* **45**, 942–951.

Roger, A.J. and Hug, L.A. (2006) The origin and diversification of eukaryotes: problems with molecular phylogenetics and molecular clock estimation. *Philosophical Transactions of the Royal Society B: Biological Sciences* **361**, 1039–1054.

Rogozin, I.B., Wolf, Y.I., Carmel, L. and Koonin, E.V. (2007a) Analysis of rare amino acid replacements supports the Coelomata clade. *Molecular Biology and Evolution* **24**, 2594–2597.

Rogozin, I.B., Wolf, Y.I., Carmel, L., and Koonin, E.V. (2007b) Ecdysozoan clade rejected by genome-wide analysis of rare amino acid replacements. *Molecular Biology and Evolution* **24**, 1080–1090.

Rogozin, I.B., Thomson, K., Csürös, M., Carmel, L., and Koonin, E.V. (2008) Homoplasy in genome-wide analysis of rare amino acid replacements: the molecular-evolutionary basis for Vavilov's law of homologous series. *Biology Direct* **3**, 7.

Rokas, A. and Carroll, S.B. (2006) Bushes in the tree of life. *PLoS Biology* **4**, e352.

Rokas, A. and Holland, P.W.H. (2000) Rare genomic changes as a tool for phylogenetics. *Trends in Ecology and Evolution* **15**, 454–459.

Rokas, A., Krüger, D., and Carroll, S.B. (2005) Animal evolution and the molecular signature of radiations compressed in time. *Science* **310**, 1933–1938.

Romano, L.A. and Wray, G.A. (2003) Conservation of *Endo16* expression in sea urchins despite evolutionary divergence in both cis and trans-acting components of transcriptional regulation. *Development* **130**, 4187–4199.

Romero, P.R., Zaidi, S., Fang, Y.Y., Uversky, V.N., Radivojac, P., Oldfield, C.J. *et al.* (2006) Alternative splicing in concert with protein intrinsic disorder enables increased functional diversity in multicellular organisms. *Proceedings of the National Academy of Sciences USA* **103**, 8390–8395.

Ronquist, F. and Huelsenbeck, J.P. (2003). MrBayes3: Bayesian phylogenetic inference under mixed models. *Bioinformatics* **19**, 1572–1574.

de Rosa, R., Grenier, J.K., Andreeva, T., Cook, C.E., Adoutte, A., Akam, M. *et al.* (1999) Hox genes in brachiopods and priapulids and protostome evolution. *Nature* **399**, 772–776.

Rota-Stabelli, O. and Telford, M.J. (2008) A multi criterion approach for the selection of optimal outgroups in phylogeny: recovering some support for Mandibulata over Myriochelata using mitogenomics. *Molecular Phylogenetics and Evolution* **48**, 103–111.

Roule, L. (1891) Considerations sur l'embranchement des Trochozoaires. *Annales des Sciences Naturelle (Zoologie) 7e série* **11**, 121–178.

Rouse, G.W. (1999) Trochophore concepts: ciliary bands and the evolution of larvae in spiralian Metazoa. *Biological Journal of the Linnean Society* **66**, 411–464.

Rouse, G.W. (2000) The epitome of hand waving? Larval feeding and the hypotheses of metazoan phylogeny. *Evolution & Development* **2**, 222–233.

Rouse, G.W. and Fauchald, K. (1995) The articulation of annelids. *Zoologica Scripta* **24**, 269–301.

Rouse, G.W. and Fauchald, K. (1997) Cladistics and polychaetes. *Zoologica Scripta* **26**, 139–204.

Rouse, G. W. and Pleijel, F. (2001) *Polychaetes*. Oxford University Press, Oxford.

Rouse, G. W. and Pleijel, F. (2007) Annelida. In: Z.-Q. Zhang and W.A. Shear (eds) *Linnaeus tercentenary: progress in invertebrate taxonomy. Zootaxa* **1668**, 245–264.

Rousset, V., Rouse, G.W., Siddall, M.E., Tillier, A., and Pleijel, F. (2004) The phylogenetic position of Siboglinidae (Annelida) inferred from 18S rRNA, 28S rRNA and morphological data. *Cladistics* **20**, 518–533.

Rousset, V., Pleijel, F., Rouse, G.W., Erséus, C., and Siddall, M.E. (2007) A molecular phylogeny of annelids. *Cladistics* **23**, 41–63.

Rowe, T. (2004) Chordate phylogeny and development. In: J. Cracraft and M.J. Donoghue (eds) *Assembling the tree of life*, pp. 384–409. Oxford University Press, Oxford.

Roy, S.W. and Irimia, M. (2008a) Rare genomic characters do not support Coelomata: intron loss/gain. *Molecular Biology and Evolution* **25**, 620–623.

Roy, S.W. and Irimia, M. (2008b) Rare genomic characters do not support Coelomata: RGC_CAMs. *Journal of Molecular Evolution* **66**, 308–315.

Rubenstein, J.L., Shimamura, K., Martinez, S., and Puelles, L. (1998) Regionalization of the prosencephalic neural plate. *Annual Review of Neuroscience* **21**, 445–477.

Ruby, J.G., Jan, C., Player, C., Axtell, M.J., Lee, W., Nusbaum, C. *et al.* (2006) Large-scale sequencing reveals 21U-RNAs and additional microRNAs and endogenous siRNAs in *C. elegans*. *Cell* **127**, 1193–1207.

Ruby, J.G., Stark, A., Johnston, W.K., Kellis, M., Bartel, D.P., and Lai, E.C. (2007) Evolution, biogenesis, expression, and target predictions of a substantially expanded set of *Drosophila* microRNAs. *Genome Research* **17**, 1850–1864.

Ruiz-Trillo, I., Riutort, M., Littlewood, D.T.J., Herniou, E.A., and Baguñà, J. (1999) Acoel flatworms: earliest extant bilaterian metazoans, not members of Platyhelminthes. *Science* **283**, 1919–1923.

Ruiz-Trillo, I., Paps, J., Loukota, M., Ribera, C., Jondelius, U., Baguñà, J. *et al.* (2002) A phylogenetic analysis of myosin heavy chain type II sequences corroborates that Acoela and Nemertodermatida are basal bilaterians. *Proceedings of the National Academy of Sciences USA* **99**, 11246–11251.

Ruiz-Trillo, I., Inagaki, Y., Davis, L.A., Sperstad, S., Landfald, B., and Roger, A.J. (2004) *Capsaspora owczarzaki* is an independent opisthokont lineage. *Current Biology* **14**, R946–R947.

Ruiz-Trillo, I., Burger, G., Holland, P.W.H., King, N., Lang, B.F., Roger, A.J. *et al.* (2007) The origins of multicellularity: a multi-taxon genome initiative. *Trends in Genetics* **23**, 113–118.

Ruiz-Trillo, I., Roger, A.J., Burger, G., Gray, M.W., and Lang, B.F. (2008) A phylogenomic investigation into the origin of Metazoa. *Molecular Biology and Evolution* **25**, 664–672.

Runnegar, B. (1982a) A molecular-clock date for the origin of the animal phyla. *Lethaia* **15**, 199–205.

Runnegar, B. (1982b) The Cambrian explosion: animals or fossils? *Journal of the Geological Society Australia* **29**, 395–411.

Runnegar, B. (1982c) Oxygen requirements, biology and phylogenetic significance of the late Precambrian worm *Dickinsonia*, and the evolution of the burrowing habit. *Alcheringa* **6**, 223–239.

Runnegar, B. (1986) Molecular palaeontology. *Palaeontology* **29**, 1–24.

Runnegar, B. (1991) Precambrian oxygen levels estimated from the biochemistry and physiology of early eukaryotes. *Palaeogeography, Palaeoclimatology, Palaeoecology* **71**, 97–111.

Runnegar, B. (1996) Early evolution of the Mollusca: the fossil record. In: J. Taylor (ed.) *Origin and evolutionary radiation of the Mollusca*, pp. 77–87. Oxford University Press, Oxford.

Runnegar, B. (2000) Loophole for snowball earth. *Nature* **405**, 403–404.

Ruppert, E.E. (1991a) Introduction to the aschelminth phyla: a consideration of mesoderm, body cavities, and cuticle. In: F.W. Harrison and E.E. Ruppert (eds) *Microscopic anatomy of invertebrates, Vol. 4. Aschelminthes*, pp. 1–17. Wiley-Liss, New York.

Ruppert, E.E. (1991b) Gastrotricha. In: F.W. Harrison and E.E. Ruppert (eds) *Microscopic anatomy of invertebrates, Vol. 4. Aschelminthes*, pp. 41–109. Wiley-Liss, New York.

Ruppert, E.E. (2005) Key characters uniting hemichordates and chordates: homologies or homoplasies? *Canadian Journal of Zoology* **83**, 8–23.

Rusten, T.E., Cantera, R., Kafatos, F.C., and Barrio, R. (2002) The role of TGF beta signaling in the formation of the dorsal nervous system is conserved between *Drosophila* and chordates. *Development* **129**, 3575–3584.

Ruta, M., Wagner, P.J., and Coates, M.I. (2006) Evolutionary patterns in early tetrapods. I. Rapid initial diversification followed by decrease in rates of character change. *Proceedings of the Royal Society B: Biological Sciences* **273**, 2107–2111.

Ruvinsky, I. and Ruvkun, G. (2003) Functional tests of enhancer conservation between distantly related species. *Development* **130**, 5133–5142.

Ruvkun, G., Wightman, B., and Ha, I. (2004) The 20 years it took to recognize the importance of tiny RNAs. *Cell* **S116**, S93–S96.

Rychel, A.L. and Swalla, B.J. (2007) Development and evolution of chordate cartilage. *Journal of Experimental Zoology (Molecular Development and Evolution)* **308B**, 325–335.

Rychel, A.L., Smith, S.E., Shimamoto, H.T., and Swalla, B.J. (2006) Evolution and development of the chordates: collagen and pharyngeal cartilage. *Molecular Biology and Evolution* **23**, 541–549.

Sacks, M. (1955) Observations on the embryology of an aquatic gastrotrich, *Lepidodermalla squamata* (Dujardin, 1841). *Journal of Morphology* **96**, 473–495.

Saenko, S.V., French, V., Brakefield, P.M., and Beldade, P. (2008) Conserved developmental processes and the formation of evolutionary novelties: examples from butterfly wings. *Philosophical Transactions of the Royal Society B: Biological Sciences* **363**, 1549–1555.

Salem, A.-H., Ray, D.A., Xing, J., Callinan, P.A., Myers, J.S., Hedges, D.J. *et al.* (2003) Alu elements and hominid phylogenetics. *Proceedings of the National Academy of Sciences USA* **100**, 12787–12791.

Salvini-Plawen, L. (1978) On the origin and evolution of the lower Metazoa. *Zeitschrift fur Zoologische Systematik und Evolutionsforschung* **16**, 40–88.

Salvini-Plawen, L. (1980) Phylogenetischer Status und Bedeutung der mesenchymaten Bilateria. *Zoologische Jahrbücher, Abteilung für Anatomie und Ontogenie der Tiere* **103**, 354–373.

Salvini-Plawen, L. v. and Steiner, G. (1996) Synapomorphies and plesiomorphies in higher classification of Mollusca. In: J.D. Taylor (ed.) *Origin and evolutionary radiation of the Mollusca*, pp. 29–51. Oxford University Press, Oxford

Sandelin, A., Bailey, P., Bruce, S., Engstrom, P.G., Klos, J.M., Wasserman, W.W. *et al.* (2004) Arrays of ultraconserved non-coding regions span the loci of key developmental genes in vertebrate genomes. *BMC Genomics* **5**, 99.

Sanderson, M.J. (1997) A nonparametric approach to estimating divergence times in the absence of rate constancy. *Molecular Biology and Evolution* **14**, 1218–1231.

Sanderson, M.J. (2002) Estimating absolute rates of molecular evolution and divergence times: a penalized likelihood approach. *Molecular Biology and Evolution* **19**, 101–109.

Sanderson, M.J. (2004) *r8s, version 1.70 user's manual.* Section of Evolution and Ecology, University of California, Davis. Available at: http://loco.biosci.arizona.edu/r8s/r8s1.7.manual.pdf.

Sanderson, M.J. and Shaffer, H.B. (2002) Troubleshooting molecular phylogenetic analyses. *Annual Review of Ecology, Evolution and Systematics* **33**, 49–72.

Sanetra, M., Begemann, G., Becker, M., and Meyer, A. (2005) Conservation and co-option in developmental programmes: the importance of homology relationships. *Frontiers in Zoology* **2**, 15.

Santos, M.A.S., Moura, G., Massey, S.E., and Tuite, M.F. (2004) Driving change: the evolution of alternative genetic codes. *Trends in Genetics* **20**, 95–102.

Sapkota, G., Alarcon, C., Spagnoli, F.M., Brivanlou, A.H., and Massague, J. (2007) Balancing BMP signaling through integrated inputs into the Smad1 linker. *Molecular Cell* **25**, 441–454.

Sasai, Y. and De Robertis, E.M. (1997) Ectodermal patterning in vertebrate embryos. *Developmental Biology* **182**, 5–20.

Sasai, Y., Lu, B., Steinbeisser, H., and De Robertis, E.M. (1995) Regulation of neural induction by the Chd and

Bmp-4 antagonistic patterning signals in *Xenopus*. *Nature* **376**, 333–336.

Sato, A. and Holland, P.W. (2008) Asymmetry in a ptero-branch hemichordate and the evolution of left-right patterning. *Developmental Dynamics* **237**, 3634–3639.

Savard, J., Tautz, D., Richards, S., Weinstock, G.M., Gibbs, R.A., Werren, J.H. *et al.* (2006) Phylogenomic analysis reveals bees and wasps (Hymenoptera) at the base of the radiation of Holometabolous insects. *Genome Research* **16**, 1334–1338.

Saville-Kent, W. (1880–82). *A manual of the Infusoria*. David Bogue, London.

Scheltema, A.H., Kerth, K., and Kuzirian, A.M. (2003) Original molluscan radula: comparisons among Aplacophora, Polyplacophora, Gastropoda, and the Cambrian fossil *Wiwaxia corrugata*. *Journal of Morphology* **257**, 219–245.

Schierenberg, E. (2005) Unusual cleavage and gastru-lation in a freshwater nematode: developmental and phylogenetic implications. *Development Genes and Evolution* **215**, 103–108.

Schierenberg, E. (2006) Embryological variation dur-ing nematode development (January 02, 2006). In: The *C. elegans* Research Community (eds) *WormBook*, doi/10.1895/wormbook.1.55.1, http://www.wormbook.org.

Schilling, T.F. and Knight, R.D. (2001) Origins of antero-posterior patterning and Hox gene regulation during chordate evolution. *Philosophical Transactions of the Royal Society B: Biological Sciences* **356**, 1599–1613.

Schinko, J.B., Kreuzer, N., Offen, N., Posnien, N., Wimmer, E.A., and Bucher, G. (2008) Divergent functions of orthodenticle, empty spiracles and buttonhead in early head patterning of the beetle *Tribolium castaneum* (Coleoptera). *Developmental Biology* **317**, 600–613.

Schleip, W. (1929) *Die Determination der Primitiventwicklung*. Leipzig: Akademische Verlagsgesellschaft.

Schlosser, G. and Ahrens, K. (2004) Molecular anatomy of placode development in *Xenopus laevis*. *Developmental Biology* **271**, 439–466.

Schmidt, B.J. and Jordan, L.M. (2000) The role of sero-tonin in reflex modulation and locomotor rhythm pro-duction in the mammalian spinal cord. *Brain Research Bulletin* **53**, 689–710.

Schmidt-Rhaesa, A. (1998) The position of the Arthropoda in the phylogenetic system. *Journal of Morphology* **238**, 263–285.

Schmidt-Rhaesa, A. (2001) Tardigrades—are they really miniaturized dwarfs? *Zoologischer Anzeiger* **240**, 549–555.

Schmidt-Rhaesa, A. (2007) *The evolution of organ systems*. Oxford University Press, Oxford.

Schmidt-Rhaesa, A. and Rothe, B.H. (2006) Postembryonic development of longitudinal musculature in *Pycnophyes kielensis* (Kinorhyncha, Homalorhagida). *Integrative and Comparative Biology* **46**, 144–150.

Schmitz, J., Ohme, M., and Zischler, H. (2001) SINE inser-tions in cladistic analyses and the phylogenetic affilia-tions of *Tarsius bancanus* to other primates. *Genetics* **157**, 777–784.

Schnabel, R., Hutter, H., Moerman, D., and Schnabel, H. (1997) Assessing normal embryogenesis in *Caenorhabditis elegans* using a 4D microscope: vari-ability of development and regional specification. *Developmental Biology* **184**, 234–265.

Schock, F. and Perrimon, N. (2002) Cellular processes associated with germ band retraction in *Drosophila*. *Developmental Biology* **248**, 29–39.

Scholtz, G. (2002) The Articulata hypothesis—or what is a segment? *Organisms Diversity and Evolution* **2**, 197–215.

Scholz, C.B. and Technau, U. (2003) The ancestral role of *Brachyury*: expression of *NemBra1* in the basal cni-darian *Nematostella vectensis* (Anthozoa). *Development Genes and Evolution* **212**, 563–570.

Scholz, G., Mittmann, B., and Gerberding, M. (1998) The pattern of *Distal-less* expression in the mouthparts of crustaceans, myriapods and insects: new evidence for a gnathobasic mandible and the common origin of the Mandibulata. *International Journal of Developmental Biology* **42**, 801–810.

Schopf, J.W. and Kudryavtsev, A.B. (2005) Three-dimensional Raman imagery of Precambrian micro-scopic organisms. *Geobiology* **3**, 1–12.

Schoppmeier, M. and Damen, W.G. (2005) Suppressor of Hairless and Presenilin phenotypes imply involve-ment of canonical Notch-signalling in segmentation of the spider *Cupiennius salei*. *Developmental Biology* **280**, 211–224.

Schoppmeier, M. and Schröder, R. (2005) Maternal *torso* signaling controls body axis elongation in a short germ insect. *Current Biology* **15**, 2131–2136.

Schröder, R. (2003) The genes *orthodenticle* and *hunchback* substitute for *bicoid* in the beetle *Tribolium*. *Nature* **422**, 621–625.

Schröder, R., Eckert, C., Wolff, C., and Tautz, D. (2000) Conserved and divergent aspects of terminal pattern-ing in the beetle *Tribolium castaneum*. *Proceedings of the National Academy of Sciences USA* **97**, 6591–6596.

Schulmeister, S. and Wheeler, W.C. (2004) Comparative and phylogenetic analysis of developmental sequences. *Evolution & Development* **6**, 50–57.

Schulz, C., Schroder, R., Hausdorf, B., Wolff, C., and Tautz, D. (1998) A caudal homologue in the short germ band beetle *Tribolium* shows similarities to both, the

Drosophila and the vertebrate caudal expression patterns. *Development Genes and Evolution* **208**, 283–289.

Schulze, A., Cutler, E.B., and Giribet, G. (2005) Reconstructing the phylogeny of the Sipuncula. *Hydrobiologia* **535/536**, 277–296.

Schulze, A., Cutler, E.B., and Giribet, G. (2007) Phylogeny of sipunculan worms: a combined analysis of four gene regions and morphology. *Molecular Phylogenetics and Evolution* **42**, 171–192.

Seaver, E.C. (2003) Segmentation: mono- or polyphyletic? *International Journal of Developmental Biology* **47**, 583–595.

Sebat, J., Lakshmi, B., Troge, J., Alexander, J., Young,J., Lundin, P. *et al.* (2004) Large-scale copy number polymorphism in the human genome. *Science* **305**, 525–528.

Sedgwick, W. (1884) On the origin of metameric segmentation and some other morphological questions. *Quarterly Journal of the Microscopical Society* **24**, 43–82.

Sedgwick, W. (1885) The development of *Peripatus capensis*. *Quarterly Journal of the Microscopical Society* **25**, 449–466.

Segawa, Y., Suga, H., Iwabe, N., Oneyama, C., Akagi, T., Miyata, T. *et al.* (2006) Functional development of Src tyrosine kinases during evolution from a unicellular ancestor to multicellular animals. *Proceedings of the National Academy of Sciences USA* **103**, 12021–12026.

Seilacher, A. (1999) Biomat-related lifestyles in the Precambrian. *Palaios* **14**, 86–93.

Selbach, M., Schwanhäusser, B., Thierfelder, N., Fang, Z., Khanin, R., and Rajewsky, N. (2008) Widespread changes in protein synthesis induced by microRNAs. *Nature* **455**, 58–63.

Selenka, E. (1876) Entwicklung der Holothurien. *Zeitschrift für Wissenschaftliche Zoologie* **27**, 155–187.

Sempere, L.F., Cole, C.N., McPeek, M.A., and Peterson, K.J. (2006) The phylogenetic distribution of metazoan microRNAs: insights into evolutionary complexity and constraint. *Journal of Experimental Zoology (Molecular Development and Evolution)* **306B**, 575–588.

Sempere, L.F., Martinez, P., Cole, C., Baguñà, J., and Peterson, K.J. (2007) Phylogenetic distribution of microRNAs supports the basal position of acoel flatworms and the polyphyly of Platyhelminthes. *Evolution & Development* **9**, 409–415.

Shalchian-Tabrizi, K., Minge, M.A., Espelund, M., Orr, R., Ruden, T., Jakobsen, K.S. *et al.* (2008) Multigene phylogeny of Choanozoa and the origin of Metazoa. *PLoS ONE* **3**(5), e2098.

Sharman, A.C. and Brand, M. (1998) Evolution and homology of the nervous system: cross-phylum rescues of *otd/Otx* genes. *Trends in Genetics* **14**, 211–214.

Sharp, A.J., Locke, D.P., McGrath, S.D., Cheng, Z., Bailey, J.A., Vallente, R.U. *et al.* (2005) Segmental duplications and copy-number variation in the human genome. *American Journal of Human Genetics* **77**, 78–88.

Shaw, P.J., Wratten, N.S., McGregor, A.P., and Dover, G.A. (2002) Coevolution in *bicoid*-dependent promoters and the inception of regulatory incompatibilities among species of higher Diptera. *Evolution & Development* **4**, 265–277.

Shen, Y.N., Canfield, D.E., and Knoll, A.H. (2002) Middle Proterozoic ocean chemistry: evidence from the McArthur Basin, northern Australia. *American Journal of Science* **302**, 81–109.

Shen, Y.N., Zhang, T.G., and Hoffman, P.F. (2008) On the coevolution of Ediacaran oceans and animals. *Proceedings of the National Academy of Sciences USA* **105**, 7376–7381.

Shendure, J. and Ji, H. (2008) Next-generation DNA sequencing. *Nature Biotechnology* **26**, 1135–1145.

Shimamura, K., Hartigan, D.J., Martinez, S., Puelles, L., and Rubenstein, J.L. (1995) Longitudinal organization of the anterior neural plate and neural tube. *Development* **121**, 3923–3933.

Shimamura, M., Yasue, H., Ohshima, K., Abe, H., Kato, H., Kishiro, T. *et al.* (1997) Molecular evidence from retroposons that whales form a clade within even-toed ungulates. *Nature* **388**, 666–670.

Shimeld, S.M., McKay, I.J., and Sharpe, P.T. (1996) The murine homeobox gene Msx-3 shows highly restricted expression in the developing neural tube. *Mechanisms of Development* **55**, 201–210.

Shoguchi, E., Satoh, N., and Maruyama, Y.K. (1999) Pattern of *Brachyury* gene expression in starfish embryos resembles that of hemichordate embryos but not of sea urchin embryos. *Mechanisms of Development* **82**, 185–189.

Shu, D., Zhang, X.-L., and Chen, L. (1996a) Reinterpretation of *Yunnanozoon* as the earliest known hemichordate. *Nature* **380**, 428–430.

Shu, D.G., Conway Morris, S., and Zhang, X.L. (1996b) A *Pikaia*-like chordate from the Lower Cambrian of China. *Nature* **384**, 157–158.

Shu, D.G., Luo, H.L., Conway Morris, S., Zhang, X.L., Hu, S. X., Chen, L. *et al.* (1999) Lower Cambrian vertebrates from South China. *Nature* **402**, 42–46.

Shu, D.G., Chen, L., Han, J., and Zhang, X.L. (2001a) An Early Cambrian tunicate from China. *Nature* **411**, 472–473.

Shu, D.G., Conway Morris, S., Han, J., Chen, L., Zhang, X.L., Zhang, Z.F. *et al.* (2001b) Primitive deuterostomes from the Chengjiang Lagerstatte (Lower Cambrian, China). *Nature* **414**, 419–424.

Shu, D.G., Conway Morris, S., Han, J., Zhang, Z.F., Yasui, K., Janvier, P. *et al.* (2003a) Head and backbone of the Early Cambrian vertebrate *Haikouichthys. Nature* **421**, 526–529.

Shu, D.G., Conway Morris, S., Zhang, Z.F., Liu, J.N., Han, J., Chen, L. *et al.* (2003b) A new species of *Yunnanozoan* with implications for deuterostome evolution. *Science* **299**, 1380–1384.

Shu, D.G., Conway Morris, S., Han, J., Zhang, Z.F., and Liu, J.N. (2004) Ancestral echinoderms from the Chengjiang deposits of China. *Nature* **430**, 422–428.

Shu, D.G., Conway Morris, S., Han, J., Li, Y., Zhang, X.L., Hua, H. *et al.* (2006) Lower Cambrian vendobionts from China and early diploblast evolution. *Science* **312**, 731–734.

Shulman, M. and Bermingham, E. (1995) Early life histories, ocean currents, and the population genetics of Caribbean reef fishes. *Evolution* **49**, 897–910.

Signor, P.W. and Vermeij, G.J. (1994) The plankton and the benthos: origins and early history of an evolving relationship. *Paleobiology* **20**, 297–319.

Sigwart, J.D. and Sutton, M.D. (2007) Deep molluscan phylogeny: synthesis of palaeontological and neontological data. *Philosophical Transactions of the Royal Society B: Biological Sciences* **274**, 2413–2419.

da Silva, F.B., Muschner, V.C., and Bonatto, S.L. (2007) Phylogenetic position of Placozoa based on large subunit (LSU) and small subunit (SSU) rRNA genes. *Genetics and Molecular Biology* **30**, 127–132.

Simpson, P. (2007) The stars and stripes of animal bodies: evolution of regulatory elements mediating pigment and bristle patterns in *Drosophila. Trends in Genetics* **23**, 350–358.

Singla, V. and Reiter, J.F. (2006) The primary cilium as the cell's antenna: signaling at a sensory organelle. *Science* **313**, 629–633.

Siveter, D.J., Sutton, M.D., Briggs, D.E.G., and Siveter, D.J. (2007) A new probable stem lineage crustacean with three-dimensionally preserved soft parts from the Herefordshire (Silurian) Lagerstatte, UK. *Proceedings of the Royal Society B: Biological Sciences* **274**, 2099–2107.

Skovsted, C.B., Brock, G.A., Paterson, J.R., Holmer, L.E., and Budd, G.E. (2008) The scleritome of *Eccentrotheca* from the Cambrian of South Australia: lophophorate affinities and implications for tommotiid phylogeny. *Geology* **36**, 171–174.

Slack, J.M., Holland, P.W.H., and Graham, C.F. (1993) The zootype and the phylotypic stage. *Nature* **361**, 490–492.

Sly, B.J., Hazel, J.C., Popodi, E.M., and Raff, R.A. (2002) Patterns of gene expression in the developing adult sea urchin central nervous system reveal multiple domains and deep-seated neural pentamery. *Evolution & Development* **4**, 189–204.

Sly, B.J., Snoke, M.S., and Raff, R.A. (2003) Who came first? Origins of bilaterian metazoan larvae. *International Journal of Developmental Biology* **47**, 623–632.

Slyusarev, G.S. and Kristensen, R.M. (2003) Fine structure of the ciliated cells and ciliary rootlets of *Intoshia variabili* (Orthonectida). *Zoomorphology* **122**, 33–39.

Smith, A.B. (1984) Classification of the Echinodermata. *Palaeontology* **27**, 431–459.

Smith, A.B. (1994) *Systematics and the fossil record—documenting evolutionary patterns.* Blackwell Scientific Publications, Oxford.

Smith, A.B. (2005) The pre-radial history of echinoderms. *Geological Journal* **40**, 255–280.

Smith, A.B. (2008) Deuterostomes in a twist: the origins of a radical new body plan. *Evolution & Development* **10**, 493–503.

Smith, A.B., Peterson, K.J., Wray, G., and Littlewood, D.T.J. (2004) From bilateral symmetry to pentaradiality: the phylogeny of hemichordates and echinoderms. In: J. Cracraft and M.J. Donoghue (eds) *Assembling the tree of life*, pp. 365–383. Oxford University Press, Oxford.

Smith, A.B., Pisani, D., Mackenzie-Dodds, J.A., Stockley, B., Webster, B.L., and Littlewood, D.T.J. (2006) Testing the molecular clock: molecular and paleontological estimates of divergence times in the Echinoidea (Echinodermata). *Molecular Biology and Evolution* **23**, 1832–1851.

Snel, B., Bork, P., and Huynen, M.A. (1999) Genome phylogeny based on gene content. *Nature Genetics* **21**, 108–110.

Snel, B., Huynen, M.A., and Dutilh, B.E. (2005) Genome trees and the nature of genome evolution. *Annual Review of Microbiology* **59**, 191–209.

Snell, E.A., Brooke, N.M., Taylor, W.R., Casane, D., Philippe, H., and Holland, P.W.H. (2006) An unusual choanoflagellate protein released by Hedgehog autocatalytic processing. *Proceedings of the Royal Society B: Biological Sciences* **273**, 401–407.

Sodergren, E., Weinstock, G.M., Davidson, E.H., Cameron, R.A., Gibbs, R.A., Angerer, R.C. *et al.* (2006) The genome of the sea urchin *Strongylocentrotus purpuratus. Science* **314**, 941–952.

Soler-Lopez, M., Petosa, C., Fukuzawa, M., Ravelli, R., Williams, J.G., and Muller, C.W. (2004) Structure of an activated *Dictyostelium* STAT in its DNA-unbound form. *Molecular Cell* **13**, 791–804.

Sommer-Knudsen, J., Bacic, A., and Clarke, A.E. (1998) Hydroxyproline-rich plant glycoproteins. *Phytochemistry* **47**, 483–497.

Sorek, R. and Ast, G. (2003) Intronic sequences flanking alternatively spliced exons are conserved between human and mouse. *Genome Research* **13**, 1631–1637.

Sørensen, M.V. (2001) *On the phylogeny and jaw evolution in Gnathifera*. PhD Thesis, University of Copenhagen, Copenhagen.

Sørensen, M.V. (2003) Further structures in the jaw apparatus of *Limnognathia maerski* (Micrognathozoa), with notes on the phylogeny of the Gnathifera. *Journal of Morphology* **255**, 131–145.

Sørensen, M.V. and Giribet, G. (2006) A modern approach to rotiferan phylogeny: combining morphological and molecular data. *Molecular Phylogenetics and Evolution* **40**, 585–608.

Sørensen, M.V., Funch, P., Willerslev, E., Hansen, A.J., and Olesen, J. (2000) On the phylogeny of Metazoa in the light of Cycliophora and Micrognathozoa. *Zoologischer Anzeiger* **239**, 297–318.

Sørensen, M.V., Sterrer, W., and Giribet, G. (2006) Gnathostomulid phylogeny inferred from a combined approach of four molecular loci and morphology. *Cladistics* **22**, 32–58.

Sørensen, M.V., Hebsgaard, M.B., Heiner, I., Glenner, H., Willerslev, E., and Kristensen, R.M. (2008) New data from an enigmatic phylum: evidence from molecular sequence data supports a sister-group relationship between Loricifera and Nematomorpha. *Journal of Zoological Systematics and Evolutionary Research* **46**, 231–239.

Sperling, E.A., Pisani, D., and Peterson, K.J. (2007) Poriferan paraphyly and its implications for Precambrian palaeobiology. In: P. Vickers-Rich and P. Komarower (eds) *The rise and fall of the Ediacaran biota. Geological Society of London, Special Publications* **286**, pp. 355–368.

Sprecher, S.G., Reichert, H., and Hartenstein, V. (2007) Gene expression patterns in primary neuronal clusters of the *Drosophila* embryonic brain. *Gene Expression Patterns* **7**, 584–595.

Sprinkle, J. and Guensberg, T.E. (1997) Early radiation of echinoderms. *Paleontological Society Papers* **3**, 205–224.

Srivastava, M., Begovic, E., Chapman, J., Putnam, N.H., Hellsten, U., Kawashima, T. *et al.* (2008) The *Trichoplax* genome and the nature of placozoans. *Nature* **454**, 955–960.

Stach, T. and Turbeville, J.M. (2002) Phylogeny of Tunicata inferred from molecular and morphological characters. *Molecular Phylogenetics and Evolution* **25**, 408–428.

Stach, T., Winter, J., Bouquet, J.-M., Chourrot, D., and Schnabel, R. (2008) Embryology of a planktonic tunicate reveals traces of sessility. *Proceedings of the National Academy of Sciences USA* **105**, 7229–7234.

Stamatakis, A., Ludwig, T., and Meier, H. (2005) RAxML-III: a fast program for maximum likelihood-based inference of large phylogenetic trees. *Bioinformatics* **21**, 456–463.

Stark, A., Lin, M.F., Kheradpour, P., Pedersen, J.S., Parts, L., Carlson, J.W. *et al.* (2007a) Discovery of functional elements in 12 *Drosophila* genomes using evolutionary signatures. *Nature* **450**, 219–232.

Stark, A., Kheradpour, P., Parts, L., Brennecke, J., Hodges, E., Hannon, G.F. *et al.* (2007b) Systematic discovery and characterization of fly microRNAs using 12 *Drosophila* genomes. *Genome Research* **17**, 1865–1879.

Stauber, M., Prell, A., and Schmidt-Ott, U. (2002) A single Hox3 gene with composite *bicoid* and *zerknüllt* expression characteristics in non-Cyclorrhaphan flies. *Proceedings of the National Academy of Science USA* **99**, 274–9.

Steenkamp, E.T., Wright, J., and Baldauf, S.L. (2006) The protistan origins of animals and fungi. *Molecular Biology and Evolution* **23**, 93–106.

Stefani, G. and Slack, F.J. (2008) Small non-coding RNAs in animal development. *Nature Reviews Molecular Cell Biology* **9**, 219–230.

Stein, L.D., Bao, Z., Blasiar, D., Blumenthal, T., Brent, M.R., Chen, N. *et al.* (2003) The genome sequence of *Caenorhabditis briggsae*: a platform for comparative genomics. *PLoS Biology* **1**, e45.

Steiner, G. and Salvini-Plawen, L. (2001) *Acaenoplax*—polychaete or mollusc? *Nature* **414**, 601–602.

Steiner, M., Zhu, M., Li, G., Qian, Y., and Erdtmann, B.-D. (2004) New Early Cambrian bilaterian embryos and larvae from China. *Geology* **32**, 833–836.

Stern, D.L. and Orgogozo, V. (2008) The loci of evolution: how predictable is genetic evolution? *Evolution* **62**, 2155–2157.

Stevens, M. (2005) The role of eyespots as anti-predator mechanisms, principally demonstrated in the Lepidoptera. *Biological Reviews*, **80**, 573–588.

St Johnston, D. and Nusslein-Volhard, C. (1992) The origin of pattern and polarity in the *Drosophila* embryo. *Cell* **68**, 201–219.

Stollewerk, A. and Simpson, P. (2005) Evolution of early development of the nervous system: a comparison between arthropods. *BioEssays* **27**, 874–883.

Stollewerk, A., Schoppmeier, M., and Damen, W.G. (2003) Involvement of *Notch* and *Delta* genes in spider segmentation. *Nature* **423**, 863–865.

Strausfeld, N.J., Strausfeld, C.M., Loesel, R., Rowell, D., and Stowe, S. (2006) Arthropod phylogeny: onychophoran brain organization suggests an archaic relationship with a chelicerate stem lineage. *Proceedings of the Royal Society B: Biological Sciences* **273**, 1857–1866.

Strong, E.E. and Lipscomb, D. (1999) Character coding and inapplicable data. *Cladistics* **15**, 363–371.

Struck, T.H. and Fisse, F. (2008) Phylogenetic position of Nemertea derived from phylogenomic data. *Molecular Biology and Evolution* **25**, 728–736.

Struck, T.H., Schult, N., Kusen, T., Hickman, E., Bleidorn, C., McHugh, D. *et al.* (2007) Annelid phylogeny and the status of Sipuncula and Echiura. *BMC Evolutionary Biology* **7**, 57.

Sulston, J.E., Schierenberg, E., White, J.G., and Thomson, J.N. (1983) The embryonic cell lineage of the nematode *Caenorhabditis elegans. Developmental Biology* **100**, 64–119.

Sundberg, P., Turbeville, J. M., and Lindh, S. (2001) Phylogenetic relationships among higher nemertean (Nemertea) taxa inferred from 18S rDNA sequences. *Molecular Phylogenetics Evolution* **20**, 327–334.

Surface, F.M. (1907) The early development of a polyclad *Planocera inquilina. Proceedings of the Academy of Natural Sciences Philadelphia* **59**, 514–559.

Sutton, M.D., Briggs, D.E.G., Siveter, D.J., and Siveter, D.J. (2001a) An exceptionally preserved vermiform mollusc from the Silurian of England. *Nature* **410**, 461–463.

Sutton, M.D., Briggs, D.E.G., and Siveter, D.J. (2001b) A three-dimensionally preserved fossil polychaete worm from the Silurian of Herefordshire, England. *Philosophical Transactions of the Royal Society B: Biological Sciences* **268**, 2355–2363.

Sutton, M.D., Briggs, D.E.G., Siveter, D.J., Siveter, D.J., and Orr, P.J. (2002) The arthropod *Offacolus kingi* (Chelicerata) from the Silurian of Herefordshire, England: computer based morphological reconstructions and phylogenetic affinities. *Proceedings of the Royal Society B: Biological Sciences* **269**, 1195–1203.

Sutton, M.D., Briggs, D.E.G., Siveter, D.J., and Siveter, D.J. (2004) Computer reconstruction and analysis of the vermiform mollusc *Acaenoplax hayae* from the Herefordshire Lagerstätte (Silurian, England), and implications for molluscan phylogeny. *Palaeontology* **47**, 293–318.

Sutton, M.D., Briggs, D.E.G., Siveter, D.J., and Siveter, D.J. (2005a) Fossilized soft tissues in a Silurian platyceratid gastropod. *Proceedings of the Royal Society B: Biological Sciences* **273**, 1039–1044.

Sutton, M.D., Briggs, D.E.G., Siveter, D.J., and Siveter, D.J. (2005b) Silurian brachiopods with soft-tissue preservation. *Nature* **436**, 1013–1015.

Sutton, M.D., Briggs, D.E.G., Siveter, D.J., Siveter, D.J., and Gladwell, D.J. (2005c) A starfish with three-dimensionally preserved soft parts from the Silurian of England. *Proceedings of the Royal Society B: Biological Sciences* **272**, 1001–1006.

Swalla, B.J. (2006) Building divergent body plans with similar genetic pathways. *Heredity* **97**, 235–243.

Swalla, B.J. (2007) New insights into vertebrate origins. In: S. Moody (ed.) *Principles of developmental genetics*, pp. 114–128. Elsevier Press, San Diego.

Swalla, B.J. and Smith, A.B. (2008) Deciphering deuterostome phylogeny: molecular, morphological and palaeontological perspectives. *Philosophical Transactions of the Royal Society B: Biological Sciences* **363**, 1557–1568.

Swalla, B.J., Cameron, C.B., Corley, L.S., and Garey, J.R. (2000) Urochordates are monophyletic within the deuterostomes. *Systematic Biology* **49**, 52–64.

Swofford, D.L. (2002) *PAUP* Phylogenetic Analysis Using Parsimony (* and other methods)*, version 4.0b10 for Macintosh. Sinauer Associates, Sunderland, MA.

Szaniawski, H. (2005) Cambrian chaetognaths recognized in Burgess Shale fossils. *Acta Palaeontologica Polonica* **50**, 1–8.

Szucsich, N.U. and Wirkner, C.S. (2007) Homology: a synthetic concept of evolutionary robustness of patterns. *Zoologica Scripta* **36**, 281–289.

Taft, R.J., Pheasant, M., and Mattick, J.S. (2007) The relationship between non-protein-coding DNA and eukaryotic complexity. *BioEssays* **29**, 288–299.

Tagawa, K., Humphreys, T., and Satoh, N. (2000) T-Brain expression in the apical organ of hemichordate tornaria larvae suggests its evolutionary link to the vertebrate forebrain. *Journal of Experimental Zoology (Molecular Development and Evolution)* **288B**, 23–31.

Tagawa, K., Satoh, N., and Humphreys, T. (2001) Molecular studies of hemichordate development: a key to understanding the evolution of bilateral animals and chordates. *Evolution & Development* **3**, 443–454.

Taguchi, S., Tagawa, K., Humphreys, T., and Satoh, N. (2002) Group B *sox* genes that contribute to specification of the vertebrate brain are expressed in the apical organ and ciliary bands of hemichordate larvae. *Zoological Science* **19**, 57–66.

Tahirov, T.H., Inoue-Bungo, T., Morii, H., Fujikawa, A., Sasaki, M., Kimura, K. *et al.* (2001) Structural analyses of DNA recognition by the AML1/Runx-1 Runt domain and its allosteric control by CBFβ. *Cell* **104**, 755–767.

Takacs, C.M., Amore, G., Oliveri, P., Poustka, A.J., Wang, D., Burke, R.D. *et al.* (2004) Expression of an NK2 homeodomain gene in the apical ectoderm defines a new territory in the early sea urchin embryo. *Develomental Biology* **269**, 152–164.

Takada, N., Goto, T., and Satoh, N. (2002) Expression pattern of the *Brachyury* gene in the arrow worm *Paraspadella gotoi* (Chaetognatha). *Genesis* **32**, 240–245.

Taneri, B., Snyder, B., Novoradovsky, A., and Gaasterland, T. (2004) Alternative splicing of mouse transcription factors affects their DNA-binding domain architecture and is tissue specific. *Genome Biology* **5**, R75.

Tang, F., Yin, C.Y., Stefan, B., Liu, Y.Q., Wang, Z.Q., Liu, P.J. *et al.* (2006) A new discovery of macroscopic fossils from the Ediacaran Doushantuo Formation in the Yangtze Gorges area. *Chinese Science Bulletin* **51**, 1487–1493.

Telford, M.J. (2004) The multimeric beta-thymosin found in nematodes and arthropods is not a synapomorphy of the Ecdysozoa. *Evolution & Development* **6**, 90–94.

Telford, M.J. (2006) Animal phylogeny. *Current Biology* **16**, R981–R985.

Telford, M.J. (2008) Resolving animal phylogeny: a sledgehammer for a tough nut? *Developmental Cell* **14**, 457–459.

Telford, M.J. and Budd, G.E. (2003) The place of phylogeny and cladistics in *Evo-Devo* research. *International Journal of Developmental Biology* **47**, 479–490.

Telford, M.J. and Thomas, R.H. (1995) Demise of the Atelocerata? *Nature* **376**, 123–124.

Telford, M.J. and Thomas, R.H. (1998) Expression of homeobox genes shows chelicerate arthropods retain their deutocerebral segment. *Proceedings of the National Academy of Sciences USA* **95**, 10671–10675.

Telford, M.J., Herniou, E.A., Russell, R.B., and Littlewood, D.T.J. (2000) Changes in mitochondrial genetic codes as phylogenetic characters: two examples from the flatworms. *Proceedings of the National Academy of Sciences Sciences USA* **97**, 11359–11364.

Telford, M.J., Lockyer, A.E., Cartwright-Finch, C., and Littlewood, D.T.J. (2003) Combined large and small subunit ribosomal RNA phylogenies support a basal position of the acoelomorph flatworms. *Proceedings of the Royal Society B: Biological Sciences* **270**, 1077–1083.

Telford, M.J., Wise, M.J., and Gowri-Shankar, V. (2005) Consideration of RNA secondary structure significantly improves likelihood-based estimates of phylogeny: examples from the Bilateria. *Molecular Biology and Evolution* **22**, 1129–1136.

Telford, M.J., Bourlat, S.J., Economou, A., Papillon, D., and Rota-Stabelli, O. (2008) The evolution of the Ecdysozoa. *Philosophical Transactions of the Royal Society B: Biological Sciences* **363**, 1529–1537.

Tessmar-Raible, K. and Arendt, D. (2003) Emerging systems: between vertebrates and arthropods, the Lophotrochozoa. *Current Opinions in Genetics and Development* **13**, 331–340.

Tessmar-Raible, K., Raible, F., Christodoulou, F., Guy, K., Rembold, M., Hausen, H. *et al.* (2007) Conserved sensory-neurosecretory cell types in annelid and fish forebrain: insights into hypothalamus evolution. *Cell* **129**, 1389–1400.

Teuchert, G. (1968) Zur Fortpflanzung und Entwicklung der Macrodasyoidea (Gastrotricha). *Zeitschrift für Morphologie Tiere* **63**, 343–418.

Theissen, U., Hoffmeister, M., Grieshaber, M., and Martin, W. (2003) Single eubacterial origin of eukaryotic sulfide: quinone oxidoreductase, a mitochondrial enzyme conserved from the early evolution of eukaryotes during anoxic and sulfidic times. *Molecular Biology and Evolution* **20**, 1564–1574.

Thollesson, M. and Norenburg, J.L. (2003) Ribbon worm relationships: a phylogeny of the phylum Nemertea. *Proceedings of the Royal Society B: Biological Sciences* **270**, 407–415.

Thomas, A.L.R. (1997) The breath of life—did increased oxygen levels trigger the Cambrian explosion? *Trends in Ecology and Evolution* **12**, 44–45.

Thomson, K.S., Sutton, M.D., and Thomas, B. (2003) A larval Devonian lungfish. *Nature* **426**, 833–834.

Thor, S. and Thomas, J. (2002) Motor neuron specification in worms, flies and mice: conserved and 'lost' mechanisms. *Current Opinions in Genetics and Development* **12**, 558–564.

Thorne, J.L. and Kishino, H. (2002) Divergence time and evolutionary rate estimation with multilocus data. *Systematic Biology* **51**, 689–702.

Thorne, J.L., Kishino, H., and Painter, I.S. (1998) Estimating the rate of evolution of the rate of molecular evolution. *Molecular Biology and Evolution* **15**, 1647–1657.

Tielens, A.G.M., Rotte, C., van Hellemond, J.J., and Martin, W. (2002) Mitochondria as we don't know them. *Trends in Biochemical Sciences* **27**, 564–572.

Ting, S.B., Caddy, J., Hislop, N., Wilanowski, T., Auden, A., Zhao, L-L. *et al.* (2005) A homolog of *Drosophila grainy head* is essential for epidermal integrity in mice. *Science*, **308**, 411–413.

Todaro, M.A., Telford, M.J., Lockyer, A.E., and Littlewood, D.T.J. (2006) Interrelationships of the Gastrotricha and their place among the Metazoa inferred from 18S rRNA genes. *Zoologica Scripta* **35**, 251–259.

Tour, E., Hittinger, C.T., and McGinnis, W. (2005) Evolutionarily conserved domains required for activation and repression functions of the *Drosophila* Hox protein Ultrabithorax. *Development* **132**, 5271–5281.

Towe, K.M. (1970) Oxygen-collagen priority and the early metazoan fossil record. *Proceedings of the National Academy of Sciences USA* **65**, 781–788.

Treisman, J., Gonczy, P., Vashishtha, M., Harris, E., and Desplan, C. (1989) A single amino acid can determine the DNA binding specificity of homeodomain proteins. *Cell* **59**, 553–562.

True, J. and Carroll, S.B. (2002) Gene co-option in physiological and morphological evolution. *Annual Reviews of Cell and Developmental Biology* **18**, 53–80.

True, J., Edwards, K.A., Yamamoto, D., and Carroll, S.B. (1999) *Drosophila* wing melanin patterns form by vein-dependent elaboration of enzymatic prepatterns. *Current Biology* **9**, 1382–1391.

Turbeville, J.M., Field, K.G., and Raff, R.A. (1992) Phylogenetic position of phylum Nemertini, inferred from 18S rRNA sequences: molecular data as a test of morphological character homology. *Molecular Biology and Evolution* **9**, 235–249.

Turbeville, J.M., Schulz, J.R., and Raff, R.A. (1994) Deuterostome phylogeny and the sister group of the chordates: evidence from molecules and morphology. *Molecular Biology and Evolution* **11**, 648–655.

Tyler, S. (2003) Epithelium—the primary building block for metazoan complexity. *Integrative and Comparative Biology* **43**, 55–63.

Ueno, K., Nagata, T., and Suzuki, Y. (1995) Roles of homeotic genes in the *Bombyx* body plan. In: M.R. Goldsmith and A.S. Wilkins (eds) *Molecular model systems in the Lepidoptera*, pp.165–180. Cambridge University Press, Cambridge.

Ulrich, W. (1951) Vorschläge zu einer Revision der Großeinteilung des Tierreichs. *Verhandlungen der Deutschen Zoologische Gesellschaft, Marburg (1950)* Suppl., 244–271.

Urbach, R. and Technau, G.M. (2003a) Molecular markers for identified neuroblasts in the developing brain of *Drosophila*. *Development* **130**, 3621–3637.

Urbach, R. and Technau, G.M. (2003b) Segment polarity and DV patterning gene expression reveals segmental organization of the *Drosophila* brain. *Development* **130**, 3607–3620.

Valentine, J.W. (2004) *On the origin of phyla*. University of Chicago Press, Chicago.

Valentine, J.W. and Collins, A.G. (2000) The significance of moulting in ecdysozoan evolution. *Evolution & Development* **2**, 152–156.

Valentine, J.W., Jablonski, D., and Erwin, D.H. (1999) Fossils, molecules and embryos: new perspectives on the Cambrian explosion. *Development* **126**, 851–859.

Van der Zee, M., Berns, N., and Roth, S. (2005) Distinct functions of the *Tribolium zerrknullt* genes in serosa specification and dorsal closure. *Current Biology* **15**, 624–636.

Vannier, J. and Chen, J.-Y. (2000) The Early Cambrian colonization of pelagic niches exemplified by *Isoxys* (Arthropoda). *Lethaia* **33**, 295–311.

Vannier, J. and Chen, J. (2005) Early Cambrian food chain: new evidence from fossil aggregates in the Maotianshan Shale biota, SW China. *Palaios* **20**, 3–26.

Vannier, J., Steiner, M., Renvoisé, E., Hu, S.-X., and Casanova, J.-P. (2007) Early Cambrian origin of modern food webs: evidence from predator arrow worms. *Proceedings of the Royal Society B: Biological Sciences* **274**, 627–633.

Vavouri, T., Walter, K., Gilks, W.R., Lehner, B., and Elgar, G. (2007) Parallel evolution of conserved non-coding elements that target a common set of developmental regulatory genes from worms to humans. *Genome Biology* **8**, R15.

Venkatesh, B., Ning, Y., and Brenner, S. (1999) Late changes in spliceosomal introns define clades in vertebrate evolution. *Proceedings of the National Academy of Sciences USA* **96**, 10267–10271.

Venter, J.C., Adams, M.D., Myers, E.W., Li, P.W., Mural, R.J., Sutton, G.G. *et al.* (2001) The sequence of the human genome. *Science* **291**, 1304–1351.

Vera, J.C., Wheat, C.W., Fescemyer, H.W., Frilander, M.J., Crawford, D.L., Hanski, I. *et al.* (2008) Rapid transcriptome characterization for a nonmodel organism using 454 pyrosequencing. *Molecular Ecology* **17**, 1636–1647.

Vermeij, G.J. (1993) *Evolution and escalation: an ecological history of life*. Princeton University Press, Princeton.

Vinther, J. and Nielsen, C. (2005) The Early Cambrian *Halkieria* is a mollusc. *Zoologica Scripta* **34**, 81–89.

Vinther, J., Van Roy, P., and Briggs, D.E. (2008) Machaeridians are Palaeozoic armoured annelids. *Nature* **451**, 185–188.

Voronov, D.A. and Panchin, Y.V. (1998) Cell lineage in marine nematode *Enoplus brevis*. *Development* **125**, 143–150.

Wada, H. and Satoh, N. (1994) Details of the evolutionary history from invertebrates to vertebrates, as deduced from sequences of 18S rDNA. *Proceedings of the National Academy of Sciences USA* **91**, 1801–1804.

Waegele, J.W. and Mayer, C. (2007) Visualizing differences in phylogenetic information content of alignments and distinction of three classes of long-branch effects. *BMC Evolutionary Biology* **7**, 147.

Wagner, G.P. (2001) What is the promise of developmental evolution? Part II: a causal explanation of evolutionary innovations may be impossible. *Journal of Experimental Zoology (Molecular Development and Evolution)* **291B**, 305–309.

Wagner, G.P. (2007) The developmental genetics of homology. *Nature Reviews Genetics* **8**, 473–479.

Wagner, G.P. and Lynch, V.J. (2008) The gene regulatory logic of transcription factor evolution. *Trends in Ecology and Evolution* **23**, 377–385.

Wagner, P.J. (2000) Exhaustion of morphologic character states among fossil taxa. *Evolution* **54**, 365–386.

Wagner, P.J. (2001) Gastropod phylogenetics: progress, problems, and implications. *Journal of Paleontology* **75**, 1128–1140.

Wagner, P.J., Ruta, M., and Coates, M.I. (2006) Evolutionary patterns in early tetrapods. II. Differing constraints on available character space among clades. *Proceedings of the Royal Society B: Biological Sciences* **273**, 2113–2118.

Wainwright, P.O., Hinkle, G., Sogin, M.L., and Stickel, S.K. (1993) Monophyletic origins of Metazoa: an evolutionary link with fungi. *Science*, 340–342.

Wallberg, A., Thollesson, M., Farris, J.S., and Jondelius, U. (2004) The phylogenetic position of the comb jellies (Ctenophora) and the importance of taxonomic sampling. *Cladistics* **20**, 558–578.

Wallberg, A., Curini-Galletti, M., Ahmadzadeh, A., and Jondelius, U. (2007) Dismissal of Acoelomorpha: Acoela and Nemertodermatida are separate early bilaterian clades. *Zoologica Scripta* **36**, 509–523.

Waloszek, D., Chen, J.Y., Maas, A., and Wang, X.Q. (2005a) Early Cambrian arthropods – new insights into arthropod head and structural evolution. *Arthropod Structure and Development* **34**, 189–205.

Waloszek, D., Repetski, J.E., and Maas, A. (2005b) A new Late Cambrian pentastomid and a review of the relationships of this parasitic group. *Transactions of the Royal Society of Edinburgh: Earth Sciences* **96**, 163–176.

Waloszek, D., Maas, A., Chen, J., and Stein, M. (2008) Evolution of cephalic feeding structures and the phylogeny of Arthropoda. *Palaeogeography, Palaeoclimatology, Palaeoecology*, **254**, 273–287.

Wang, D.Y., Kumar, S., and Hedges, S.B. (1999) Divergence time estimates for the early history of animal phyla and the origin of plants, animals and fungi. *Proceedings of the Royal Society B: Biological Sciences* **266**, 163–171.

Wang, X. and Lavrov, D.V. (2008) Seventeen new complete mtDNA sequences reveal extensive mitochondrial genome evolution within the Demospongiae. *PloS ONE* **3**(7), e2723.

Wanninger, A., Fuchs, J., and Haszprunar, G. (2007) Anatomy of the serotonergic nervous system of an entoproct creeping-type larva and its phylogenetic implications. *Invertebrate Biology* **126**, 268–278.

Webster, B.L., Copley, R.R., Jenner, R.A., Mackenzie-Dodds, J.A., Bourlat, S.J., Rota-Stabelli, O. *et al.* (2006) Mitogenomics and phylogenomics reveal priapulid worms as extant models of the ancestral Ecdysozoan. *Evolution & Development* **8**, 502–510.

Wennberg, S.A., Janssen, R., and Budd, G.E. (2008) Early embryonic development of the priapulid worm *Priapulus caudatus*. *Evolution & Development* **10**, 326–338.

Westblad, E. (1949) *Xenoturbella bocki* n. g., n. sp., a peculiar, primitive Turbellarian type. *Arkiv för Zoologi* **1**, 3–29.

Westfall, J.A., Elliott, C.F., and Carlin, R.W. (2002) Ultrastructural evidence for two-cell and three-cell neural pathways in the tentacle epidermis of the sea anemone *Aiptasia pallida*. *Journal of Morphology* **251**, 83–92.

Weygoldt, P. (1960) Embryologische Untersuchungen an Ostracoden: die Entwicklung von *Cyprideis litoralis* (G. S. Brandy). Ostracoda, Podocopa; Cytheridae. *Zoologische Jahrbücher, Abteilung für Anatomie und Ontogenie der Tiere* **78**, 369–426.

Wheeler, B.M., Heimberg, A.M., Moy, V.N., Sperling, E.A., Holstein, T.W., Heber, S. *et al.* (2009) The deep evolution of metazoan microRNAs. *Evolution & Development* **11**, 50–68.

Wheeler, W.C., Giribet, G., and Edgecombe, G.D. (2004) Arthropod systematics. The comparative study of genomic, anatomical, and paleontological information. In:. J. Cracraft and M.J. Donoghue (eds) *Assembling the tree of life*, pp. 281–295. Oxford University Press, New York.

Wheeler, W., Aagesen, L., Arango, C.P., Faivovich, J., Grant, T., D'Haese, C. *et al.* (2006) *Dynamic homology and phylogenetic systematics: a unified approach using POY*. American Museum of Natural History, New York.

Whelan, S. and Goldman, N. (2001) A general empirical model of protein evolution derived from multiple protein families using a maximum-likelihood approach. *Molecular Biology and Evolution* **18**, 691–699.

Whitfield, J.B. and Lockhart, P.J. (2007) Deciphering ancient rapid radiations. *Trends in Ecology and Evolution* **22**, 258–265.

Whiting, M.F. (2002) Phylogeny of the holometabolous insect orders: molecular evidence. *Zoologica Scripta* **31**, 3–15.

Whittington, H.B. (1978) The lobopod animal *Aysheaia pedunculata* Walcott, Middle Cambrian, Burgess Shale, British Columbia. *Philosophical Transactions of the Royal Society B: Biological Sciences* **284**, 165–197.

Wickens, M. and Takayama, K. (1994) Deviants—or emissaries. *Nature* **367**, 17–18.

Widersten, B. (1973) On the morphology of actinarian larvae. *Zoologica Scripta* **2**, 119–124.

Wiens, J.J. (2005) Can incomplete taxa rescue phylogenetic analyses from long-branch attraction? *Systematic Biology* **54**, 731–742.

Wiens, J.J. (2006) Missing data and the design of phylogenetic analyses. *Journal of Biomedical Informatics* **39**, 34–42.

Wiens, J.J., Bonett, R.M., and Chippindale, P.T. (2005a) Ontogeny discombobulates phylogeny: paedomorphosis and higher-level salamander relationships. *Systematic Biology* **54**, 91–110.

Wiens, J.J., Fetzner, J.W. Jr, Parkinson, C.L., and Reeder, T.W. (2005b) Hylid frog phylogeny and sampling strategies for speciose clades. *Systematic Biology* **54**, 719–748.

Wiens, J.J., Engstrom, T.N., and Chippindale, P.T. (2006) Rapid diversification, incomplete isolation, and the 'speciation clock' in North American salamanders (genus *Plethodon*): testing the hybrid swarm hypothesis of rapid radiation. *Evolution* **60**, 2585–2603.

Wightman, B., Ha, L., and Ruvkun, G. (1993) Posttranscriptional regulation of the heterochronic gene *lin-14* by *lin-4* mediates temporal pattern formation in *C. elegans*. *Cell* **75**, 855–862.

Wignall, P.B. and Twitchett, R.J. (1996) Oceanic anoxia and the End Permian mass extinction. *Science* **272**, 1155–1158.

Wills, M.A. (1998) Cambrian and recent disparity: the picture from priapulids. *Paleobiology* **24**, 177–199.

Winchell, C.J., Sullivan, J., Cameron, C.B., Swalla, B.J., and Mallatt, J. (2002) Evaluating hypotheses of deuterostome phylogeny and chordate evolution with new LSU and SSU ribosomal DNA data. *Molecular Biology and Evolution* **19**, 762–776.

Wolf, Y.I., Rogozin, I.B., and Koonin, E.V. (2004) Coelomata and not Ecdysozoa: evidence from genome-wide phylogenetic analysis. *Genome Research* **14**, 29–36.

Wolff, C., Sommer, R., Schröder, R., Glaser, G., and Tautz, D. (1995) Conserved and divergent expression aspects of the *Drosophila* segmentation gene *hunchback* in the short germ band embryo of the flour beetle *Tribolium*. *Development* **121**, 4227–4236.

Wolff, C., Schröder, R., Schulz, C., Tautz, D., and Klingler, M. (1998) Regulation of the *Tribolium* homologues of *caudal* and *hunchback* in *Drosophila*: evidence for maternal gradient systems in a short germ embryo. *Development* **125**, 3645–3654.

Woltereck, R. (1904) Beiträge zur praktischen Analyse der *Polygordius*–Entwicklung nach dem 'Nordsee-' und dem 'Mittelmeer-Typus'. *Archiv für Entwicklungsmechanik der Organismen* **18**, 377–403.

Wong, K.M., Suchard, M.A., and Huelsenbeck, J.P. (2008) Alignment uncertainty and genomic analysis. *Science* **319**, 473–476.

Woolfe, A., Goodson, M., Goode, D.K., Snell, P., McEwen, G.K., Vavouri, T. *et al.* (2005) Highly conserved non-coding sequences are associated with vertebrate development. *PLoS Biology* **3**, e7.

Woolley, K. and Martin, P. (2000) Conserved mechanisms of repair: from damaged single cells to wounds in multicellular tissues. *BioEssays* **22**, 911–919.

Worsaae, K. and Kristensen, R.M. (2003) A new species of *Paranerilla* (Polychaeta : Nerillidae) from Northeast Greenland waters, Arctic Ocean. *Cahiers de Biologie Marine* **44**, 23–39.

Worsaae, K. and Rouse, G.W. (2008) Is *Diurodrilus* an annelid? *Journal of Morphology* **269**, 1426–1455.

Wray, G.A. (2007) The evolutionary significance of *cis*-regulatory mutations. *Nature Reviews Genetics* **8**, 206–216.

Wray, G.A., Levinton, J.S., and Shapiro, L.H. (1996) Molecular evidence for deep Precambrian divergences among metazoan phyla. *Science* **274**, 568–573.

Wray, G.A., Hahn, M.W., Abouheif, E., Balhoff, J.P., Pizer, M., Rockman, M.V. *et al.* (2003) The evolution of transcriptional regulation in eukaryotes. *Molecular Biology and Evolution* **20**, 1377–1419.

Xian-Guang, H., Aldridge, R.J., Siveter, D.J., Siveter, D.J., and Xiang-Hong, F. (2002) New evidence on the anatomy and phylogeny of the earliest vertebrates. *Proceedings of the Royal Society B: Biological Sciences* **269**, 1865–1869.

Xian-Guang, H., Bergström, J., Xiao-Ya, M., and Jie, Z.G. (2006) The Lower Cambrian *Phlogites* Luo & Hu re-considered. *GFF* **128**, 47–51.

Xiao, S.H. and Knoll, A.H. (2000) Phosphatized animal embryos from the Neoproterozoic Doushantuo Formation at Weng'An, Guizhou, South China. *Journal of Paleontology* **74**, 767–788.

Xiao, S., Zhang, Y., and Knoll, A.H. (1998) Three-dimensional preservation of algae and animal embryos in a Neoproterozoic phosphorite. *Nature* **391**, 553–558.

Xiao, S.H., Yuan, X.L., and Knoll, A.H. (2000) Eumetazoan fossils in terminal Proterozoic phosphorites? *Proceedings of the National Academy of Sciences USA* **97**, 13684–13689.

Xiao, S.H., Hagadorn, J.W., Zhou, C.M., and Yuan, X.L. (2007a) Rare helical spheroidal fossils from the Doushantuo Lagerstätte: Ediacaran animal embryos come of age? *Geology* **35**, 115–118.

Xiao, S.H., Zhou, C.M., and Yuan, X.L. (2007b) Palaeontology – undressing and redressing Ediacaran embryos. *Nature* **446**, E9–E10.

Yamada, A., Pang, K., Martindale, M.Q., and Tochinai, S. (2007) Surprisingly complex T-box gene complement in diploblastic metazoans. *Evolution & Development* **9**, 220–230.

Yang, Z. and Rannala, B. (2006) Bayesian estimation of species divergence times under a molecular clock using multiple fossil calibrations with soft bounds. *Molecular Biology and Evolution* **23**, 212–226.

Yap, A.S., Brieher, W.M., and Gumbiner, B.M. (1997) Molecular and functional analysis of cadherin-based adherens junctions. *Annual Review of Cell and Developmental Biology* **13**, 119–146.

Yeates, D.K. (1995) Groundplans and exemplars—paths to the tree of life. *Cladistics*, **11**, 343–357.

Yeo, G.W., Van Nostrand, E., Holste, D., Poggio, T., and Burge, C.B. (2005) Identification and analysis of alternative splicing events conserved in human and mouse. *Proceedings of the National Academy of Sciences USA* **102**, 2850–2855.

Yin, C.Y., Yue, Z., and Gao, L.Z. (2001) Discovery of phosphatized gastrula fossils from the Doushantuo Formation, Weng'an, Guizhou Province, China. *Chinese Science Bulletin* **46**, 1713–1716.

Yin, C.Y., Bengtson, S., and Yue, Z. (2004) Silicified and phosphatized *Tianzhushania*, spheroidal microfossils of possible animal origin from the Neoproterozoic of South China. *Acta Palaeontologica Polonica* **49**, 1–12.

Yin, L.M., Zhu, M.Y., Knoll, A.H., Yuan, X.L., Zhang, J.M., and Hu, J. (2007) Doushantuo embryos preserved inside diapause egg cysts. *Nature* **446**, 661–663.

Yochelson, E.L. (1991) Problematica/Incertae Sedis. In: A.M. Simonetta and S. Conway Morris (eds) *The early evolution of Metazoa and the significance of problematic taxa. Proceedings of an international symposium held at the University of Camerino 27–31 March 1989*, pp. 287–296. Cambridge University Press, Cambridge.

Yoder, J.H. and Carroll, S.B. (2006) The evolution of abdominal reduction and the recent origin of distinct *Abdominal-B* transcript classes in Diptera. *Evolution & Development* **8**, 241–251.

Yokobori, S.-I., Iseto, T., Asakawa, S., Sasaki, T., Shimizu, N., Yamagishi, A. *et al.* (2008) Complete nucleotide sequences of mitochondrial genomes of two solitary entoprocts, *Loxocorone allax* and *Loxosomella aloxiata*: implications for lophotrochozoan phylogeny. *Molecular Phylogenetics and Evolution* **47**, 612–628.

Yu, J.K., Satou, Y., Holland, N.D., Shin, I.T., Kohara, Y., Satoh, N. *et al.* (2007) Axial patterning in cephalochordates and the evolution of the organizer. *Nature* **445**, 613–617.

Zaffran, S., Das, G., and Frasch, M. (2000) The NK-2 homeobox gene scarecrow (scro) is expressed in pharynx, ventral nerve cord and brain of *Drosophila* embryos. *Mechanisms of Development* **94**, 237–241.

Zdobnov, E.M., von Mering, C., Letunic, I., Torrents, D., Suyama, M., Copley, R.R. *et al.* (2002) Comparative genome and proteome analysis of *Anopheles gambiae* and *Drosophila melanogaster*. *Science* **298**, 149–159.

Zelinka, C. (1891) Studien über Räderthiere. III. Zur Entwicklungsgeschichte der Räderthiere nebst Bemerkungen über ihre Anatomie und Biologie. *Zeitschrift für Wissenschaftliche Zoologie* **53**, 1–159.

Zeng, L. and Swalla, B.J. (2005) Molecular phylogeny of the protochordates: chordate evolution. *Canadian Journal of Zoology* **83**, 24–33.

Zeng, L., Jacobs, M.W., and Swalla, B.J. (2006) Coloniality has evolved once in stolidobranch ascidians. *Integrative and Comparative Biology* **46**, 255–268.

Zhang, X. and Briggs, D.E.G. (2007) The nature and significance of the appendages of *Opabinia* from the Middle Cambrian Burgess Shale. *Lethaia* **40**, 161–173.

Zhang, X.G. and Xian-Guang, H. (2004) Evidence for a single median fin-fold and tail in the Lower Cambrian vertebrate, *Haikouichthys ercaicunensis*. *Journal of Evolutionary Biology* **17**, 1162–1166.

Zhang, X. L. and Reitner, J. (2006) A fresh look at *Dickinsonia*: removing it from Vendobionta. *Acta Geologica Sinica – English Edition* **80**, 636–642.

Zhao, Y. and Srivastava, D. (2007) A developmental view of microRNA function. *Trends in Biochemical Sciences* **32**, 189–197.

Zhou, C.M., Tucker, R., Xiao, S.H., Peng, Z.X., Yuan, X.L., and Chen, Z. (2004) New constraints on the ages of Neoproterozoic glaciations in south China. *Geology* **32**, 437–440.

Zigler, K.S., Raff, E.C., Popodi, E., Raff, R.A., and Lessios, H.E. (2003) Adaptive evolution of bindin in the genus *Heliocidaris* is correlated with the shift to direct development. *Evolution* **57**, 2293–2302.

Zrzavý, J. (2001) The interrelationships of metazoan parasites: a review of phylum- and higher-level hypotheses from recent morphological and molecular phylogenetic analyses. *Folia Parasitologica* **48**, 81–103.

Zrzavý, J. (2003) Gastrotricha and metazoan phylogeny. *Zoologica Scripta* **32**, 61–81.

Zrzavý, J., Hypsa, V., and Tietz, D.F. (2001) Myzostomida are not annelids: molecular and morphological support for a clade of animals with anterior sperm flagella. *Cladistics* **17**, 170–198.

Zrzavý, J., Mihulka, S., Kepka, P., Bezdek, A., and Tietz, D. (1998) Phylogeny of the Metazoa based on morphological and 18S ribosomal DNA evidence. *Cladistics* **14**, 249–285.

Zuckerkandl, E. and Pauling, L. (1965) Molecules as documents of evolutionary history. *Journal of Theoretical Biology* **8**, 357–366.

Zuker, M., Mathews, D.H., and Turner, D.H. (1999) Algorithms and thermodynamics for RNA secondary structure prediction: a practical guide. In: J. Barciszewski and B.F.C. Clark (eds) *RNA biochemistry and biotechnology*, pp. 11–43. Kluwer Academic Publishers, Boston.

Index